Thermal Management in Electronic Devices

Thermal Management in Electronic Devices

Prof. Dr. R.K. Singh

ARCLER
PRESS

www.arclerpress.com

Thermal Management in Electronic Devices

Prof. Dr. R.K. Singh

Arcler Press
4164 Lakeshore Road
Burlington, ON L7L 1A4
Canada
www.arclerpress.com
Email: orders@arclereducation.com

ISBN: 978-1-77956-976-9 (Hardcover)

ABOUT THE AUTHOR

Dr. R. K. Singh, born on 11th June, 1969, is a Professor and worked as Vice Chancellor, Patliputra University, Patna, Bihar, a state university of Bihar. Having throughout first division career, he obtained his Ph.D. from Allahabad University and M. E. from BITS, Pilani. He is recipient of numerous awards and honours. He is engaged in teaching, research, administration and consultancy for the last 28 years. Apart from this, he has worked as Director, B.T. Kumaon Institute of Technology, Dwarahat, Almora, Uttarakhand an autonomous Institution of Government of Uttarakhand and appointed as Director since 01/10/2015 at B.T. Kumaon Institute of Technology, Dwarahat, Almora,

Dr. Singh started his esteemed career as teaching assistant at Birla Institute of Technology and Science (BITS), Pilani (one and half years, Oct. 4, 1990 - July 1993) and as RA & Scientist(P) at CEERI, Pilani (more than one and half years, July 1993 – March 20, 1995), as an Assistant Professor March 21, 1995 – July 30, 1998) at BT Kumaon Institute of Technology (BTKIT), Dwarahat, formerly known as Kumaon Engineering College (KEC), Dwarahat, Almora (more than 3 years) and Associate Professor July 31, 1998 – July 30, 2006) at BT Kumaon Institute of Technology, Dwarahat (8 years). He became professor (July 31, 2006 – till date) at the age of 37 years, at BT Kumaon Institute of Technology, Dwarahat. He held the positions of HOD ECE, Principal, BT Kumaon Institute of Technology, Dwarahat, Officer on Special Duty, Registrar and Controller of Examination, at Uttarakhand Technical University, Dehradun.

Prof. Singh has received felicitations and awards by professional bodies such as Rastriya Samman Purskar- presented by Indian Society for Intellectual Development in presence of His Excellency, ShriSatya Narayan Reddy, Ex. Governor, U.P. and His Excellency, Shri Bhism Narayan Singh, Ex. Governor, Bihar, Jewel of India award- presented by International Institute of Education and Management in presence of His Excellency, Shri Bhism Narayan Singh, Ex. Governor, Bihar and Hon. Minister, Health, Govt. of Delhi, Shri Yoga Nand, Rastriya Ekta Award- presented by Global Economic Council, New Delhi, in presence of His Excellency, Shri Madan Lal Khurana, Governor, Rajasthan on 28th April, 2004 and Arch of Excellence Award-2004- presented in National Seminar-All India Achievers Conference in presence of His Excellency, Shri Satya Narayan

Reddy, Ex. Governor, and U.P. He is member of institution of Engineers (India). He has worked as Programme Officer of National Service Scheme at BTKIT, Dwarahat.

Prof. Singh has held positions of responsibility such as - Member, various committees of UTU University and Govt. of Uttarakhand, First Appellate Authority, BTKIT, Dwarahat (for Right 2 to Information Act, 2005); Dealing World Bank assisted Technical Education Quality Improvement Program for BTKIT; Expert Member, of various committees of several Institutions and organizations of national and international repute. Technical Expert for AICTE, NBA, and UGC of Government of India

Prof. Singh has guided 22 Ph.Ds. and about 15 M.Tech.thesis. He has participated in various National and International conferences, and published more than 120 research papers in national and international journals of repute. He has also authored eight books. Besides these, he has undertaken numerous consultancies for state government and various PSUs. He has provided technical assistance as member to groups of expert set up for various tasks.

Prof. Singh is member of the various prestigious bodies like Institution of Engineers (India), Kolkata, IEEE, etc.

Prof. Singh has visited various International countries as key note speaker and session chair-man. Some of visited countries are Peris France, Hong Kong, Zurich, Switzerland, Kuala lumpur, Malaysia and Bangkok, Thailand where he Delivered Key note speech and chaired the session of International Conference.

TABLE OF CONTENTS

LIST OF FIGURES

PREFACE

Electronics are the support columns of modern technology these days, powering everything from household appliances and personal devices to factory equipment and satellites. As electronic devices get more powerful, compact, and power-hungry, thermal management has become an integral engineering challenge. Electronic systems degrade in performance, dependability, and even fail if there is no proper thermal management. Overheating not only shortens the life of components but also limits the efficiency and safety of newer technologies. Thermal management is, therefore, not an accessory issue—it is a key to the success of modern electronics.

The need for effective heat control has never been more intense. With miniaturization, increased density of integration, increased switching speed, and increased power requirements, electronic devices operate in environments where tiny increases in temperature can lead to gigantic problems. In order to make sure that thermal challenges are anticipated and addressed, engineers therefore must combine knowledge of device physics, material science, heat transfer, and system design. The evolution of cooling systems, simulation software, and new materials is transforming the field, at once highly technical and deeply practical.

Thermal Management in Electronic Devices is designed to provide a comprehensive exploration of this subject. The book introduces fundamental concepts, analyzes the mechanisms of heat generation and transfer, and presents strategies for modeling, simulation, and cooling. It integrates theory with application, ensuring readers understand not only the scientific principles but also their relevance to real-world systems. From passive and active cooling methods to reliability considerations, this text offers both breadth and depth, preparing students, researchers, and practitioners to address the pressing challenges of thermal management in modern electronics.

Chapter 1 establishes what Thermal Management in Electronics is. The chapter explains why thermal control is central to electronic device design and function. The chapter establishes the effects of inadequate thermal control, reviews historic achievements in thermal management, and underscores growing applicability to emerging technologies such as 5G, electric vehicles, and high-end computing.

Chapter 2 addresses Fundamentals of Heat Transfer. Knowledge of conduction, convection, and radiation is required in thermal performance analysis. The chapter provides the theoretical foundation on which heat flow is predicted and computed in electronic systems. Examples show how the rules govern device temperature and guide design decisions.

Chapter 3 discusses the Thermal Properties of Materials. Since materials define how heat is stored and transported, their properties play a central role in thermal management. This chapter covers thermal conductivity, specific heat, thermal diffusivity, and expansion. A focus is put on advanced materials such as graphene, phase-change materials, and ceramics that are transforming cooling strategies.

Chapter 4 covers Heat Generation in Electronic Components. Electrical mechanisms such as resistive losses, leakage currents, and switching action generate heat. This chapter looks at ways these mechanisms vary across components like microprocessors, LEDs, batteries, and power modules. Quantifying heat sources enables designers to accurately forecast thermal loads and needs.

Chapter 5 introduces Thermal Modeling and Simulation. Thermal behavior is predicted by engineers using analytical and computational techniques prior to implementation of solutions. This chapter explains lumped parameter models, finite element analysis, and computational fluid dynamics in illustrating how simulation optimizes device configurations as well as cooling schemes.

Chapter 6 covers Active Cooling Methods. Active techniques such as fans, liquid cooling, thermoelectric coolers, and phase-change systems are discussed in this chapter. It deals with their principles of operation, advantages, and disadvantages along with their use in high-performance and mission-critical electronics.

Chapter 7 is on Passive Cooling Methods. In the majority of applications, passive solutions are effective and solid answers. This chapter presents natural convection, radiation cooling, heat pipes, and the role of design geometry. It also indicates the way passive solutions reduce energy use and maintenance requirements.

Chapter 8 addresses Heat Sinks and Thermal Interface Materials (TIMs). Heat sinks are the most common thermal management device, while TIMs enable efficient thermal transfer between surfaces. This chapter considers their design, materials, performance factors, and applications, including comparisons of innovations in fin design, surface treatment, and nanomaterials.

Chapter 9 analyzes Reliability and Thermal Stress in Electronics. Temperature influences the life and performance of devices, often inducing stress, fatigue, and failure. The chapter treats mechanisms such as thermal cycling, electromigration, and material degradation. It concentrates on how thermal reliability testing guides product design and enables safe operation.

Chapter 10 concludes with Thermal Management in Power Electronics. Power electronics exhibit some of the most demanding thermal challenges due to high currents and high rates of switching. The chapter addresses cooling of devices such as inverters, converters, and power modules. It associates thermal management with renewable power systems, electric drives, and industrial power applications.

This book has been written for a broad audience consisting of undergraduate and graduate students in electrical engineering, mechanical engineering, and materials science. It is also intended for researchers and professionals in the domains of electronics, semiconductor fabrication, renewable energy, and power systems. Students can take advantage of a rational progression from basic principles to advanced applications. Professionals can avail themselves of a reference book that is scientifically rigorous as well as practically relevant.

Educators may find the book valuable as a teaching resource for thermal engineering courses, while designers and engineers can apply its principles directly to product development and testing.

CHAPTER 1

INTRODUCTION TO THERMAL MANAGEMENT

Contents

INTRODUCTION

Thermal management is a critical science that deals with the techniques and technologies involved in controlling and managing the thermal conditions of systems and environments. It is particularly critical in aerospace, automotive, and electronic systems where temperature fluctuations can be exceedingly harmful to performance and safety. Thermal management seeks to keep components within their optimal temperature range to prevent overheating and enhance reliability. This involves a chain of heat generation, transfer, and dissipation processes that must be properly balanced. As OAST highlights, effective thermal management is among the eight key technological needs for the enhancement of national space programs and must be extensively validated in various environments, presenting the importance of well-tested thermal management strategies for future technological endeavors (Beck et al., 2025). The need for thermal control extends beyond avoiding failures, as it is also a key factor in attaining maximum energy efficiency and comfort for users. In dwellings, for instance, thermal conditions need to be controlled in the conservation of energy while maintaining appropriate living conditions. Different strategies involving balancing comfort and energy requirements have been discussed in the recent literature.

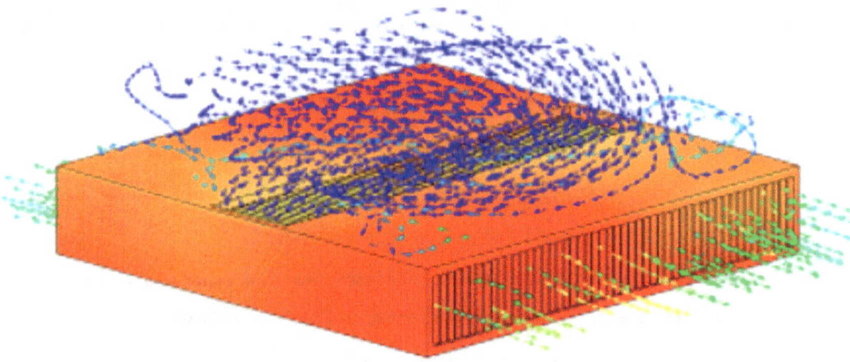

Figure 1.1. Thermal management (electronics).

Source: https://upload.wikimedia.org/wikipedia/commons/9/99/CFD_Forced_Convection_Heat_Sink_v4.gif.

These methods apply different optimization algorithms that focus on maintaining preferred comfort levels at minimum energy consumption. This is an important challenge of optimization since technological developments tend to bring with them new complications that their current counterparts cannot fully handle. Aside from this, the development of new algorithms is capable of enhancing accuracy in the management of energy consumption but yet to be implemented on a wide scale in practical applications, even in smart home environments (Fayaz et al., 2019). For the most part, the premise of thermal management can be extended to not only the physical and mechanical control over heat moderation but also the strategic implications for saving energy and enhancing comfort in different industries. Where precise temperature control is imperative, for example, in aerospace and home environments, thermal management is a complex job. As technology improves, the methods utilized in thermal management will need to adapt more and more in accordance with emerging needs and conventions. Hence, a deep understanding of thermal management is critical to the researchers and practitioners since they seek to design solutions that effectively address the complexities of heat management with particular focus on sustainability and efficiency in different applications (Figure 1.1).

Explanation of thermal management concepts

Thermal management is an engineering science that deals with the understanding and application of heat transfer principles, dissipation of energy, and temperature control in various systems. The main to effective thermal management is maintaining optimum working temperatures in good condition to enhance performance without overheating. This can broadly be classified into passive and active mechanisms. Passive thermal management relies on materials and design characteristics, such as thermal insulation and conduction behavior, to control temperature without the use of energy. Active thermal management uses hardware such as fans, pumps, and heat exchangers to actively control temperature so that systems can dynamically respond to changing operation requirements. Thermal management also extends beyond temperature control to influence energy efficiency and environmental performance. For instance, in office buildings, CO_2 emissions primarily stem from poorly controlled thermal conditions, which indicate a direct relationship between effective thermal control and environmental stewardship. Technological breakthroughs can enable interventions not only to address energy efficiency but also to incorporate the behavior of the occupants into the calculation. Examples of such initiatives include the Empower program that explores user-centric design practices in order to modulate energy usage among building occupants. Through interactions with users to understand their behavior towards heating, ventilation, and air conditioning (HVAC) systems, solutions can be designed that will optimally trigger energy-saving behavior, thereby decreasing the carbon footprint of workplace spaces (Cain et al., 2011). Principles of thermal management are also of utmost significance in specialized industries such as aerospace engineering, where temperature control is necessary for equipment performance and safety. Historical data, such as that available at the Lewis Research Center, establish the need for more advanced thermal management technology within spacecraft fluid management systems. Numerical modeling and experimental analysis projects for radiators and heat exchangers establish the sophisticated complexities found in managing heat in space missions. While spacecraft operators make efforts to ensure the unimpeded functioning of spacecraft under extremum conditions, a combination of theoretical developments and real application in thermal management is called for, revealing its general importance in engineering fields (Aydelott et al., 1985).

Importance in various industries

Thermal management is essential across numerous industries and has a major impact on the performance and safety of products. Thermal control in applications such as aerospace and automotive is especially important to maintain systems in their original shape and not to overheat and trigger catastrophic failure. Thermal management practices are not only used to enhance operation efficiency but also to prolong the lifespan of crucial parts. For example, in aerospace, thermal management of electronics is critical because fluctuating temperatures could potentially affect avionics and propulsion systems. With advancing technologies and materials, thermal performance has been maximized, resulting in lighter and more fuel-efficient aircraft. Such advancements are transferred to total fuel efficiency, which makes the economic and environmental importance of good thermal management practices even more apparent. In manufacturing, especially with the onset of additive manufacturing (AM), thermal management has become a vital part of the manufacturing process. Controlling the temperature while printing materials guarantees that parts have the necessary structural integrity and performance parameters. As highlighted by recent studies, AM facilitates advanced designs with reduced waste, an important issue in the wake of sustainability in manufacturing systems (Dehghanghadikolaei et al., 2019). Manufacturing can be greatly reduced in terms of energy consumption and cost, which fits into industry 4.0's greater goals of facilitating efficiency and technological implementation. Effective application of thermal management in AM not only streamlines manufacturing processes but also promotes innovation in product design, solidifying its central role in determining the future direction of industry standards. Moreover, industries such as oil and gas and renewable energy have comprehended the significance of thermal management to enhance safety and operating efficiency. In oil production and refining, instantaneous temperature conditions monitoring is extremely critical to ensure safe shipping and processing of volatile materials. The use of non-destructive testing (NDT) techniques has emerged as an effective way of monitoring temperature conditions and ensuring the cleanliness of equipment applied in these industries. The techniques support real-time testing that can prevent costly shutdowns and overall safety improvement during operations (Addepalli et al., 2018). In addition, in renewable energy systems, thermal control of solar panels and wind turbines is imperative to maximize energy output and system reliability. As industries evolve, the role of thermal management remains increasingly important in pushing the boundaries of performance, sustainability, and security in various industries.

Overview of thermal management systems

Thermal management systems play a crucial role in the control of optimum temperatures in most environments, particularly industrial environments. The systems control heat produced by processes, equipment, and external factors in order to maintain components within defined temperature ranges. Accurate thermal management is not only essential for maximizing equipment life and reducing failures but also for improving overall efficiency. Using various technologies, such as heat exchangers, insulations, and active cool systems, industry can reduce energy consumption effectively and increase performance. As discussed in literature sources, screening for energy-efficient technologies that are specifically made to retrofit industrial buildings can be beneficial for sustainable operations and cost saving (Fisher et al., 2011). Therefore, understanding how this thermal management systems function is the foundation for enhancing energy efficiency in countless industrial processes. Among countless innovations in thermal management, advances in materials science in recent times have given rise to a new generation of heat exchanger technology. Particularly noteworthy is the development of metal foams and metal matrix composites (MMCs) that has proven to hold great promise for efficient thermal control. These materials can be engineered to possess optimally tailored properties for a particular use, thereby enhancing their performance in heat exchange operations. Metal foams, in particular, possess an extraordinary contrast between light structure and high thermal conductivity, making them suitable to be applied in various applications as heat exchangers, varying from liquid-liquid, liquid-gas, and gas-gas systems. The current study is concerned with addressing the manufacturing problems of such materials in a way to make them more suitable for real-world applications (Han et al., 2012). Since industries continue to seek new innovations, these high-tech materials lead the way in establishing dynamic thermal management systems. The way forward for thermal management systems is through the integration of smart technologies to enhance the dynamic control of heat across all applications. The application of real-time sensing and predictive analytics can lead to more dynamic thermal management that allows immediate response to variations in temperature and load conditions. This pro-active approach reduces wastage of energy and improves system reliability, establishing a more environmentally friendly operational environment. Moreover, the technology of Internet of Things (IoT) enables easy amalgamation of communication between all the elements of a thermal management system so that data-driven decision-making is enabled to achieve optimal performance. Creating a complete picture of thermal management systems, one can observe that the confluence of innovation and efficiency not only solves the problems of the present but also lays the groundwork for the advancements of the future in industrial processes. All in all, continuous research on thermal management technologies is critical towards realizing greater sustainability and operational excellence in the long term.

1.1. THERMAL MANAGEMENT TECHNIQUES

Effective thermal energy management is necessary in many industries, especially for increasing the operational efficiency of systems. Of the numerous thermal management practices, surface modification is a significant approach. With the utilization of specially tailored surface structures, such as the electron beam enhanced configurations discussed in current investigations, the heat transfer rates can be radically improved.

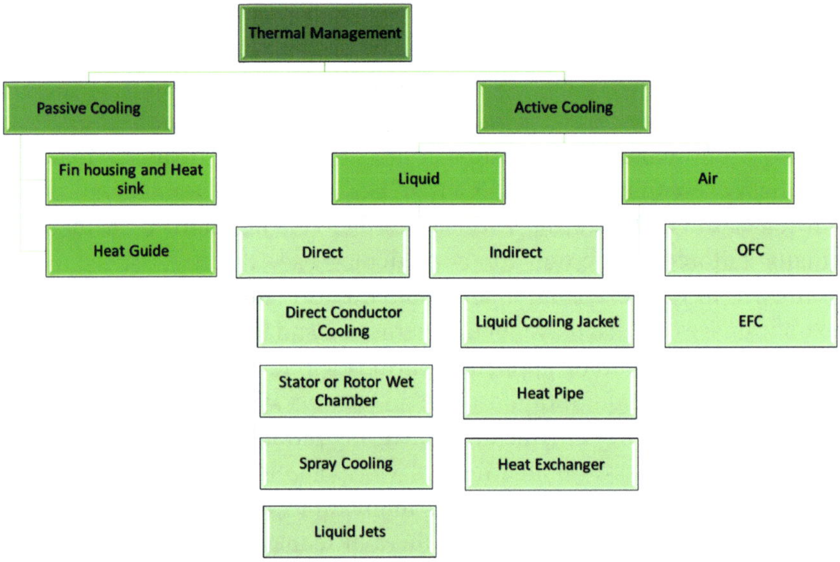

Figure 1.2. Various thermal management methods for electrical machines.

Source: https://www.researchgate.net/figure/arious-thermal-management-methods-for-electrical-machines_fig2_349651442.

A single experimental study has demonstrated that pyramid-shaped protrusions on surfaces led to a dramatic enhancement of the average Nusselt number to up to three times that of a smooth surface. This is because of surface imperfections, which increase the area for heat transfer in addition to permitting greater mixing within the fluid flow, significantly disrupting boundary layers for velocity and thermal. While the enhancements in the heat transfer rates are convincing, designers must weigh them against the ensuing pressure drop increases, which could influence total system performance and efficiency (Ferhati et al., 2015). In home energy use, the application of thermal management strategies can provide dramatic optimization outcomes. With sustainability coming into focus, maintaining user comfort while lowering energy use is most critical. Researchers have tried different optimization algorithms to balance this tradeoff, highlighting the challenge of the dynamic conditions of thermal environments. Successful methods have to account for different considerations such as thermal comfort, visual comfort, and air quality that all combine to form the overall user experience in smart homes. Interestingly, literature reveals the need to include novel cutting-edge optimization algorithms that are in their inception phases, some of which have shown better accuracy during benchmarking scenarios but whose performance must be validated in real-world contexts. The disparity emphasizes the need to further examine the application of such algorithms to real-world energy consumption scenarios in smart homes (Fayaz et al., 2019) (Figure 1.2).

In addition, novel materials also play an essential role in today's thermal management systems. Continuous developments in material with high thermal conductivity can now facilitate successful heat dissipation in a wide range of applications, from electronics and air conditioning systems to other advanced technologies. These materials, including some such as graphene, carbon nanotubes, and metal matrix composites, are being researched presently for the unique properties that enhance thermal functionality at minimal weight and flexibility. The use of these kinds of materials not only facilitates the effective management of thermal loads but also contributes to miniaturization and power efficiency of devices, which is of utmost importance in a world where power utilization and space are of prime importance. As researchers continue to discover more about the properties of these new materials, their impact on thermal management approaches will have the capability to transform design and operational approaches across numerous industries, leading to enhanced performance and sustainability practices.

1.1.1. Passive thermal management strategies

Good thermal management is critical in contemporary architecture, particularly as the requirements for energy efficiency become tighter. Passive thermal management methods are a starting point in achieving optimal building performance with lower energy consumption. They incorporate a variety of design elements to harness natural climatic conditions to manage temperatures indoors. For instance, the orientation of a building can be designed to maximize winter sunlight intake and minimize summer, thereby reducing heating and cooling requirements. Additionally, materials used in buildings with high thermal mass, such as concrete or brick, can absorb heat during the day and dissipate heat at night, maintaining an inside climate that is consistent. By focusing on these passive approaches, architects can create energy-conserving spaces that resonate with sustainability concepts in building design. Passive thermal control techniques become particularly pertinent in multifamily dwelling, where collective energy consumption can significantly impact design. One of the best examples of this is found in older buildings, where retrofitting for greater energy efficiency can yield more economic benefits. As has been demonstrated through current studies, energy savings resulting from retrofit can vary considerably based on occupant energy usage patterns, potentially as much as 80% in situations where applicable energy efficiency measures are implemented. Not only do such measures maximize thermal comfort but also minimize the need for active heating and cooling, enabling more sustainable energy use in the housing sector. Emphasizing this feature, the use of evidence-based tests in decision-making can make property owners and architects use the most appropriate passive strategies within their specific contexts (Padura et al., 2019). Furthermore, building design development has increasingly embraced the values of net zero energy buildings (NZEBs), particularly in regions with hot and humid climates. In these situations, passive design approaches such as natural ventilation and daylighting are critical to reducing energy use while ensuring occupant comfort. Studies have shown that most successful NZEBs have annual energy use intensities of less than 100 kWh per square meter, proving that passive methods can be combined effectively with renewable energy technologies. But it is natural that all buildings become energy-efficient by implementation of renewable energy alone; there are some which stay short even after being NZEB compliant. Such an insight indicates the importance of an all-round strategy by incorporating passive thermal control along with cutting-edge technology solutions in a bid to unlock the complete extent of energy-efficient buildings (Feng et al., 2019).

1.1.2. Active thermal management systems

Active thermal management systems play a vital role in today's technology since they control temperature in various applications, particularly data centers. With the demand for processing and storing data, the energy usage of these systems also grows. A good active thermal management system enhances energy efficiency, which allows for optimized performance and conserves excessive power consumption, important towards effective sustainable operation in a big-scale system. The application of newer techniques, such as server consolidation and dynamic voltage and frequency scaling (DVFS), is indicative of the novel approach to solving the issue of high heat generation in data centers with high density.

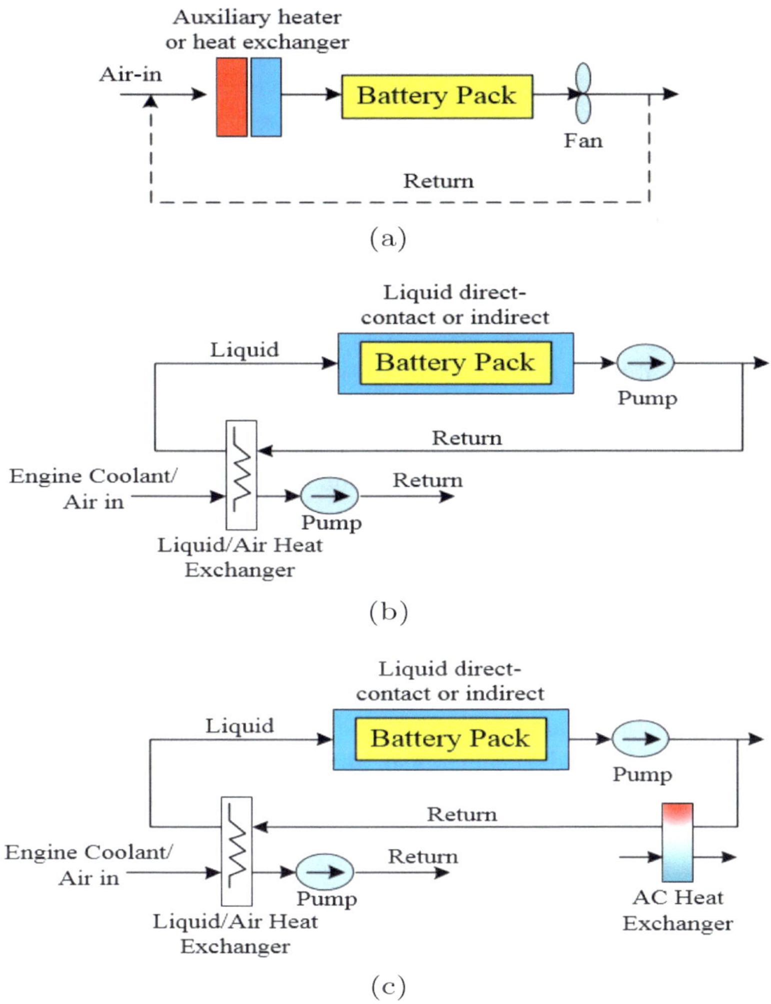

Figure 1.3. Schematic of active thermal management systems: (a) Active cooling/heating using outside or cabin air; (b) active moderate cooling/heating using liquid circulation; and (c) active cooling (high temperature) and heating (cold temperature) using liquid circulation.

Source: https://www.researchgate.net/publication/283842696_Integrated_Battery_Management_System.

These systems not only offer improved performance but also serve to maintain the operational cost low and environmental issues at a distance, making them fundamental in the big data era (Mittal et al., 2014). In aerospace applications, active thermal control systems are also significant as they offer solutions to the issue posed by the harsh thermal environment of space. High priority has been assigned by the Office of Aeronautics and Space Technology (OAST) to the development of these systems to deliver effective thermal management in spacecraft where passive techniques may fall short. New algorithms of control and sensors are applied to control heat exchange processes dynamically and thus enhance the efficiency and longevity of onboard systems. These technologies must be used in the fulfillment of the key demands of national space missions because this integrates high thermal management technique and successful space missions (Beck et al., 2025) (Figure 1.3).

Therefore, active thermal management technologies are essential, not only because they enable operational effectiveness but also because they guarantee the reliability and safety of advanced aerospace systems. Moreover, the application of active thermal management systems is not only confined to data centers and aerospace technology but includes a broad range of applications in a wide range of industries including automotive and consumer electronics. For applications in the automotive field, for instance, thermal management is essential to offer engine efficiency and prolong component life. With the use of active cooling methods like liquid cooling systems, vehicle manufacturers are able to effectively optimize performance and minimize risk of overheating. In consumer electronics, active thermal management offers devices operation in wanted temperature ranges, thereby enhancing user experience and product reliability. Continued advancements in materials and technology have the potential for a more sophisticated method of heat management across many platforms, similar to the increasing demand for proper thermal management in more complex applications (Beck et al., 2025; Mittal et al., 2014).

1.1.3. Hybrid approaches to thermal management

In thermal management, hybrid technologies have emerged as an important strategy for optimizing energy efficiency and reducing environmental impacts. Hybrid strategies combine various technologies in order to overcome the restrictions inherent in traditional thermal management systems, particularly in city areas where transportation-related pollution is a matter of urgent concern. Hybrid electric vehicles (HEVs), for instance, leverage the power of internal combustion engines and electric motors, harnessing greater flexibility in the energy consumption. Beyond reducing harmful emissions—accounting for some 20% of city transport-related air pollution, as indicated by recent studies—this twin system also meets improved fuel efficiency, thus satisfying environmental as well as economic needs (Capata et al., 2018). Besides, the development of hybrid technology plays a central role in making thermal management systems more efficient as a major catalyst for sustainable practices in most industries. The unification of diverse energy sources and management technologies is the foundation of hybrid thermal management systems. A case in point is the establishment of microgrids, which demonstrates decentralized energy distribution wherein multiple energy sources like renewables and traditional generators can be utilized to their optimum capacity. This evolution to a more sophisticated and efficient energy model is crucial when it comes to urbanization and climate change because it facilitates improved thermal load control and energy usage patterns. Sophisticated energy management strategies (EMSs) can be implemented by operators for dynamical adjustment of energy flows so as to realize efficient thermal performance as well as lower emissions (Alba et al., 2016) (Figure 1.4).

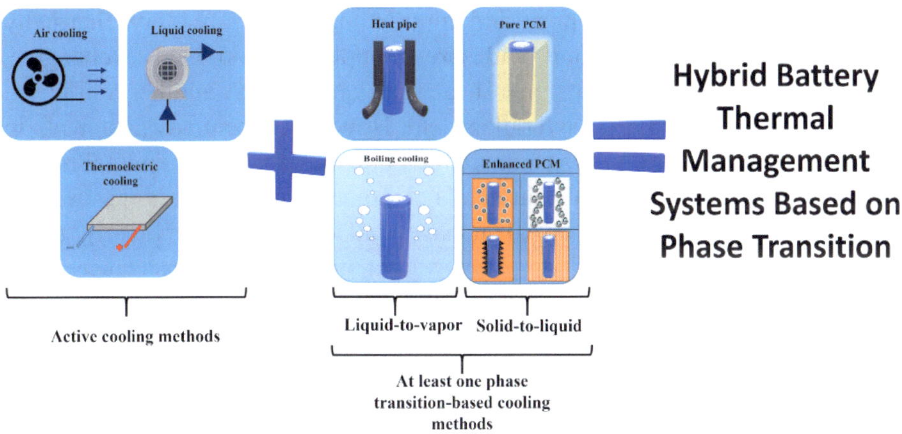

Figure 1.4. Hybrid battery thermal management systems based on phase-transition processes.

Source: https://ars.els-cdn.com/content/image/1-s2.0-S2352152X24008119-ga1_lrg.jpg.

This adaptability not only ensures legal compliance with environmental regulations but also supports growing demands for energy resilience amid climate fluctuation. The dividends of efficiency achieved through hybrid solutions to thermal management have a reach far broader than single systems and influence larger-scale infrastructural construction. As cities become further constrained by stringent environmental regulations, hybrid thermal management systems can serve to secure compliance through optimizing energy efficiency in transportation systems and building systems. Technologies such as Model Predictive Control, when integrated with hybrid systems, enhance operational efficiency by predicting thermal loads and then controlling energy inputs. Effective implementation of such technologies demonstrates that hybrid approaches can help in meeting energy goals and also foster economic benefits through reduced operating costs (Alba et al., 2016). The constant research and development of these hybrid strategies will be vital to sustainable urban development, as they represent an integral solution to the multifaceted issue of modern thermal management.

1.2. MATERIALS USED IN THERMAL MANAGEMENT

Material selection for thermal management is critical in the realization of maximum energy efficiency in various applications. Such optimal materials include metals such as copper and aluminum, which are very thermally conductive. Such metals find widespread application in heat exchangers, where they facilitate the transfer of heat from one fluid to another and hence resulting in a more efficient overall system. However, recent advancements in technology have allowed new-age materials like metal foams and metal matrix composites (MMCs) to be introduced. They possess engineered properties that can significantly enhance thermal management performance. Notably, metal foams have a high permeability and low weight, hence ideal for use where there are effective heat dissipation and minimal weight, which is vitally necessary in fields like automotive and aerospace (Han et al., 2012). In addition to metals, polymers with special thermal properties are also being used more in thermal management. Polymers are now being formulated with additives that enhance their conductivity so that they can be utilized effectively as insulators or as part of thermal interface materials (TIMs). Such multifunctionality renders them effective for application in most electronic devices and components where heat generation must be controlled for reliability and performance. Thermal insulating materials such as thermally conductive silicones or epoxies are employed for retarding heat transfer, thereby maintaining sensitive components at optimal operating temperatures. Moreover, the formulation flexibility of polymers enables manufacturers to adjust not only thermal performance but also mechanical and chemical resistance, hence broadening their application in thermal management products (Ali et al., 2015). The future of thermal management materials also seems promising with the introduction of nanomaterials, which possess great thermal properties at reduced weights. Incorporation of nanostructures such as carbon nanotubes or graphene into traditional materials can significantly increase their thermal conductivity without reducing mechanical strength. This feature makes them an extremely suitable option for upcoming applications like high-performance computing and electric vehicles where efficient heat transfer is a key requirement. Yet, the scalability of such materials and their integration in conventional manufacturing processes remains a challenge. Solving such challenges includes advancing research on the characteristics of nanomaterials and their behavior in composite systems, thus potentially revolutionizing thermal management. Solving these complexities will set the industry for increased efficiency in energy consumption and heat control (Han et al., 2012).

1.2.1. Conductive materials: metals and composites

Conductive materials, particularly metal and composite materials, play key roles in the context of thermal management since they are capable of conducting heat efficiently.

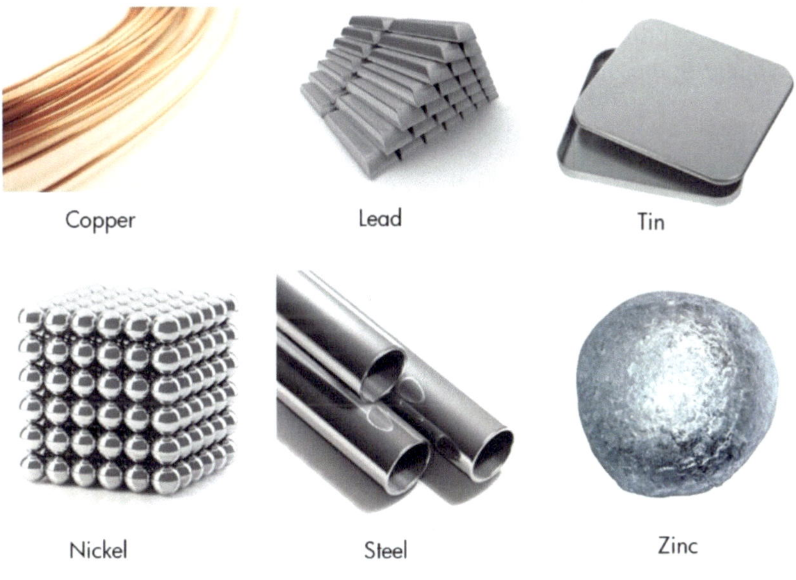

Figure 1.5. Conductive metals.

Source: https://external-content.duckduckgo.com/iu/?u=https%3A%2F%2Fglobal scraptrading.org%2Fwp-content%2Fuploads%2F2022%2F09%2Fmetals-min.jpg &f=1&nofb=1&ipt=e2f9c320f25f66d8a940f5e351aa59e9a2fb36511b711eaaec21 2a1f9bd926f0.

Metals such as copper and aluminum have been traditional choices for thermal applications due to their high thermal conductivity, thus making them the material of choice for heat exchangers and heat sinks. These materials enable efficient removal of heat from susceptible elements, thus prolonging their lifespan and enhancing operation efficiency. The use of these metals with high conductivity in various thermal management systems has been recognized to be fundamental because their properties can form a significant determining factor for the performance of electronic and mechanical systems. Along with this, new advancements in manufacturing technologies introduced new types of metals, i.e., metal foams and metal matrix composites, which possess improved thermal characteristics particular to specific use areas, demonstrating the relentless rate of development in material science to meet thermal management requirements (Han et al., 2012) (Figure 1.5).

In addition to traditional metals, composite materials have turned out to be essential components in thermal management systems. Composites, particularly metal matrix composites (MMCs), balance the beneficial properties of metals with the lightness of polymers or ceramics. The synergy not only enhances higher thermal conductivity but also enhances structural integrity under changing thermal stresses (Figure 1.6).

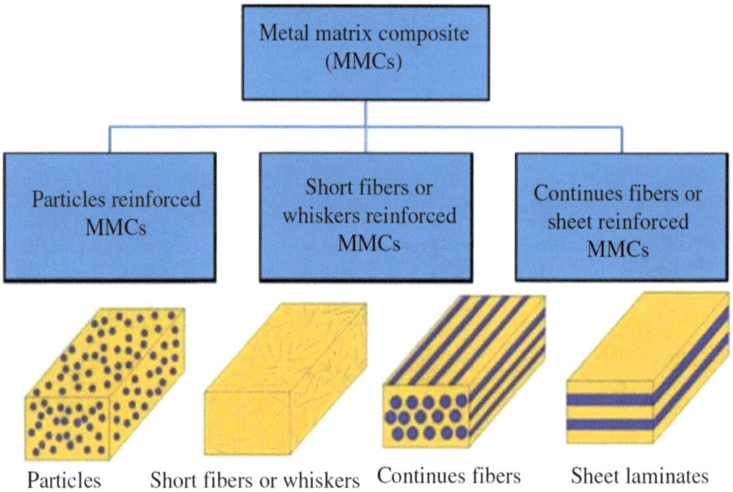

Figure 1.6. Types of metal matrix composite.

Source: https://www.researchgate.net/figure/Types-of-metal-matrix-composite_fig1_317171331.

Use of MMCs is particularly advantageous in applications such as aerospace and automobile industries, where saving weight is the most important parameter without sacrificing thermal efficiency. Research on the optimization of the composite material indicates potential benefits such as improved heat transfer and tailored thermal behaviors for specific application.

Enhanced utilization of these composites is opening the door to new heat exchanger technologies that fulfill the pressing needs for more effective thermal management technologies in a variety of engineering applications (Han et al., 2012). Furthermore, ionomeric polymer-metal composite (IPMC) is the next breakthrough in conductive material technology that brings the scientific community together under mechanical actuation and thermal management. These electromechanically active soft material systems can be designed to react to very small electrical inputs, providing large displacement and high-speed motion, which is particularly beneficial in biomedical applications.

Although so far restricted to bending motion, new advances in freeform fabrication techniques hold promise for creating more complex actuating systems that incorporate specialized geometries and are paired with other active elements. By encouraging innovation in IPMC design and operation, researchers are not just attempting to enhance their actuation performance but also their potential thermal management uses. Such innovations would be welcoming to universal usability that facilitates the efficiency of devices requiring meticulous thermal control, hence facilitating the essential relationship between conductive materials and thermal management systems.

1.2.2. Insulating materials: types and properties

The selection of insulating material has a significant role to play in thermal management system efficiency. Insulation materials have various characteristics that fit specific applications, balancing among thermal resistance, cost, durability, and environmental impacts. Fiberglass, foam, cellulose, and mineral wool are some of the most common insulating materials with given thermal conductivity and moisture resistance. For instance, fiberglass is well thermally resistant and relatively inexpensive, whereas foam insulation, particularly closed-cell varieties, possesses superior water resistance along with thermal properties.

All such diverse materials enable designers and engineers to make thermal management systems adjustable, with structures having optimal internal temperatures and low energy consumption, which eventually contributes to sustainability goals. In terms of building technology, the use of high-performance insulating material increases in increasing significance. Existing research is indicative of the development of sophisticated insulating systems that promise a substantial saving in energy needs. For example, technology like the R-5, 1-inch-thick triple-pane insulating glass unit, as explained in local studies, offers impressive saving in heating and cooling loads, particularly useful during periods of maximum energy utilization (Breshears et al., 2020) (Figure 1.7).

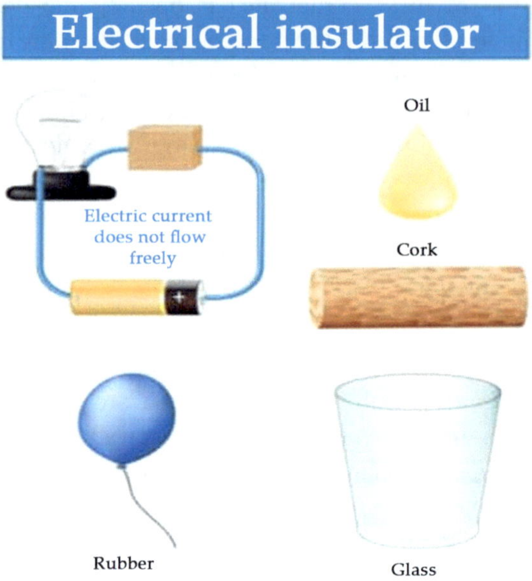

Figure 1.7. Insulating materials.

Source: https://external-content.duckduckgo.com/iu/?u=https%3A%2F%2Fcache.careers360.mobi%2Fmedia%2Farticle_images%2F2022%2F11%2F29%2FElectric_conductors.jpg&f=1&nofb=1&ipt=2f55bb53ae0a8f6ea38a07cb8bc56af56d59adf267da24759f8f16bd3700d554.

The success of these new materials is boosted further by the fact that they can be used in conjunction with other green technologies, for instance, energy recovery devices. Apart from their application improving thermal performance, they also help address regulations in strict regulation systems intended to avert environmental impacts as well as reducing greenhouse gas emissions. Despite the use of newer materials being associated with challenges, particularly concerning safety and fire risks, it is still worth considering. The growing application of photovoltaic (PV) solar systems, which can be integrated with insulating technologies, has also brought about concerns over electrical component fire risks (Smith, 2015). Low knowledge or fire threat awareness from such systems is found to prevent safety measures from being adopted. Therefore, ongoing research on the thermal stability and longevity of insulating materials remains essential, particularly for high-temperature applications and electrical insulation. By addressing such safety issues and improving material characteristics, the construction and energy industries can continuously improve their practices in thermal management, eventually building safer and more efficient building conditions.

1.2.3. Phase change materials and their applications

Phase change materials (PCMs) are materials that absorb and release heat during phase changes, such as melting or solidification, and are extremely valuable to thermal management systems. These materials utilize the latent heat of phase change to offer effective thermal energy storage (TES), which can be used for a vast variety of applications, such as building temperature regulation to advanced aerospace systems. The ability of PCMs to produce steady thermal output during phase transitions provides a method of controlling energy flows efficiently, particularly in situations where temperature regulation is crucial. With the industries of the world looking to boost energy efficiency and reduce costs, the use of PCMs in thermal regulation becomes not only desirable but imperative for sustainability and enhanced performance. A principal application of PCMs is to thermal energy storage systems, and these are crucial in regulating the excess heat produced by solar energy systems. By utilizing the heat of fusion of PCMs, systems are constructed to store heat energy during periods of high generation and release it when needed. This capability is particularly nuanced in the context of space power generation systems, where the lightweight character of PCMs can be used to reduce overall system mass while optimizing for energy efficiency (Figure 1.8).

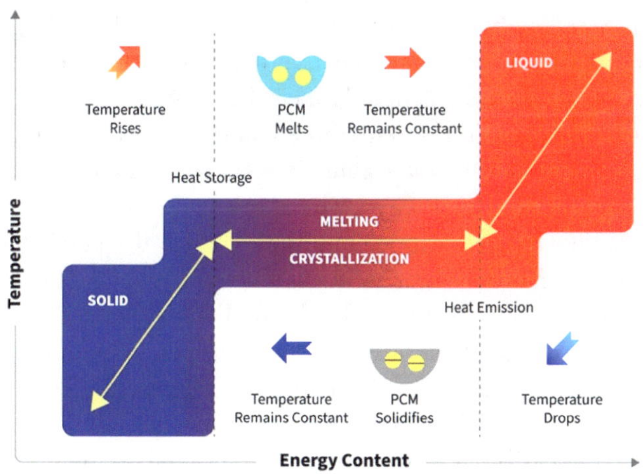

Figure 1.8. Working of phase change materials.

Source: https://thermtest.com/phase-change-material-pcm.

For instance, research has established that by minimizing the volume of space held by TES, along with the high energy density contributed by PCMs, design and operation of solar dynamic systems can be optimized. This innovation renders PCMs a key component in creating new technologies that are capable of enhancing thermal management applications' performances (Migra et al., 2025). Besides their role in thermal energy storage, the thermal transfer behavior of PCMs can be significantly enhanced by the addition of materials such as aluminum foams that enhance the thermal conductivity of the system. This approach serves to offset one of the greatest challenges faced in the utilization of PCMs—namely, their inherent low thermal transfer rates. Experimental studies have also shown that the integration of aluminum foams with paraffin wax, one of the most commonly used PCMs, led to improved rates of loading and unloading in thermal storage. Applying foams of various pore densities (PPI) made it possible for scientists to streamline the thermal transfer operations involved in PCMs, demonstrating the potential of increased system performance in heating and cooling uses. This study underscores the versatility of PCMs and their ability to serve varied functions relative to various technologies that aim at improving energy management (Lazzarin et al., 2018).

1.3. THERMAL MANAGEMENT IN ELECTRONICS

Effective thermal management is important in electronics, particularly as devices decrease in size and demand more processing. Miniaturization of components in devices like smartphones necessitates sophisticated cooling systems to achieve optimal performance without overheating. While software becomes more complex, performance demands have been moving toward a matching increase in power dissipation. Even with advancements in cooling technology, thermal management failure may cause catastrophic events, such as battery explosion due to thermal runaways. These occur due to insufficient analysis of thermal behavior under various conditions, hence the need to comprehend device thermal dynamics properly. Thermal management that includes safe application of electronics is not only thermal management optimization in cooling but also an overall thermal design approach considering all aspects, including the heat generated by the battery (Chen et al., 2019), which has remained underrepresented in thermal management literature. Processing infrastructure, especially in data centers, is seriously under pressure when it comes to thermal management. The resultant exponential growth in data consumption has necessitated an increase in the size and power consumption of data centers. This creates immense power usage, generating massive amounts of heat, which makes it increasingly difficult to maintain it at optimum operating temperatures. Power efficiency under such conditions is therefore an important concern for designers and engineers at present. Various architectural strategies have been developed to manage power consumption within data centers, with consideration for how to optimize thermal management without degrading system performance. Strategies such as dynamic voltage and frequency scaling (DVFS) and server consolidation are created to optimize power efficiency and reduce thermal output. By taking these steps, data centers can effectively offset thermal risks without sacrificing operational efficiency (Mittal et al., 2014) (Figure 1.9).

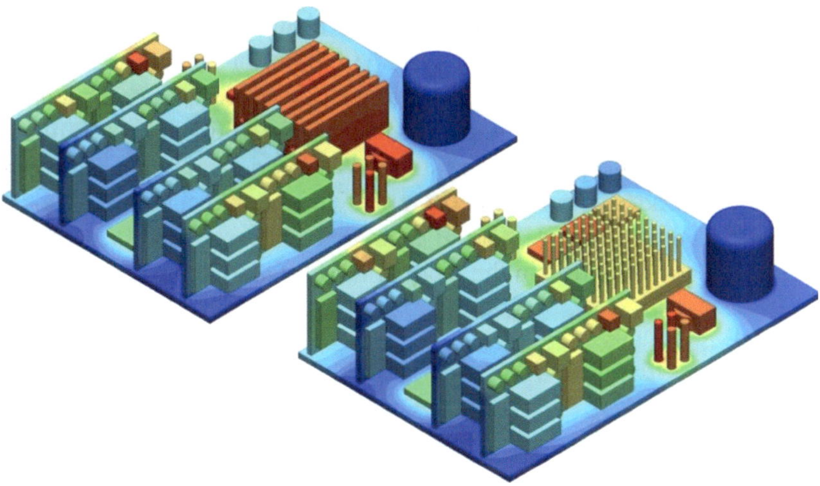

Figure 1.9. Thermal management in electronics.

Source: https://www.simscale.com/wp-content/uploads/2021/02/thermal-management-simulation-with-heat-sinks-e1623794701229.jpg.

Furthermore, the interplay of thermal management and technological advancements in microelectronics is a critical area to be studied. Developments in materials science, for instance, new TIMs, can potentially play a huge role in heat dissipation. They can become a component of key importance in electronic design as they potentially could increase the thermal conductivity between components and heat sinks, hence offering better heat management in reduced designs. In addition, the integration of intelligent thermal management algorithms through real-time inputs has the potential to optimize dynamically cooling solutions in accordance with fluctuating loads and thermal patterns. Not only does this strategic move extend the life of electronic devices, but it also highly optimizes safety by preventing overheating. Continuous investment in research and development of thermal management technology is the future of electronics and guaranteeing sustained sustainable performance in every field (Chen et al., 2019; Mittal et al., 2014).

1.3.1. Challenges in electronic thermal management

Thermal management in electronic devices is a significant challenge that has emerged in the context of increased sophistication and performance demands of modern technologies. The emergence of electric vehicles (EVs), for instance, requires efficient heat dissipation solutions for ensuring maximum performance and longevity of key components such as electric traction motors, power electronic components (PECs), and batteries. Traditional cooling technologies such as active and passive cooling systems might not be sufficient to handle the growing thermal loads, leading to inefficiency and system breakdown. The need for new solutions is evident, with temperature stability directly influencing the reliability and safety of these systems. With various thermal management techniques, the trade-off between cost, effectiveness, and complexity in design is crucial to forming good thermal solutions that are capable of critically addressing electronic issues in the modern world (Gharehghani et al., 2025).

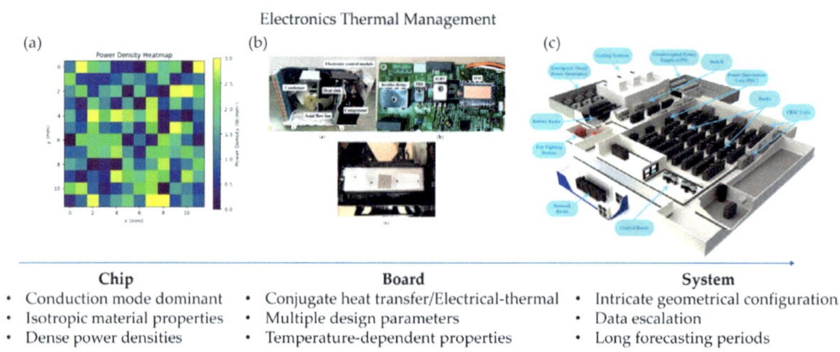

Figure 1.10. Electronics thermal management challenges at the (a) chip; (b) board; and (c) system levels.

Source: https://www.researchgate.net/publication/392010634/figure/fig2/AS:11431281515842108@1750870624645/Electronics-thermal-management-challenges-on-a-Chip-b-Board-and-c-System-Image.tif.

Another future challenge in electronic thermal management stems from the reality that traditional cooling technologies do not have ample ability to reasonably manage the localized heating effects in highly dense electronic systems. The trend of miniaturization in device structure makes thermal resistance greater, and it is not easy to control heat dissipation. In response to these challenges, scientists are looking into newer cooling techniques, including directly injected cooling techniques, which offer a less expensive and scalable alternative. This new approach makes it possible to cool multiple components within electronic packages simultaneously and evenly, leading to a greatly enhanced overall performance. Moreover, the integration of new materials, such as phase change materials (PCMs) and heat pipes, into existing designs can potentially surpass these thermal management limitations, improve energy efficiency, and enhance electronic system life (Brok et al., 2008) (Figure 1.10).

As the demand for high-performance electronics increases, so too will the challenge to dissipate heat. Strategic research and development investment must be made to propel the field and stimulate innovations that can meet these complex thermal challenges. Most importantly, interdisciplinarity in collaboration between other fields of engineering will be necessary in developing complete thermal management solutions that allow sophisticated electronic devices to run safely. Mutual collaboration in such situations can lead to identification of synergies that can maximize the production processes as well as enhance the thermal performance of the products. Solutions to these thermal management problems ultimately play a vital role in sustainable technological development, particularly in industries such as electric mobility where efficient cooling is expected to have a significant impact on performance and safety results (Brok et al., 2008).

1.3.2. Cooling solutions for high-Performance devices

Effective thermal management is crucial to the functionality and longevity of high-performance devices, since high temperatures can significantly impact their operating efficiency and reliability. Among the best-known solutions, the application of advanced cooling techniques stands out, which have progressed a long way to meet the unique thermal requirements of today's electronic packages. An approach that holds great potential is the implementation of directly injected cooling for common package types, including ball grid arrays, so that the process can reduce production stages and component sourcing by a considerable margin. Not only does this novel process simplify manufacturing, but it also improves cooling efficacy, which makes it simple to control heat dissipation on many components at once. As noted by Brok et al. (2008), such flexibility in design offers greater performance, volume, and manufacturing efficiency, positioning directly injected cooling in a position to become a potential competitor to traditional thermal management technologies. In addition, high-performance applications increasingly call for integrated thermal management systems that are adjustable to dynamic operating conditions. Following this demand, recent research has outlined the development of multifunctional façade systems addressing cooling as well as energy efficiency, which reflects the convergence between thermal management and sustainable design. For instance, a novel insulating glass unit based on low-conductance materials offers a radical reduction of residential heating and cooling loads, significantly affecting the energy demand in peak load hours. These new solutions highlight the importance of integrating energy-efficient design into thermal management practices because they provide twin advantages in terms of maintaining device functionality and minimizing environmental impact, thus aligning with broader regulatory and sustainability goals, as outlined by Breshears et al. (2020). Finally, a forward-looking strategy for thermal management includes both active and passive techniques employed in combination to optimize device cooling. Adaptive control systems, for instance, enable real-time adjustment of shading and ventilation based on environmental conditions, resulting in energy saving and cost reduction. Field testing has shown that the systems can save electricity costs significantly, especially on hot summer days. As found in research findings, these trends indicate a paradigm shift towards smart, adaptable design for cooling systems, which are required to meet the growing demands of high-performance devices. Through their transition from fixed parts to integrated solutions acting on instantaneous conditions, as described by Breshears et al. (2020), manufacturers are not only enhancing performance but also ensuring compliance with current energy standards, thus driving thermal management innovation.

1.3.3. Future trends in electronic thermal management

As digital hardware increases in complexity, the requirement for effective thermal management solutions becomes crucial. New trends in electronic thermal management are strongly reflected in new materials and approaches to combat heat problems in high-performance systems. As one of the numerous notable developments, the application of phase change materials (PCMs) is quickly becoming a research priority area. These materials, which store and discharge heat energy in the form of phase transition, have much potential for enhancing electronic device thermal stability. By means of heat dissipation optimization, PCMs have the capacity to enable devices to maintain levels of performance with less risks of overheating, thereby contributing towards overall system reliability and lifespan. Ongoing research into these materials presents the manner in which electronic systems can be configured in the future through predictive thermal management methods. Another major trend is the application of machine learning and advanced data analytics to electronic thermal management. The ability to predict thermal behavior using past history is a paradigm shift from traditional reactive approaches to more proactive, predictive strategies. For instance, as shown by recent research, implementation of machine learning models can make dynamic voltage and frequency scaling within multicore processors optimal and thus boost performance with power conservation (Ababei et al., 2019). This forecasting aspect allows for smart control that can prevent overheating potential through reacting to environmental conditions in real time. In due turn, the integration of these cutting-edge technologies into electronic thermal management systems forms a critical framework for enhanced efficiency and sustainability in the operation of devices in the future. In addition, the shifting paradigm of green thermal management strategies cannot be overlooked. As more profound concerns toward environmental sustainability emerge, the development of thermal management systems in tandem with green technologies has become inevitable. Research indicates the implementation of hybrid-electric engine technologies significantly reduces the emission rate of transport systems, thus proving to be a paradigm shift towards green design practices (Capata et al., 2018). By focusing on minimizing carbon prints and improving energy efficiency in electronic parts through cutting-edge thermal management solutions, manufacturers can remain compliant with regulation as well as appealing to increasingly environmentally conscious consumers. In all, future electronic thermal management trends will definitely drive significant advancement in performance ability as well as sustainability, with a twin focus on technological and environmental progress.

1.4. THERMAL MANAGEMENT IN AUTOMOTIVE APPLICATIONS

Effective thermal management is critical in automotive use, particularly in maintaining internal combustion engines and electric cars at their optimal operating temperatures. As cars become increasingly more performance-capable, the demand for sophisticated thermal management systems has become more urgent. Aside from governing engine temperatures, these systems also help the vehicle efficiency overall by minimizing energy loss and preventing overheating, which results in component failure. Incorporation of thermal management methods is synergistic with various engine management systems, necessitating advanced algorithms for calibration and optimization. Approaches such as meta-heuristic algorithms offer positive new advancements in engine management and design, allowing for real-time adjustment with the ability to improve thermal efficiency while reducing emissions, pointing towards the key milestone between temperature control and environmental issues in modern automotive engineering (Tayarani et al., 2015). The technological advancement in automobiles has undoubtedly directed focus towards electrification and hybrid technology, revealing a growing demand for sophisticated thermal management methods. As electric vehicles (EVs) lack traditional internal combustion engines, thermal management becomes the key to regulating battery temperatures, thus impacting vehicle performance, safety, and longevity. Good heat dissipating techniques, like liquid cooling and phase change materials, are needed to enable batteries to operate within the appropriate ranges of temperature. Moreover, efficient thermal management can improve energy efficiency by preventing loss of energy in terms of overheating or overcooling. Here, research and innovation are continuously evolving, centering on topics such as weight reduction and integration of the system to enhance the performance of EVs and hybrids (Addepalli et al., 2018) (Figure 1.11).

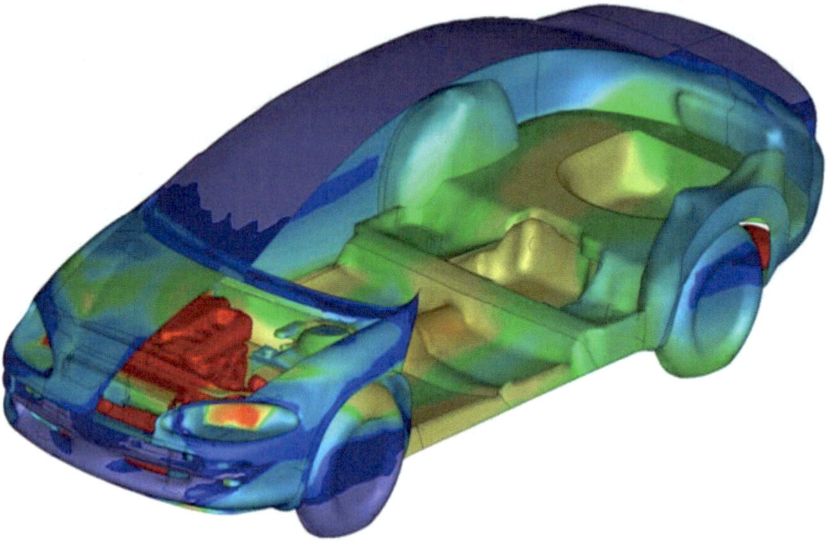

Figure 1.11. Automotive thermal management.

Source: https://econtroldevices.com/wp-content/uploads/2022/02/AutomotiveUnderhoodThermalManagement.jpg.

Non-destructive test (NDT) techniques are presently very relevant in the context of automotive thermal management. With growing complexity and integration in thermal management systems, quality assurance through NDT techniques becomes a matter of great significance. NDT techniques facilitate efficient assessment of thermal management components without compromising on reliability and performance while not impacting structural integrity. With such systems becoming miniaturized, NDT methods also fall under weight and size limits, thus necessitating the development of new inspection techniques. Development in NDT not only guarantees enhanced levels of safety but also averts failure in thermal management components that will lead to catastrophic malfunctioning. Incorporating NDT into common practices is a testament to the commitment of the automotive industry to quality and safety with continued development of thermal management technology (Addepalli et al., 2018; Tayarani et al., 2015) (Figure 1.12).

Figure 1.12. Non-destructive testing in automobiles.

Source: https://assets.onestopndt.com/p/uploads/ndt-articles/original/automotive-industry.jpg.

1.4.1. Importance in vehicle performance and safety

Effective thermal management is vital in optimizing vehicle performance and safety because it has a direct impact on the most critical operating parameters. Cars, especially cars employed in aerospace activities, experience severe temperature conditions, and this can affect not only the mechanical system but also the comfort and health of the occupants. The integration of advanced thermal management systems enables better control of heat dissipation, which directly correlates to enhanced engine efficiency and reliability. This is particularly important in launch vehicles, where the structures must be flexible enough to accommodate a large range of payloads while maintaining performance under rigorous thermal environments. By good thermal load control, designers can ensure critical components are at optimal temperature levels, which reduces the likelihood of failures that could compromise mission success. For instance, when a launch vehicle core stages overheat due to inadequate thermal control, it results in catastrophic flight failures, thereby the importance of thermal management for safety and performance (Burton et al., 2025). In addition to maximizing performance, thermal management heavily influences making vehicles safe from overheating and component degradation. As vehicles become more advanced with complex systems and materials, the need for effective thermal control is even more pertinent. Current research estimates that upcoming technologies focused on thermal management can bring dramatic gains in safety and operating efficiency. For example, the development of advanced cooling techniques and materials that withstand elevated temperatures can reduce mechanical failure in critical components. Furthermore, such advancements are not only for improving the vehicles' safety but also some of the designs that allow for more efficient operations, thereby contributing positively to the general environmental footprint of transport systems. Safety components are, therefore, incorporated with efficient thermal management, supporting the contention that efficient thermal systems play an important role in the reliability of modern vehicles (Albers et al., 2025). With the automobile and aviation industries continually raising the bar on innovation, the application of thermal management becomes increasingly vital. The introduction of new technologies will make the vehicles safer as well as more efficient, with enhanced performance capabilities. New designs focused on sustainability and environmental compatibility are indicative of the industry's determination to maintain high levels of safety along with heightened performance. By investing in innovative thermal management technology, producers are in a position to launch vehicles that better respond to the challenges of modern transportation requirements while preventing potential risks from temperature fluctuations. Both safety and efficiency are enhanced in contributing to competitiveness in the aerospace and automotive industries, placing them at the vanguard of progress. Its use in future vehicle application and design and functionality will largely be based on the thermal management principles to deliver unprecedented performance and safety and thereby become increasingly central to vehicle design and functionality development (Albers et al., 2025; Burton et al., 2025).

1.4.2. Techniques used in automotive thermal management

Improved thermal control in the automotive sector is crucial to attain maximum vehicle performance, enhance safety, and prolong the life of components. Among the most renowned solutions is the use of advanced materials with superior thermal conductivity and thermal resistance. High-temperature polymers, aluminum alloys, and thermally resistant composites are just a few examples of materials employed, for instance, to design components such as engine blocks, exhaust systems, and heat exchangers. Not only do the materials help in dissipating heat more effectively but also reduce the vehicle's weight, which serves to enhance fuel efficiency. Using the materials in the design of automobiles underscores the importance of selecting the appropriate material as a fundamental technique in managing a vehicle's heat. By proper engineering and selection of materials, car manufacturers can minimize the risks of overheating and run optimally even under varying conditions. Active cooling systems are another significant technique employed in thermal management of automobiles, relying on various mechanisms such as liquid cooling and vapor cycle systems to get rid of excess heat from critical components, especially in high-performance cars. For instance, liquid cooling systems circulate coolant through hoses and radiators, basically staying hot by having the heat created by the engine transferred and released through the radiator (Figure 1.13).

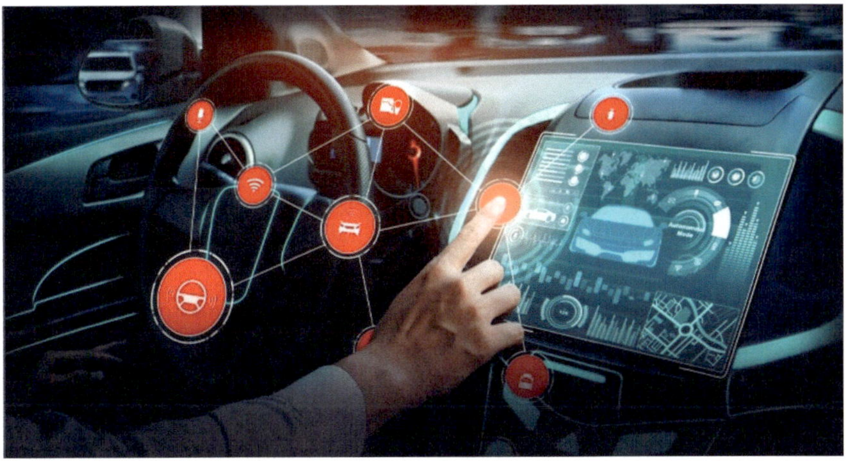

Figure 1.13. Effective thermal management in the automotive industry is critical for optimizing vehicle performance, enhancing safety, and prolonging component lifespan.

Source: https://mgchemicals.com/wp-content/uploads/2021/10/Thermal-Management-In-Automobiles-1.jpg.

Moreover, advancements in coolant formulations, which involve the typical addition of additives to enhance thermal properties, further improve the efficiency of the cooling systems. Not only does this active temperature control maintain the engine within tolerance levels of temperature but also enhances the longevity of components. With increasing complexity of automotive systems, active cooling technologies have to be incorporated for the purposes of performance and reliability, creating constant innovations in this critical aspect of thermal management. Recycling and reuse of materials used in automotive thermal management is finally gaining prominence, and this has environmental as well as practical benefits. It is a strategy that is consistent with legislative policies aimed at reducing waste and sustainability in the automotive sector. Recyclable products, particularly composite and polymer-derived components, may be recycled and reused, lowering the environmental profile of car manufacturing. Such ventures of creating efficient recycling schemes are necessary as they can help overcome the issue of separating combined polymer waste into individual materials. Migrating towards more eco-friendly processes in thermal management not only fulfills environmental requirements but also enhances overall efficiency in car manufacture. In total, the methods of material innovation, active cooling technologies, and recycling are a major factor in determining the direction of thermal management in the auto sector into the future, such that it is optimally aligned with performance and environmental requirements.

1.4.3. Regulatory standards and environmental considerations

Standards in regulations play a crucial role in setting the design and operation of thermal systems, particularly in those industries with considerable impacts on the environment. Regulatory frameworks guarantee that thermal management technologies are not only compliant with the law but also with sustainable processes. In California, for instance, power plant siting is an integrated analysis reflective of the state's energy efficiency and environmental responsibility. Such requirements do usually develop through convergence by government agencies, developers, and public interest groups toward the aim of an equitable plan that protects both public health and environmental wholeness (Paul et al., 2002). Consequently, regulatory compliance impacts design choice, operational procedures, and overall efficiency in thermal management solutions. Additionally, environmental factors are being more commonly incorporated into the regulatory standards framework, in particular, energy-efficient, and sustainable practices. One of the exemplary cases of this in action is the screening of energy-efficient retrofit technologies for use in industrial buildings. By giving highest priority to the installation of new thermal management technology, policymakers aim to save energy and reduce greenhouse gas emissions. Top-priority energy-efficient retrofitting aligns with broader environmental goals, prioritizing reducing the carbon intensity of production processes (Fisher et al., 2011). This intersection of regulation and environmental protection speaks to the need for organizations to keep up with evolving standards that govern thermal management practice and adopt innovative solutions that align with such standards. As regulatory bodies continue to develop and implement standards in coordination with environmental needs, thermal management-practicing organizations must walk the line between compliance and innovation. The interplay between technology and regulation is a prime industry practice determinant. Companies that prepare for these standards not only prevent risks of non-compliance but are also at the forefront of sustainability. To do so frequently requires investing in research and development of efficient thermal systems and retrofitting existing plants to exceed or meet regulatory levels. By merging business objectives and regulatory compliance and environmental considerations, businesses can develop a sustainability culture without decreasing their thermal management practices, ensuring compliance and corporate responsibility are maintained in the evolving arena of energy management.

1.5. THERMAL MANAGEMENT IN BUILDING DESIGN

Ideal building design is based on effective thermal management to achieve maximum energy efficiency and warm indoor conditions. Thermal management is based on the heat manipulation principle to achieve maximum efficiency while reducing heating and cooling loads. Knowing thermodynamics is the foundation, for it educates architects and engineers on the materials' nature and climatic interaction. Various techniques, including insulation, ventilation, and thermal mass, can be employed to optimize the energy efficiency of a building. Improved insulation minimizes heat transfer, and deliberate window placement can facilitate natural ventilation and hence induce passive cooling. Therefore, a holistic approach that includes these must be prioritized at the design phase to ensure that buildings meet not only short-term thermal comfort requirements but also contribute to long-term energy efficiency goals. Thermal control practice development has more recently focused on net zero energy buildings (NZEBs) as the norm for sustainable building. NZEBs try to produce as much energy as they consume over one year, offering a viable alternative to the problems of conventional building energy consumption. It is a comprehensive mix of energy-saving technologies and renewable energy installations, tailored to the regional climatic conditions, which makes a building an NZEB. Passively designed features, such as daylight and natural ventilation, are particularly important for buildings located in regions of hot and humid climates. By reducing the energy intensity to less than 100 kW-hours per square meter of floor space, many NZEBs can demonstrate that there is efficient thermal management, achieving better energy performance while some buildings gain a "net-positive energy" rating. As such, NZEBs development focuses on the fundamental importance of thermal management to modern building design (Feng et al., 2019). Heat management is also essential in retrofitting existing buildings to meet contemporary energy levels. The practice of retrofitting, aside from maximization of energy efficiency, also addresses issues of historic preservation in order to preserve the aesthetic and functional nuances of existing old buildings. Through the screening of energy-efficient solutions, stakeholders are able to pick suitable interventions that minimize energy consumption while maximizing occupant comfort. In addition, new technologies such as smart systems that monitor energy consumption in real-time can be integrated into retrofitted buildings. Such technological innovation not only enhances the efficiency of thermal control systems but also encourages green culture in the construction sector. Therefore, retrofitting measures are crucial in reducing overall carbon footprints and establishing a culture of sustainability in the urban environment, emphasizing the need to make commitments to advanced thermal management practices in new buildings as well as existing constructions (Fisher et al., 2011) (Figure 1.14).

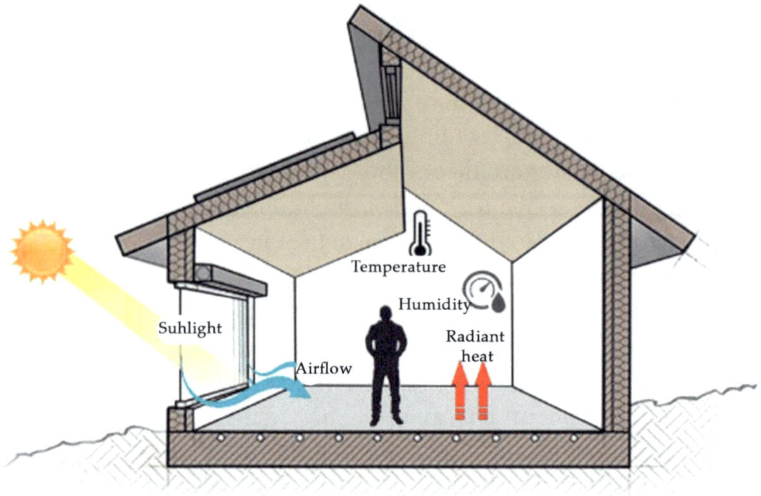

Figure 1.14. Factors influencing residential building thermal comfort.

Source: https://www.researchgate.net/figure/Factors-that-influence-residential-building-thermal-comfort_fig1_356656452.

1.5.1. Role of thermal management in energy efficiency

Thermal management is the most significant factor in enhancing energy efficiency across many applications, particularly building systems. Effective temperature regulation not only delivers occupant comfort but also reduces energy usage. As buildings contribute to the majority of global energy use, the use of advanced thermal management methods becomes unavoidable. Techniques such as predictive modeling, which incorporate data-driven methods such as support vector machines, are pivotal as far as optimal heating, ventilation, and air conditioning (HVAC) system optimization is concerned. An instance involves the utilization of relevant data selection methods, which can result in more accurate energy consumption predictions, thereby enabling better control measures, such as presented in research findings from Ecole des

Mines de Nantes (Corre et al., 2015). Thus, improving thermal management measures culminates in lowering operational expenses, highlighting its crucial role towards the attainment of energy-efficient building designs. The connection between thermal management and energy efficiency is not solely within individual buildings but also affects greater energy systems, such as multicarrier energy grids. Within such grids, thermally coordinated energy hubs optimize use of diverse carriers of energy in such a manner as to achieve optimal operation under varying demands and resources. By utilizing cutting-edge concepts like the virtual energy hub (VEH), operators can participate optimally in local thermal energy markets, enhancing efficiency and income. The hubs learn from existing data in real time, optimizing thermal management to address uncertainties, particularly from renewable power sources. Application of strategies that are designed for both risk-averse and risk-seeker conditions enables energy suppliers to manage their thermal resources in an adaptive way, significantly improving overall energy efficiency (Jadidbonab et al., 2021). Therefore, thermal management development in these disparate systems holds a great promise for energy savings and increased sustainability. Further, efficient thermal management techniques enable the use of renewable sources of energy, a necessary component in maintaining energy efficiency in contemporary systems. With augmented use of fluctuating renewable resources like solar and wind, the necessity for a responsive thermal management technique is even more paramount. With the use of novel storage technologies, for instance, compressed air energy storage, thermal management systems can buffer the impact of fluctuating energy production. This responsiveness allows energy systems to capitalize on periods of high renewable output, hence lowering fossil fuel usage and improving overall efficiency. Thermal system optimization also ensures the maximum utilization of available power, as is seen with the innovation in local energy markets. Optimal thermal operation, therefore, not only supports building efficiency but also forms a greener energy scenario through strategic optimization in both energy markets and configurations (Jadidbonab et al., 2021).

1.5.2. Passive design strategies for buildings

In building design, thermal control is addressed effectively with the aid of passive strategies. They reduce a building's reliance on mechanical heating and cooling by utilizing natural phenomena. Using features such as thermal mass, sufficient insulation, and properly placed windows achieves optimal temperature control. Thermal mass buildings can acquire heat throughout the day and release it at night when there are lower temperatures, reducing the temperature fluctuation and enhancing comfort. Moreover, the window orientation and utilization of solar shading devices are extremely crucial elements to manage the solar heat gain, particularly in hot and humid climates. These methods not only cut down energy consumption but also adhere to the contemporary demand for sustainable building practices, proving passive design to be effective and viable within energy-efficient design (Feng et al., 2019). Additionally, close examination of case studies indicates that advanced passive design approaches have the potential to make significant impacts on thermal performance in climates (Figure 1.15).

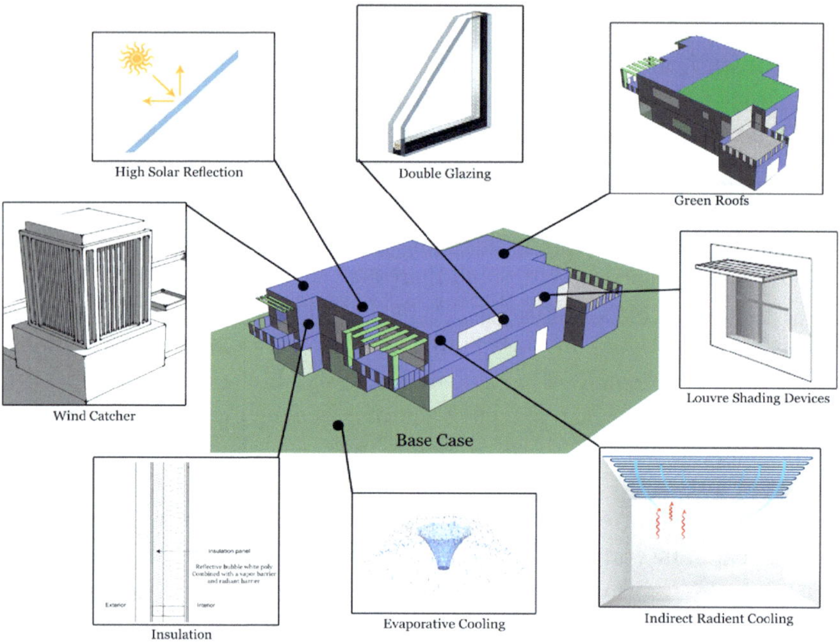

Figure 1.15. Passive design techniques support effective building design.

Source: https://archiroots.com/powerful-passive-design-strategies/

For instance, net zero energy buildings (NZEBs) frequently utilize passive strategies such as natural ventilation and daylighting, both contributing to the reduction of overall energy consumption. The evaluation of 34 NZEB case studies identifies the importance of climate-appropriate passive technology in modern building practice today. These strategies enable designers to maximize energy use while still being environmentally compliant. However, under the analysis, while a majority of NZEBs have great energy performance, not all NZEBs maintain high energy efficiency. Rather, this gap indicates a strong imperative to conduct careful analysis of in-use energy use and also the implementation of design principles (Feng et al., 2019). The symbiosis of architecture and climatic influences dictates the paradigm shift towards low-energy design. Allowing for the fact that high levels of air conditioning are viewed as inevitable, it is then necessary to return to simple principles of passive design. This renaissance centers on taking advantage of architectural elements that resist reliance on mechanical systems, constructing an environment that harmonizes energy conservation and comfort for occupants. By assigning priority to primitive environmental thinking, architects are prompted towards embracing easy passive strategies that can contribute significantly to enhancing thermal control. Advanced environmental modeling techniques also make it possible to correctly analyze sunlight, temperature, and airflow early in the design process, making it possible to consider wisely passive approaches. Thus, the architectural environment can get remolded to integrate functionality along with sustainability, giving a solid base for future building structures.

1.5.3. Technologies for active thermal control in buildings

Technologies for active building thermal control continue to advance, and their developments significantly contribute to energy efficiency and comfort of occupants. Of all these technologies, heat pumps have become a leading technology in addressing indoor temperature moderation as well as enhancing energy consumption optimization. Heat pump functioning can be tactically adjusted to suit the variable nature of energy systems, specifically when combined with renewable energy systems. One study emphasized the potential for control manipulation of these systems, citing that properly executed thermal control will provide a more stable and secure supply of electrical energy. This flexibility not only addresses energy grid reliability but also optimizes overall building system performance. Therefore, the convergence of heat-driven technologies reflects a revolutionary method in the area of thermal management that meets environmental as well as user requirements, as the harmony between comfort and efficiency becomes increasingly important. Along with heat pumps, new insulation materials and thermal buffer technologies have become key elements to active thermal management. Enhanced insulation reduces heat loss and gain and accordingly diminishes the strain on heating and cooling systems. The integration of thermal storage systems provides considerable flexibility in energy use patterns. Emerging research has suggested that increased levels of insulation coupled with the application of sophisticated thermal buffering measures can result in astronomical improvements to energy performance. This integration allows for considerable alterations in operation time for heating equipment, including heat pumps, without compromising comfort. For example, it is possible to achieve up to six-hour time shifts in heating operations, thus enabling additional energy savings and operational efficiency. These approaches demonstrate the dynamic synergy between building design and energy technology, setting the stage for more intelligent, more sustainable buildings. Even with the promising developments, the performance of active thermal control technologies is impacted heavily by the building stock in place and its thermal properties. Previous studies find that there are constraints on how much heating operating times can be changed without affecting occupant comfort, particularly without major improvements to insulation and efficiency of the system. The results of a simulation-based method reported time shifts of merely one to two hours in traditional buildings, highlighting the need for retrofitting and improving thermal characteristics to provide more flexibility. The challenge lies not only in deploying new technologies but also in screening and retrofitting existing buildings for optimal performance. This reflects the importance of comprehensive energy-efficient technologies within retrofit strategies, as explored in related literature, which emphasizes targeted improvements for industrial and residential environments (Fisher et al., 2011). Therefore, the future course in thermal management will rely on a systems understanding of these technologies and their application within the built environment.

SUMMARY

In summarizing thermal management, it is evident that effective thermal control is essential to the performance and longevity of electronic devices. The integration of engineering disciplines offers creative solutions to cooling technology, particularly in high-performance electronic devices. The integration of direct cooling technologies, as is noted in the case study, offers the opportunity for cost-effective innovations that minimize manufacturing complexity. Through the reduction of the number of necessary components and production processes, designers can enhance flexibility in design, leading to products with better performance, reduced weight, and increased production efficiency. This emerging paradigm emphasizes the importance of multidisciplinary integration towards achieving paradigm jumps in thermal management. In addition, the industrial symbiosis principle is essential in order to achieve optimized thermal management practices. By creating synergies between industries, organizations can facilitate the exchange of resources that enable more sustainable performance. This "whole-system approach" not only minimizes waste but also encourages reuse and recycling of material, energy, and water, thereby enabling enhanced thermal management outcomes. By agglomerating geographically and strategically, industries can form networks geared to emissions and resource minimization. The thermal management innovation is achievable when industries make use of by-products and waste products, making the overall process more efficient and good for the environment. Finally, technological innovations in thermal management are necessary to ensure continued growth of the vast electronics industry. The case studies and research demonstrate a clear trajectory, emphasizing interdisciplinary collaborative efforts. Through continuous innovation in cooling technologies and research into new ways of resource efficiency, such as the type outlined in industrial symbiosis, the industry can respond to current and future demands. Enhanced thermal management needs to be a continued emphasis, with a vision that will enable innovation to thrive as well as sustainable initiatives. As we move forward, it is imperative that we share a common vision for integrating these innovations into holistic industrial approaches so that thermal management remains a cornerstone of technology innovation.

Thermal management vocabulary has dramatically shifted, a reflection of increasing sensitivity to energy efficiency and environmental sustainability in building design. A notable undertaking in this regard is the St Nicholas Court Project, its aim being to enhance energy performance requirements for new homes, particularly timber-framed dwellings. The project, funded by the Joseph Rowntree Foundation, was significant in creating qualitative and quantitative evidence to guide Part L updates of the Building Regulations. All phases of the project, including design and communication, were significant in addressing the thermal performance of domestic buildings in urban areas, with a focus on the thin line between innovative design and compliance. This type of activity emphasizes the need to integrate thermal management designs at early project phases of planning and construction, thus making sustainable housing options available that align with current energy performance requirements (Bell et al., 2003). In modern technology, especially in multiprocessor SoCs, thermal management plays a contrasting but equally critical role. The use of multiple heterogenous cores in such systems puts a priority on power use and consequent heat production as issues. With increasing temperature, so increases leakage power, creating a complicate feedback cycle that can destabilize system operation. Current research indicates needed conditions for thermal different and security, providing an equation to compute maximum run-time thermally safe power utilization. This innovation is particularly vital in ensuring dependable operation, as demonstrated by hardware measurements on contemporary cell phone processors, which showed a satisfactory predictive accuracy of the stable temperature point (Bhat et al., 2017). The confluence of thermal management and computational efficiency portrays a new field that requires continuous research and development. Moreover, the impact of thermal control not only is within individual buildings and equipment but also has wider implications for city planning and energy networks. The experience of such projects as the St

Nicholas Court Development underlines the value of an integrative approach to thermal performance, combining architectural, regulatory, and technological systems. With urban areas still expanding and the demand for energy-efficient measures increasingly present, the understanding gained from both local and computational milieus will be of pivotal importance. Effective thermal management not only yields improved performance for buildings but also works to reduce overall carbon footprints, and with this making the argument for its importance in future developments impossible to resist. Ensuring thermal strategies are at the forefront of technology and design will be a key element in designing an energy-conscious society, which will ultimately bring benefits to not just people but also nature (Bell et al., 2003; Bhat et al., 2017).

REFERENCES

1. Ababei, C., & Moghaddam, M. G. (2019). A survey of prediction and classification techniques in multicore processor systems. https://core.ac.uk/download/322825320.pdf (accessed on 23rd October 2025).
2. Addepalli, S., Bilik, M., Broberg, D., Ditchburn, J., Garnier, B., Jolly, M., Li, X., et al. (2018). A review of miniaturized non-destructive testing technologies for *in-situ* inspections. https://core.ac.uk/download/323306202.pdf (accessed on 23rd October 2025).
3. Albers, J. A., & Zuk, J. (2025). Aircraft technology opportunities for the 21st century. https://core.ac.uk/download/pdf/42830883.pdf (accessed on 23rd October 2025).
4. Ali, M. (2015). Developing the knowledge-based human resources that support the implementation of the National Dual Training System (NDTS): Evaluation of TVET teacher's competency at MARA Training Institutions. https://core.ac.uk/download/12006712.pdf (accessed on 23rd October 2025).
5. Andrews, D. L. (2008). Energy harvesting: A review of the interplay between structure and mechanism. https://core.ac.uk/download/2761160.pdf (accessed on 23rd October 2025).
6. Aydelott, J. C., Carney, M. J., & Hochstein, J. I. (1985). NASA Lewis Research Center low-gravity fluid management technology program. https://core.ac.uk/download/pdf/42843059.pdf (accessed on 23rd October 2025).
7. Barrios Padura, Á., Lizana Moral, F. J., Molina Huelva, M., Serrano Jiménez, J., et al. (2019). Decision-support method for profitable residential energy retrofitting based on energy-related occupant behaviour. https://core.ac.uk/download/286563277.pdf (accessed on 23rd October 2025).
8. Beck, S. M., Breckenridge, R. A., Clark, L. G., Collier, D., et al. (2025). Technology for the future: In-space technology experiments program, part 2. https://core.ac.uk/download/pdf/42817270.pdf (accessed on 23rd October 2025).
9. Bell, M., Lowe, R. J., & Roberts, D. (2003). Developing future UK energy performance standards: The St Nicholas Court project, final report. https://core.ac.uk/download/29018940.pdf (accessed on 23rd October 2025).
10. Bhat, G., Gumussoy, S., & Ogras, U. Y. (2017). Power-temperature stability and safety analysis for multiprocessor systems. http://arxiv.org/abs/1806.06180 (accessed on 23rd October 2025).

11. Boardman, B., Clarke, J. A., de Dear, R., Fanger, P. O., Kelly, N. J., et al. (2013). Assessing heat pumps as flexible load. https://core.ac.uk/download/9843648.pdf (accessed on 23rd October 2025).
12. Bordons Alba, C., Guerra Macho, J. J., Rosa Iglesias, M. F., Valverde Isorna, A., et al. (2016). Energy management strategies in hydrogen smart-grids: A laboratory experience. https://core.ac.uk/download/288003657.pdf (accessed on 23rd October 2025).
13. Brataas, A., Brataas, A., Einstein, A., Slachter, A., Slachter, A., Sugihara, A., Kovalev, A. A., et al. (2011). Spin caloritronic. https://core.ac.uk/download/pdf/148239576.pdf (accessed on 23rd October 2025).
14. Breshears, J., Curcija, C., Deisler-Moroder, D., Fernandes, L., et al. (2020). High-performance integrated window and façade solutions for California. https://core.ac.uk/download/286728862.pdf (accessed on 23rd October 2025).
15. Brok, G. J. H. M., Legtenberg, R., Mannak, J. H., Wits, W., et al. (2008). Novel cooling strategy for electronic packages: Directly injected cooling. https://core.ac.uk/download/pdf/11480087.pdf (accessed on 23rd October 2025).
16. Burton, L., Drake, K., Kubinski, R., Mann, R., et al. (2025). Final design report of a personnel launch system and a family of heavy lift launch vehicles. https://core.ac.uk/download/pdf/42813721.pdf (accessed on 23rd October 2025).
17. Cain, R., Giudice, S., Harrison, D. J., Jennings, P., et al. (2011). Behaviour changes at work: Empowering energy efficiency in the workplace through user-centered design. https://core.ac.uk/download/2733853.pdf (accessed on 23rd October 2025).
18. Capata, R. (2018). Urban and extra-urban hybrid vehicles: A technological review. https://core.ac.uk/download/188826162.pdf (accessed on 23rd October 2025).
19. Chen, Y., Egan, D., Feng, X., Grimes, C., Gurrum, S. P., Huang, H., Jaworski, C. M., et al. (2019). Thermal analysis of lithium-ion battery-equipped smartphone explosions. https://core.ac.uk/download/289095013.pdf (accessed on 23rd October 2025).
20. Dehghanghadikolaei, A., Emamian, S. S., Fotovvati, B., Gisario, A., et al. (2019). The potential of additive manufacturing in the smart factory industrial 4.0: A review. https://core.ac.uk/download/231748826.pdf (accessed on 23rd October 2025).
21. Fayaz, M., Lajis, A., Nasir, H., Shah, L. A., et al. (2019). A review on energy consumption optimization techniques in IoT based smart building

environments. https://core.ac.uk/download/300473894.pdf (accessed on 23rd October 2025).

22. Feng, W., Hao, B., Ji, H., Lau, A. K., et al. (2019). A review of net zero energy buildings in hot and humid climates: Experience learned from 34 case study buildings. https://core.ac.uk/download/323310631.pdf (accessed on 23rd October 2025).
23. Ferhati, A., Bejan, A., Bergles, A. E., Reay, D. A., Lewis, J. S., McGlen, R. J., Shah, R. K., et al. (2015). Single-phase laminar flow heat transfer from confined electron beams enhanced surfaces. https://core.ac.uk/download/29204056.pdf (accessed on 23rd October 2025).
24. Fisher, D., Hand, J., Liesen, R., Wilson, E., et al. (2011). Screening of energy efficient technologies for industrial buildings' retrofit. https://core.ac.uk/download/9033399.pdf (accessed on 23rd October 2025).
25. Gharehghani, A., Könnö, J., Mahmoudzadeh Andwari, A., Moradi, R., et al. (2025). Advanced thermal management strategies for electric vehicles: Enhancing efficiency, reliability, and performance. https://core.ac.uk/download/653750538.pdf (accessed on 23rd October 2025).
26. Han, X. H., Jacobi, A., Park, Y. G., Sommers, A., et al. (2012). A review of metal foam and metal matrix composites for heat exchangers and heat sinks. https://core.ac.uk/download/55763987.pdf (accessed on 23rd October 2025).
27. He, J., & Tritt, T. M. (2017). Advances in thermoelectric materials research: Looking back and moving forward. *Science, 357*. https://doi.org/10.1126/science.aak9997 (accessed on 23rd October 2025).
28. Hibbard, P. J., & Tierney, S. F. (2002). Siting power plants: Recent experience in California and best practices in other states. https://core.ac.uk/download/71354895.pdf (accessed on 23rd October 2025).
29. Jadidbonab, M., Marzband, M., Mohammadi-Ivatloo, B., Siano, P., et al. (2021). Short-term self-scheduling of virtual energy hub plant within thermal energy market. https://core.ac.uk/download/288430500.pdf (accessed on 23rd October 2025).
30. Joos, J., Poelman, D., Sijbom, H., Smet, P., et al. (2017). K_2SiF_6: Mn^{4+} as a red phosphor for displays and warm-white LEDs: A review of properties and perspectives.
31. Kelsall, G., Olaleye, A. K., & Wang, M. (2015). Steady state simulation and exergy analysis of supercritical coal-fired power plant with CO_2 capture. https://core.ac.uk/download/151159412.pdf (accessed on 23rd October 2025).

32. La Marca, F. (2015). Treatment and valorization plants in materials recovery supply chain. https://core.ac.uk/download/54525197.pdf (accessed on 23rd October 2025).

33. Lazzarin, R. M., Mancin, S., Noro, M., Righetti, G., et al. (2018). Hybrid PCM–aluminum foams' thermal storages: An experimental study. https://core.ac.uk/download/168402984.pdf (accessed on 23rd October 2025).

34. Le Corre, O., Elmitri, M., Kling, W. L., Lacarrière, B., et al. (2015). Support vector machine in prediction of building energy demand using pseudo dynamic approach. https://core.ac.uk/download/50614878.pdf (accessed on 23rd October 2025).

35. Lipson, H., & Malone, E. (2005). Freeform fabrication of ionomeric polymer-metal composite actuators. https://core.ac.uk/download/287648532.pdf (accessed on 23rd October 2025).

36. Migra, R. P. (2025). Microgravity fluid management requirements of advanced solar dynamic power systems. https://core.ac.uk/download/pdf/42837294.pdf (accessed on 23rd October 2025).

37. Mittal, S. (2014). Power management techniques for data centers: A survey. http://arxiv.org/abs/1404.6681 (accessed on 23rd October 2025).

38. Napier, J. D. (2015). Climate based facade design for business buildings with examples from central London. https://core.ac.uk/download/29176762.pdf (accessed on 23rd October 2025).

39. Smith, S. P. (2015). The recognition of fires originating from photovoltaic (PV) solar systems. https://core.ac.uk/download/30713291.pdf (accessed on 23rd October 2025).

40. Tarverdi, K. (2009). Improving the mechanical recycling and reuse of mixed plastics and polymer composites. https://core.ac.uk/download/337221.pdf (accessed on 23rd October 2025).

41. Tayarani-N, M. H., Xu, H., & Yao, X. (2015). Meta-heuristic algorithms in car engine design: A literature survey. https://core.ac.uk/download/132585049.pdf (accessed on 23rd October 2025).

42. Valera-Medina, A., Xiao, H., Owen-Jones, M., David, W. I. F., & Bowen, P. (2018). Ammonia for power. *Progress in Energy and Combustion Science, 69*, 63–102. https://doi.org/10.1016/j.pecs.2018.07.001 (accessed on 23rd October 2025).

CHAPTER 2

FUNDAMENTALS OF HEAT TRANSFER

Contents

INTRODUCTION

Heat transfer theory forms the basis of understanding numerous physical phenomena with extensive application domains over various disciplines, including engineering, physics, and materials science. Heat is transferred in three primary modes: conduction, convection, and radiation. Each mode operates under distinct mechanisms that govern the way thermal energy is transferred within or between materials. Conduction, for instance, is the passing of heat via direct contact of material based on Fourier's law. Convection is fluid transport that takes away thermal energy from or adds it to surfaces. The third form of heat transfer is by radiation, which is transmission of heat through electromagnetic waves with no medium involved that enables the transmission of heat through a vacuum. A detailed comprehension of these mechanisms is required by scientists and engineers in order to design efficient systems for thermal energy management. In the majority of industrial processes, the application of efficient heat transfer principles is most important to increase energy efficiency and maximize the process. For example, heat recovery from exhaust gases is an excellent avenue for energy conservation in many industries. Issues such as gases' inability to withstand high resistance to heat transfer and particulates restrict the effectiveness of traditional heat recovery systems (Atkins et al., 2011). Solutions such as advances in the design of heat exchangers have been created to overcome the constraints. By employing various geometries of tubes, researchers have examined how variations influence heat transfer efficiency with reduced fouling. The findings not only pave the way for enhanced industrial processes but also promote environmental sustainability by conserving energy in energy-intensive processes. Furthermore, understanding the heat transfer science has key implications in the design of thermal management technologies. Chemical manufacturing processes, for example, comprise manufacturing processes of various thermally driven processes such as chromium trioxide. Regulation of heat transfer processes is therefore a prerequisite. This specific chemical process is one that demands rigorous thermal regulation due to the fact that it is a continuous process and high thermal charges in the reactions involved within it (Engg et al., 2010). Proper thermal management can optimize production expenditures as well as product quality, and this points out the important role of heat transfer principles in process engineering. With the advancements in technology, engineers must be competent in the application of heat transfer principles to enhance system performance across applications, thus facilitating innovations that drive the industry forward.

Definition of Heat Transfer

Heat transfer is a fundamental physical phenomenon that accounts for the movement of thermal energy from a system to another due to a temperature gradient. It is a prerequisite process in a majority of scientific and engineering applications, particularly where thermal management becomes an extremely crucial aspect. Heat transfer phenomena can be categorized into three primary modes: conduction, convection, and radiation. Conduction is the transfer of heat by solid materials through direct contact of particles, generally described by Fourier's law. Convection is the transport of fluid, the accompanying heat as fluid flow, and can further be classified as forced and natural convection. Lastly, radiation is a unique kind of heat transfer without a medium and by electromagnetic waves, characterized by the Stefan-Boltzmann law. Understanding these mechanisms is very important in managing thermal energy efficiently in practical applications. The theory of heat transfer extends beyond the determination of its processes; it entails their application and impact in practical conditions.

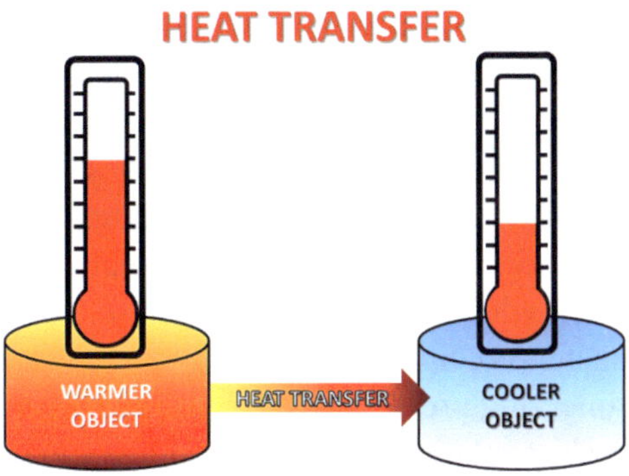

Figure 2.1. Heat transfer.

Source: https://o.quizlet.com/o1FkGmqd6O.fSHml.BPtcQ_b.jpg.

For instance, the emergence of energy-conserving technologies, particularly for industrial structures, is highly reliant on comprehension of the principles of heat transfer. Insulation, effective ventilation, and thermal energy recovery systems are just a few of the steps employed to conserve heat and maximize energy efficiency. Comprehension of the way thermal energy transports itself is the foundation on which engineers develop solutions to improve comfort as well as energy usage and consumption. As pointed out in the study of energy-saving technologies (Fisher et al., 2011), determining the most effective way to regulate heat transfer can lead to colossal savings in operational costs and environmental impact. Beyond conventional know-how, optimization of heat transfer mechanisms plays a fundamental role in the creation of new technologies. One such example is in thermoelectric devices, for which the relation between entropy production and heat transport governs performance optimization. By analyzing irreversibility associated with heat flow and thermodynamic laws, researchers have been able to determine conditions under which heat transfer is optimal. The thermoelectric performance assessment emphasizes the necessity for minimizing entropy production in attaining optimum levels of thermal efficiency (Goupil et al., 2011). Such ideas are invaluable in generating new materials and devices that utilize heat transfer principles to achieve benefits, demonstrating the enduring relevance and usability of heat transfer theory in current engineering challenges (Figure 2.1).

Importance in Engineering and Science

In engineering and scientific applications, heat transfer mechanisms are the focus of system design and analysis. Understanding mechanisms of conduction, convection, and radiation enables engineers to optimize thermal performance in a wide range of applications, from thermal building insulation of homes to industrial process heat exchangers. The ability to transfer heat effectively not only enhances energy efficiency but also contributes to the sustainability of engineering endeavors. As the push for green technology and climate-aware design continues, knowledge of heat transfer is critical to engineers tasked with optimizing performance while being environmentally conscious. Therefore, a solid grasp of heat transfer principles is crucial in developing innovative solutions to contemporary engineering problems. Furthermore, the use of computational methods, particularly Computational Fluid Dynamics (CFD), has revolutionized thermal engineering. Utilizing high-performance simulations, one can observe and predict heat transfer processes with unprecedented accuracy, allowing for the optimization of designs before physical models are built. This capability is invaluable in numerous applications, like energy generation, combustion systems, and air conditioning systems, where thermal control efficiency is essential for reliability and efficacy. As highlighted in recent reviews of CFD contributions, the two decades past have witnessed immense boosts in software modeling functionality and computational capacity, capturing the imagination of scientists and engineers equally (Paricio et al., 2019). The focus, therefore, on heat transfer as part of engineering curricula is gaining prominence, preparing future engineers to use these rich tools effectively. Finally, project-based problem-based learning (POPBL) programs are being introduced into engineering education increasingly, which highlight the applicability of heat transfer concepts. Such an approach engages students in realistic design projects involving a clear understanding of thermal dynamics and convergence of all the engineering disciplines. As shown through recent sets of engineering design projects, such types of educational paradigms not only enhance critical thinking and higher-level problem-solving capacity but also build the capacity in students to take on problems that have sustainability and cost considerations (N/A, 2017). Through hands-on experiments, the significance of heat transfer is emphasized as a crucial aspect of engineering practice to foster the development of the next generation of well-informed engineers who can design novel solutions based on fundamental principles of science.

Overview of Heat Transfer Mechanisms

There are three primary mechanisms governing heat transfer in heat transfer: conduction, convection, and radiation. Conduction, or the transfer of heat from a body through a material without the material itself moving, occurs at the molecular level. It is the collision of oscillating particles; one particle gains energy and collides with the surrounding particles, thus passing along thermal energy. High thermal conductive materials such as metals facilitate efficient heat conduction, while insulators such as rubber or glass wool possess extremely low heat transfer characteristics. An understanding of concepts of conduction is critical for thermal design, insulation systems, and material selection in engineering. Convection, however, refers to the conveyance of fluid and the transfer of heat through mass movement. It is classified as forced and natural convection; the former is caused by external means such as pumps or fans, while the latter is due to density gradients because of temperature differences. The efficiencies of the processes are defined by fluid velocity, viscosity, and temperature difference considerations. For instance, in heating and cooling systems, the concept of convection should be understood to facilitate effective energy consumption and balance heating under different environments. Convection in practical uses is better illustrated in micro-scale systems where forced boiling phenomena exhibit certain thermo-hydrodynamic characteristics that enhance the heat transport characteristics as presented by Emerson et al. (2009). Radiation is the third fundamental mode of heat transfer that does not rely on a medium. It is distinct from conduction and convection in that it transmits energy through electromagnetic waves, primarily infrared radiation. All objects emit and absorb radiation based on their emissivity and temperature. Thermal radiation is a crucial concept in numerous applications including solar energy collection and thermal insulation. It is crucial that scientists and engineers be well conversant with the laws of radiation, particularly Wiens Displacement Law and the Stefan-Boltzmann Law. These laws aid in designing more effective energy-saving systems and designs of radiant heat, particularly building designs with the aim of thermal comfort and energy saving, as addressed in the efforts submitted by Clarke et al. (2009).

2.1. MODES OF HEAT TRANSFER

The phenomena of heat transfer, primarily falling in the category of conduction, convection, and radiation, collectively define the fundamental phenomena of thermal energy transport. Conduction is where heat is passed on through a solid material by virtue of temperature difference, precisely moving the energy from hot spots to cold spots. The mode is determined by the thermal conductivity of the material, which determines how easily the heat would be able to flow through it. For instance, metals are normally good conductors of heat with high heat conduction rates for easy transfer of energy, while insulators like rubber or wood do not want any such thing. Understanding the role played by conduction is particularly crucial in temperature regulation applications requiring high accuracy, such as heat exchangers and thermal insulation systems. Convection, however, represents the passing of heat through the motion of fluids—gases or liquids—that creates the union of hot and cold areas.

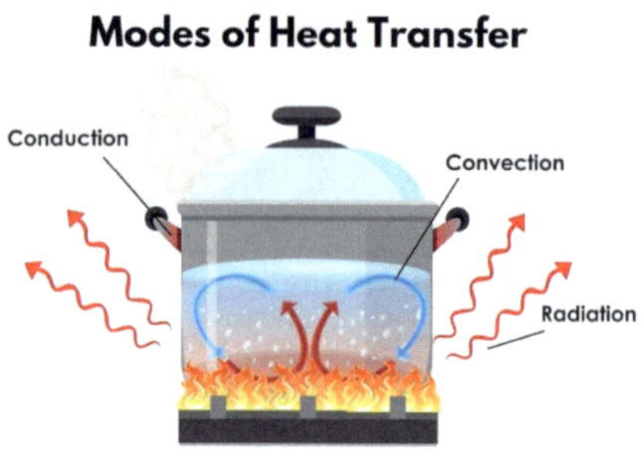

Figure 2.2. Modes of heat transfer.

Source: https://www.careerpower.in/blog/wp-content/uploads/sites/2/2024/11/14180922/Modes-of-Heat-Transfer.jpg.

This process may be classified as natural and forced convection, with natural convection resulting from the buoyancy effects of temperature gradients and forced convection due to forces from outside, e.g., fans or pumps, which cause the fluid flow. A typical example of natural convection is atmospheric heating in which warmer air moves upwards and colder air moves downwards, creating air currents. In man-made systems, the understanding of convection complexities, such as in the effective design of solar dryers, is instrumental, among them being the exploitation of convection in tandem with thermal mass to ensure optimal retention and delivery of heat, as seen in the effective drying of pineapple slices using solar and biomass power sources (Madhlopa et al., 2007). Lastly, radiation, very different from conduction and convection, is heat transfer by electromagnetic waves, allowing for energy to travel without a medium. This method of transfer does not require a medium, such as solar energy that reaches the Earth. The Stefan-Boltzmann Law quantifies radiative heat transfer with the assertion that the radiated power of a body varies directly with the fourth power of the absolute temperature of the body. This simple principle is the foundation of the study of thermal radiation in various applications from passive solar heating systems to advanced technologies in aeronautics. In addition, the coupling between conduction, convection, and radiation exists in various thermal systems, where an in-depth understanding of these modes is necessary for the invention of thermal management techniques and improving overall energy efficiency (Chibisov et al., 2008) (Figure 2.2).

2.1.1. Conduction: Mechanism and Examples

Conduction is a fundamental process of heat transfer by direct contact of materials where energy is transferred from one particle to another without the movement of material as a whole. This is caused by temperature differences; when a portion of a material is heated, its particles will move more turbulently, colliding with other adjacent particles and transferring kinetic energy. The rate of conduction varies with the thermal conductivity of the materials, which is vastly different from one substance to another. Metals, for instance, are highly conductive thermally due to free electrons facilitating easy transfer of energy, whereas insulating material such as rubber or wood exhibits significantly lower conductivity. Understanding the mechanisms of conduction is crucial in fields ranging from engineering to ecology, since it is crucial to thermal management and energy efficiency. Conduction is applied in real life through various instances, making it clear just how vital it is in everyday life. For instance, if a metal spoon is inserted into a hot pot of soup, the temperature of the soup is transferred through the spoon and ends up warming its handle. Such instances depict how conduction applies to cooking techniques and devices in the kitchen. Additionally, thermal conduction is vital in electronics too, where parts must let go of heat safely so that they are not overheated and they operate effectively. Effective thermal management is at times realized through utilization of materials with specially crafted conductive properties in an effort to enhance heat dissipation for high-performance systems. With advancing technology, the development of new materials, particularly in the case of two-dimensional (2D) crystals and heterostructure devices, is creating new avenues for innovation in heat transfer technologies that utilize conduction more optimally (Kostya et al., 2016). Furthermore, recent research has emphasized the significance of conduction in energy application, particularly photocatalysis, where the efficiency of heat transfer may influence chemical reactions. The integration of semiconductor heterojunction enhances photocatalytic activity, eliminating issues with charge recombination and light absorption. Optimization of conductive pathways in such materials is the aim of scientists, who intend to increase the rates of energy transfer and thereby enable more efficient degradation of contaminants and hydrogen generation (Wang et al., 2014). Such coupling of heat conduction with photocatalytic activity highlights the need for acknowledging conduction as not a fundamental physical phenomenon but instead a critical component in the development of sustainable technology. As materials science and energy fields extend, conduction mechanisms will continue to play a pivotal role in driving innovation and efficiency across many applications (Figure 2.3).

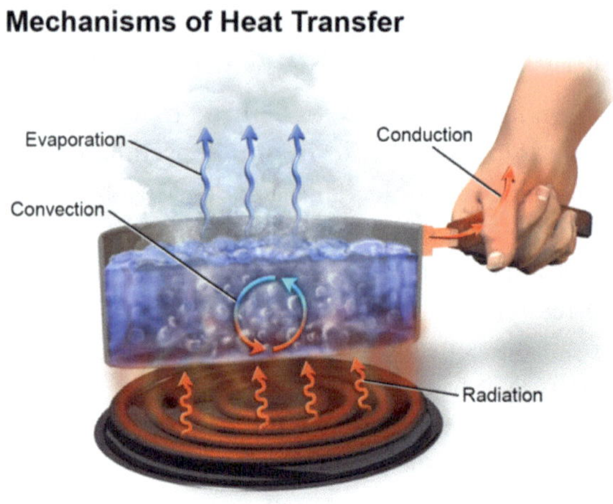

Figure 2.3. Mechanisms of heat transfer diagram.

Source: https://thermtest.com/heat-transfer-through-conduction-examples-in-everyday-life.

2.1.2. Convection: Types and Applications

Convection, which is one of the primary modes of heat transfer, consists of a broad variety of types that play an important role in natural as well as engineered systems. Predominantly, convection is categorized as natural and forced convection. Natural convection results from fluid density differences caused by temperature variations, and spontaneous fluid movement takes place. This phenomenon is typically observed in the patterns of weather, where hot gases rise and cold gases descend and consequently give rise to wind patterns. Forced convection, however, employs external forces, such as pumps or fans, to compel fluid flow, thereby enhancing the rate of heat transport. Such applications are crucial in the design of heating and cooling systems in buildings and refrigeration systems where temperature control is critical to efficiency and comfort. The distinctions between such convection are key to understanding their corresponding functions in thermal management systems, being basic principles of heat transfer. Convection has applications that are common beyond climatic control and refrigeration, influencing various industrial processes. In the example of heat exchangers, natural and forced convection are crucial in ensuring maximum thermal efficiency in liquid and gas transfers. These units utilize convection principles to facilitate heating, cooling, or phase change of fluids by producing the maximum possible contact surface area between the fluids and heat transfer surfaces (Figure 2.4).

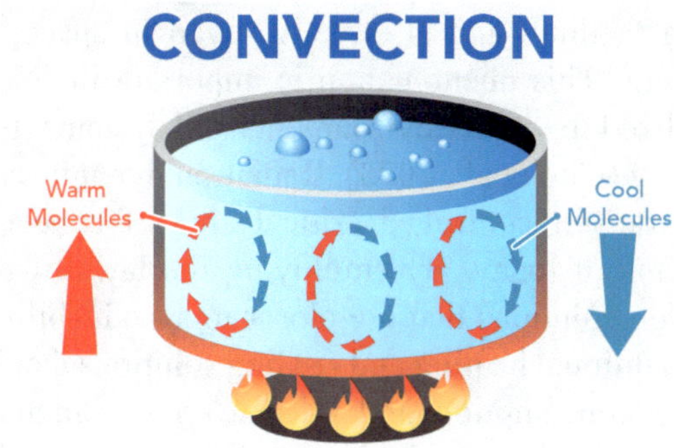

Figure 2.4. Convection refers to motion in a gas or liquid caused by temperature differences.

Source: https://letstalkscience.ca/sites/default/files/2021-01/convection_currents.png.

One significant application of forced convection is seen in high-performance electronic cooling systems, where component heat has to be removed at high rates to prevent overheating. In these applications, enhanced convection—realized through fins or liquid-based cooling systems—is a direct application of the idea to maximize heat transfer rates, reinforcing convection's vital role in thermal management across technologies and manufacturing processes. Understanding the nuances of convection also involves studying more complex situations, e.g., multiphase flows and the ensuing heat transfer concerns. Numerical analysis of such flows, often studied through Computational Fluid Dynamics (CFD), provides examples of the interaction of immiscible fluids in oil recovery and chemical processing, among others. Through the application of sophisticated simulations, researchers can model complex convection dynamics in multiphase systems, thus gaining insight critical for optimizing processes. Chapter analyses in contemporary research illustrate the way enhanced low-dissipation convection schemes improve predictive capacity in situations of turbulence, making them a necessity for formulating engineering solutions within multiphase flow scenarios. This intricate interaction between convection modes and applications ultimately highlights the complex nature and importance of effective heat transport mechanisms in natural phenomena as well as engineered devices, guiding upcoming thermal management developments (Dombrovsky, 2022; Schillaci, 2017).

2.1.3. Radiation: Principles and Significance

Radiation heat transfer is the most important to learn in terms of thermal management applications. Radiation, electromagnetic wave energy transfer, does not require a medium and is effective even in space, where there is a vacuum environment. This phenomenon is important in solar heating, where solar energy is utilized to warm environments and systems, thereby enhancing the energy efficiency (Guo et al., 2023). Radiation dynamics encompass active and passive heat transfer control, a critical area of concern in the thermal management systems of today. By employing the laws of thermal radiation, engineers can design solutions that are closer to sustainability goals, reducing the utilization of traditional heating and cooling solutions that have a propensity to increase energy consumption and adverse environmental consequences. In contemporary applications, radiative cooling has gained popularity as a groundbreaking means of heat dissipation. Radiative cooling is achieved through the natural emission of infrared radiation by surfaces into the cold outer universe, dissipating whatever object is emitting heat without energy input (Guo et al., 2023) (Figure 2.5).

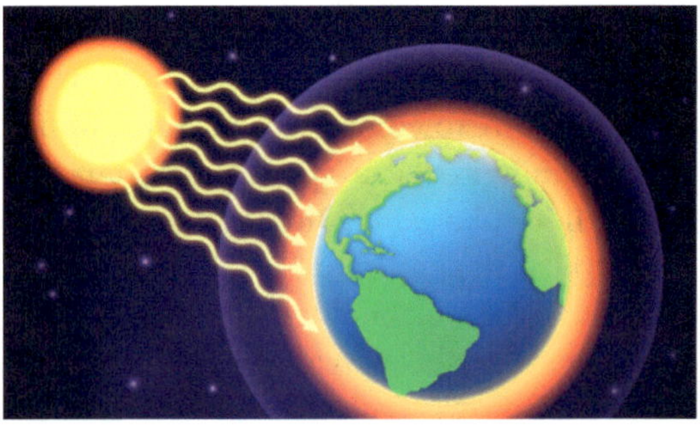

Figure 2.5. On Earth, thermal radiation originates from the Sun.

Source: https://letstalkscience.ca/educational-resources/backgrounders/introduction-heat-transfer.

The significance of the process is particularly evident in cities where heat entrapment can enhance the urban heat island effect. Through the implementation of radiative cooling practices, not only residential and commercial establishments but also buildings can assist in lessening reliance on energy-consuming air conditioning systems. More significantly, advances in material science have witnessed the development of surfaces designed for the explicit purpose of optimizing radiative cooling, thus allowing sustainable thermal management and total energy conservation. The implications of radiation extend far from its potential use for thermal management, with deep effects on climate prediction and science. Physical principles of radiative transfer provide the key to understanding greenhouse gas effects on global temperatures in terms of climate projections. Robust models of these principles allow scientists to forecast future warming trends based upon recent emissions trends and thereby guide policy and energy management strategies. The application of radiative transfer physics in Earth System Model frameworks enhances understanding of feedback processes, making deeper implications of climate change. This is why progress and advancement in study and application of radiative principles are essential in the development of effective countermeasures to mitigate climate effect and ensure sustainable energy use for future generations (Figure 2.6).

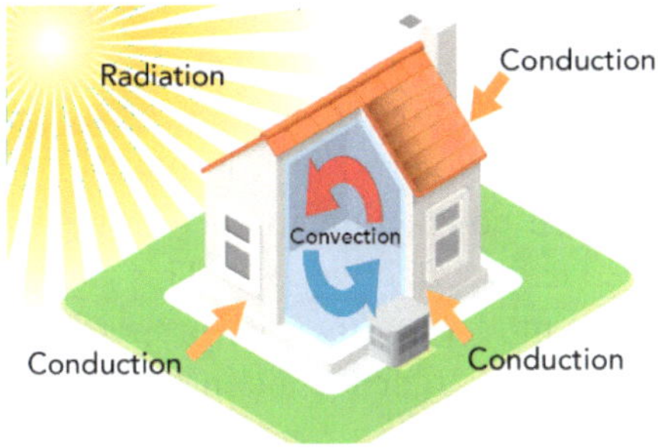

Figure 2.6. Conduction, convection, and thermal radiation in a house.

Source: https://letstalkscience.ca/educational-resources/backgrounders/introduction-heat-transfer.

2.2. CONDUCTION

Conduction is an integral process of heat transfer in solid materials and is described as direct transfer of thermal energy through a medium without material movement. It occurs as kinetic energy is transferred between adjacent molecules or atoms to facilitate heat transfer because of temperature gradients. Material properties such as thermal conductivity, a measurement of the heat-conducting ability of a substance, have a significant impact on the efficiency of conduction. High-value thermal conductivity materials such as metals facilitate efficient heat transfer, while insulators such as wood or rubber conduct less and consequently minimize heat loss. The understanding of conduction is critical in most applications, from the design of successful thermal systems to improving energy conservation methods in building materials, and its foundational function within the broader study of heat transfer principles. The recent advancement of nanotechnology has also expanded the reach of conduction, particularly through the study of nanofluids. Nanoparticle suspensions in conventional heat transfer fluids have been of great interest, largely on the basis of their ability for anomalous thermal conductivity increase even at low solid particle concentrations. Research indicates that such anomalous increase is caused by nanoparticle structuring, which alters fluid flows and thermal characteristics of the base liquid. However, the increasing viscosity due to nanoparticle agglomeration is an issue regarding the utilitarian utilization of nanofluids in real-world engineering systems. Thus, a wise evaluation of structure-property relationships is critical in applying nanofluids to practical usage in cooling systems and heat exchangers, underlining the delicate balance between enhancement and practicality in thermal management (Ding et al., 2011). Furthermore, conduction plays an important role in the performance of heat exchangers, particularly those involving heat pipes or thermosyphons. In a recent experimental study of an air-water heat exchanger with thermosyphons, thermal performance was analyzed with varying air mass flow rates and inlet temperatures. Effective conduction in such systems is indispensable in attaining optimal heat transfer effectiveness. This study validates the need to accurately measure and model the thermal properties, such as the thermal resistance of the heat pipes, in order to obtain the best system performance. Notably, results indicated satisfactory agreement between the experimental data and computational fluid dynamics (CFD) solutions with the constant thermal conductivity assumption. This alludes to the importance of conduction understanding in designing and optimizing thermal devices to facilitate greater energy efficiency and stable operation under practical application (Bai et al., 2015) (Figure 2.7).

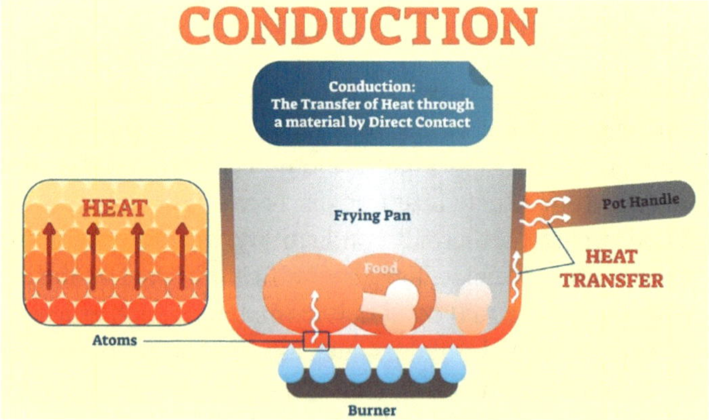

Figure 2.7. Conduction mode of heat transfer.

Source: https://www.study-ground.com/storage/app/user_blog_thumb_images/Title__Understanding_Conduction_Mode_of_Heat_Transfer_451744.webp.

2.2.1. Fourier's Law of Heat Conduction

Fourier's Law of Heat Conduction is a fundamental principle employed to learn about heat transfer and, to a great degree, accounts for the rate of thermal energy transfer in a substance according to the gradients of temperature. The law defines the rate of heat transfer in quantitative terms across a material as directly proportional to the negative gradient of temperature and the conducting area employed in transmitting the heat. Mathematically, it is expressed as $Q = -kA(dT/dx)$, where Q is the heat transfer rate; k is the thermal conductivity of the material; A is the cross-sectional area; and dT/dx is the temperature gradient. This fundamental relationship expresses the performance of heat conduction in solids and its performance in a broad spectrum of applications, ranging from industrial processes to simple thermal management systems. The use of Fourier's Law transcends mere numerical calculations and impacts various scientific and engineering disciplines.

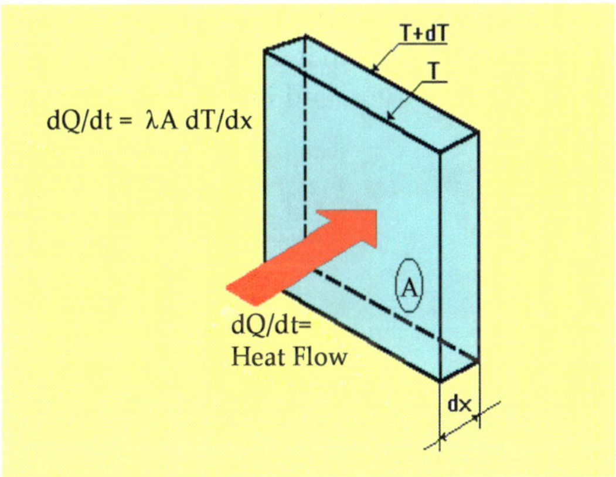

Figure 2.8. Fourier's law of conduction.

Source: https://encrypted-tbn0.gstatic.com/images?q=tbn:ANd9GcQPrrAar9XSKhx3hAE0ShbykEf0duA0UfigUw&s.

For instance, the law enables one to make an analysis of heat transfer in systems involving several interacting elements, as described in modern thermodynamic studies. As indicated in recent reactive systems analysis, understanding the forces of interaction as a consequence of heat conduction requires the merging of Fourier's laws with the chemical reaction phenomena in mixtures, as described by Bothe et al. (2013). The integration of chemical kinetics and heat conduction broadens the portfolio of analysis tools available to engineers, enabling more sophisticated modeling of processes involving heat transfer and chemical interaction. Besides, creating an understanding of

Fourier's Law in the context of nonlinear multicomponent diffusion reveals the rise in complexity of heat transfer processes. As highlighted by Gorban et al. (2011), multicomponent system dynamics often involve deviating from linear conditions and incorporating nonlinearities with significant effects on thermal conduction. The combination of thermal with concentration gradients in multicomponent systems requires more in-depth exploration of the thermodynamic theory behind such phenomena. By combining the understanding from both nonlinear heat conduction and diffusion, scientists can build more accurate models that capture the intricate behaviors of materials at various thermal and compositional conditions, enabling us to forecast more realistic heat transfer processes (Figure 2.8).

2.2.2. Thermal Conductivity and Materials

The concept of thermal conductivity is central to understanding how material influences heat transfer phenomena. Thermal conductivity quantifies a material's ability to conduct heat, with a profound impact on many application areas in construction, manufacturing, and in thermal management systems.

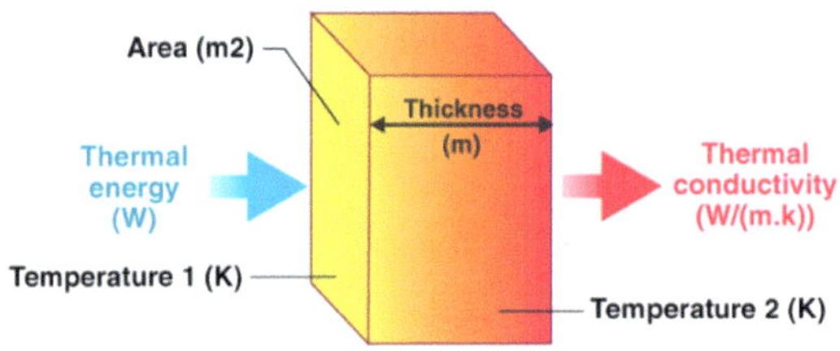

Figure 2.9. The thermal conductivity of a material in terms of the flow of heat.

Source: https://byjus.com/chemistry/thermal-conductivity/

Highly conductive materials such as metals are the model where there is high demand for heat dissipation, and insulators such as foam or wood possess low conductivity, as needed to restrict the transfer of heat. The choice of material based on thermal properties can regulate energy efficiency and overall performance. Material temperature, phase, and structure can also further change thermal conductivity, requiring proper choice in some applications in thermal engineering (Clarke et al., 2009). The advancement of material characterization techniques in recent years has made possible precise determinations of thermal conductivity. For instance, examination of accessible data sets of thermophysical properties allows researchers and engineers to make informed choices in building and construction activities. Enhanced test methods have been developed so that accurate results are assured, making it easy to classify materials required to build energy-efficient environments. The vetting processes and data integration works, as reported by the Building Research Establishment, represent significant developments in generating reliable databases to facilitate improved selection based on materials according to thermal properties, ultimately resulting in sustainable behavior in the built environment (Clarke et al., 2009) (Figure 2.9).

Also, the material interactions among materials under heat transfer can be complex, particularly in engineered systems. Conduction becomes the dominant mode of heat transfer in some applications, such as linered aluminum engine blocks. The treatment of interfaces and their role in controlling thermal conductance illustrates how minute surface modifications can become decisive for overall system efficiency. Factors such as pressure, roughness, and contact area are primary factors in determining effective thermal conductivity. Such findings, from finite element analysis, point out the need to correct for imperfections in material interfaces to attain optimal performance in thermal systems. A deep understanding of these factors is vital to advancing material science in comparison to applications in heat transfer.

2.2.3. Applications in Insulation and Building Design

In design, insulation plays a pivotal role in enhancing energy efficiency and thermal comfort. The deliberate application of various insulating materials can significantly reduce heat flow, making it crucial to manage heating and cooling loads in a building. For instance, low thermal conductivity materials such as foam boards, fiberglass, and mineral wool effectively prevent the transfer of heat. Such resistance to heat transfer not only guarantees a more stable indoor climate but also diminishes the reliance upon energy-thirsty heating and cooling systems. Consequently, structures equipped with high-quality insulation systems exhibit impressive decreases in energy consumption, testifying to the importance of selecting appropriate insulation types during the design process. These developments reflect the vital connection between insulation and energy performance in the context of sustainable architecture as part of contemporary building practice. Additionally, technological developments in insulation have expanded performance with broader application within modern building design. The use of energy-saving insulating materials, such as aerogels and phase change materials (PCMs), is becoming increasingly prevalent in innovative architectural designs. These materials have great thermal resistance but offer a light, space-saving profile that can be hugely beneficial where every square foot counts in city locations. The University of Strathclyde's Energy Systems Research Unit has emphasized the importance of screening and integrating existing datasets of the thermo-physical properties of building materials, thereby having a more reliable framework for designers and engineers to select appropriate insulation options (Clarke et al., 2009). Such advances not only make the energy efficiency better but also improve the overall sustainability of building design, visually confirming the continuing advancement of insulation technologies. Furthermore, effective insulation also plays a significant role in the reduction of greenhouse gas emissions associated with building energy use. By preventing thermal bridging and air leakage, high-performance insulation techniques can optimize building envelope's lifespan and durability and reduce environmental impact. The strict screening of retrofitting technologies for energy efficiency in industrial facilities has been explored extensively, allowing for valuable lessons towards sustainable development in insulation applications (Fisher et al., 2011). These practices acquire particular relevance in older structures that fall short of current codes of energy efficiency. Besides this, enforcement of rigorous testing and performance standards for insulation products ensures conformity of the products to the desired energy targets set by building codes and regulatory programs. Correspondingly, the use of new insulation practices enhances building performance while adhering to global ambitions for sustainability and energy conservation.

2.3. CONVECTION

Convection is a secondary mode of heat transfer, independent of conduction and radiation, relying on the transfer of fluid particles to transport energy. Its process hinges on the differences in temperature within a fluid, which gives rise to circulation patterns. When a fluid is heated, its density tends to be less, thus it rises, while denser cold fluid falls. This flow that is induced by buoyancy is of significance in many natural and man-made systems, ranging from atmospheric systems to domestic heating. For instance, in solar collectors designed for the preparation of hot water, convection mechanisms play a crucial role in attaining high energy efficiency. Through such proper simulation of thermal response of the systems, engineers can optimize design and performance, as emphasized in recent numerical analysis undertaking heat transfer kinetics using the Finite Element Method (Allegrini et al., 2016). Convection is of special significance in wildland fire behavior applications where heat transfer dynamics directly affect fire spread. Recent advances in model strategy, utilizing enhanced computational capability and spatial data analysis, aim at understanding and predicting the behavior of wildfires based on the physics of heat transfer. Convection in this case allows for the rapid transfer of heat, leading to the kindling of surrounding vegetation and the propagation of fire. Physical and quasi-physical model generation after the 1990s emphasizes rigorous understanding of convections role in wildfire conditions. With enhanced models, researchers can forecast more precisely the dynamics of fires so that effective management practices and resource allocation are possible (Albini et al., 2007). Convection is hence a keystone process with relevance across a range of scientific disciplines. In engineering and environmental use, the principles of convection also inform the construction of efficient cooling and heating systems. For example, in architecture, traditional heating can be enhanced by understanding the way convective currents spread heat throughout a room. Similarly, in geothermal systems, natural convection can enhance processes of heat exchange, resulting in an enormous increase in efficiency. Research into various modes of heat transfer indicates that convection efficiency varies with fluid properties, flow regimes, and surface interactions. With increasing concern about sustainable resources, knowledge of convection opens doors for the development of new technology, such as solar heating systems efficient in their use of natural processes to deliver effective heat production (Allegrini et al., 2016). Convection interaction with energy systems continues to be a crucial area of study within the broad overall discipline of heat transfer (Figure 2.10).

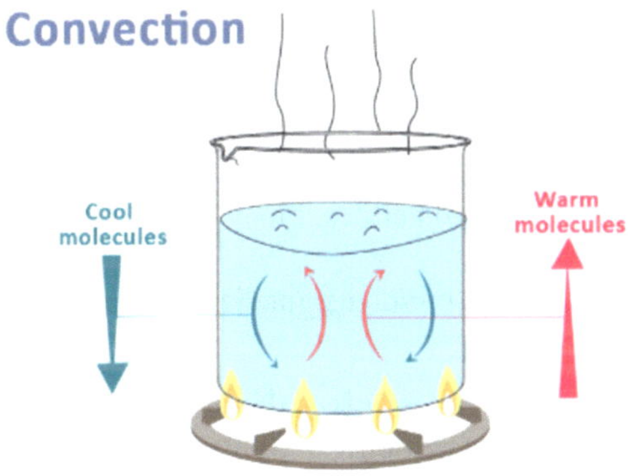

Figure 2.10. Convection process.

Source: https://www.vedantu.com/evs/convection-of-heat-in-water.

2.3.1. Natural versus Forced Convection

The fundamental heat transfer concepts entail numerous mechanisms, with natural and forced convection being paramount among them. Natural convection is characterized by fluid movement as a result of density gradients created via temperature variations. When a fluid is heated, it flows upward, and cooler fluid moves down to sustain a circulating pattern that favors the transfer of heat. This is a slow process by nature and heavily reliant on environmental conditions that surround it, which gives rise to lower heat transfer coefficients than forced convection. It is a concept that is worth knowing about, particularly in usage such as building heating systems and passive cooling measures, where it can presumably provide adequate thermal control without the help of mechanical means. According to studies, a detailed analysis of natural convection-induced steam condensation reveals heat transfer inefficiencies with the slower dynamic movement that is characteristic of natural flow (Samah et al., 2024) (Figure 2.11).

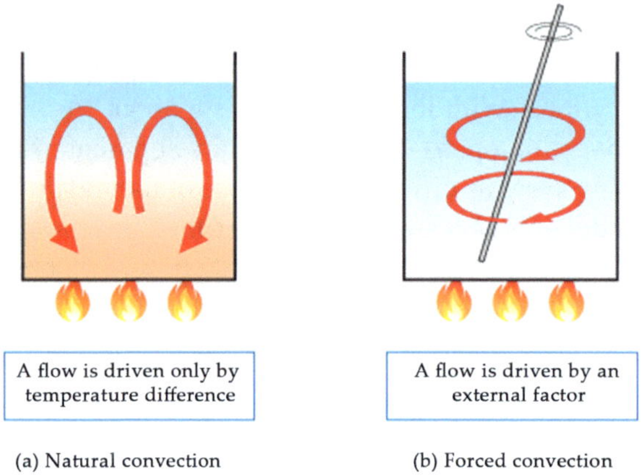

Figure 2.11. Natural versus forced convection.

Source: https://www.cradle-cfd.com/dcms_media/image/en_column_basic_fig4-4.jpg.

In contrast, forced convection involves the intentional movement of fluid by external sources, typically fans or pumps, significantly enhancing heat transfer rates. Forced convection can efficiently maximize the heat transfer coefficient by elevating the fluid velocity, which is useful in applications of high thermal efficiency. Forced convection is a common mechanism employed in industrial processes and heat exchangers where the thermal performance is maximized. Secondly, the ability to control flow rates and heat transfer conditions gives greater versatility in designing systems that must deal with a specific thermal load. Recent advances, particularly in computational fluid dynamics (CFD) modeling, have provided a lot of insight into how forced convection systems can be optimized. This modeling enables an understanding of the effects of design parameter changes on the performance of heat exchangers, ultimately leading to improved operational efficacy across industries (Kubakh et al., 2023). The convection mechanisms are both important to the complete understanding of heat transfer principles, each of which has its own advantages and limitations. The choice between forced and natural convection is determined to a large extent by the specific requirements of an application, including cost, efficiency, and environmental considerations. Natural convection is suitable for low energy input requirement passive systems, while forced convection provides potential for maximum performance in varying conditions. In this understanding, engineers and designers can design their approach on an ad-hoc basis, modifying factors including energy consumption and thermal efficiency. Lastly, the comprehension of both kinds of convection is not only useful for theoretical knowledge but also for real-world engineering applications that improve the control of heat in many areas, ranging from construction to industrial applications.

2.3.2. Governing Equations and Principles

The principles of governance and equations of heat transfer form the basis for understanding thermal systems. The fundamental laws, such as conduction, convection, and radiation, dictate the flow of heat within and between materials. Fourier's law states the heat transfer directly in proportion to the temperature gradient, where the material properties such as thermal conductivity remain a crucial parameter. This concept comes to be crucial in design applications ranging from insulation design to cooling electronic devices. Numerical and analytical models are formed differently so that such equations can be solved for actual conditions, enabling engineers and physicists to predict thermal behavior accurately. Such a deep knowledge of these governing equations hence lies at the core of designing efficient thermal systems as well as optimum energy utilization in engineering practice. Convection, the other crucial heat transfer mechanism, involves the movement of a fluid and the resulting heat transfer. Momentum and energy transfer within a fluid are explained by the Navier-Stokes equations along with energy equations. In this case, the Nusselt number plays an especially important role as it explains the convective heat transfer in comparison to conductive heat transfer. Knowledge of these principles enables consideration of convective heat transfer in a wide variety of situations, ranging from industrial operations to natural phenomena. The interactions of fluid dynamics with thermal energy are complex and often need computational models to solve them, adding significance to the governing equations to practical applications in many various fields, from HVAC systems to aerodynamics. Radiation heat transfer theory further complicates thermal analysis. Subordinated to the Stefan-Boltzmann law, radiation is qualitatively distinct from conduction and convection because it involves the emission and absorption of energy by electromagnetic waves. Attention has to be given to incorporating the contribution of the Dufour and Soret effects into applications of drying, with a focus in recent work, which states that their neglect can lead to significant errors in model predictions for heat and mass transfer (Alves-Filho et al., 2020). It is a view that integrates more comprehensive governing equations that consider such coupled phenomena, particularly vital in applications requiring precise thermal management. The interaction of various heat transfer modes emphasizes the need for governing equations to simulate and analyze thermal systems accurately, advocating their precious place in the realm of heat transfer.

2.3.3. Role in Heat Exchangers and Cooling Systems

Heat exchangers are essential units in various engineering applications, particularly in cooling systems, where thermal management is of key importance. The main role of a heat exchanger is to transfer heat from one or more fluids without fluid contact, hence maintaining efficiency and consistency in the process. In production settings, say chemical processing or power generation, heat exchangers enable waste heat to be recovered, significantly improving resource optimization. By effectively managing thermal energy, they can reduce operating costs and mitigate environmental damage. Moreover, the integration of heat exchanger technology in cooling systems is the foundation for the stabilizing of operating temperatures, which is critical to the operation and life expectancy of industrial equipment, again affirming their relevance within the broader framework of heat transfer principles and applications to engineering design (N/A, 2017). Heat exchangers are utilized most prominently in food processing industries, where correct temperatures are important for product quality and safety. For instance, in a milk powder factory, challenges in recovering heat from spray dryer exhausts have been investigated to improve energy efficiency. Such an application highlights the significance of innovative designs and effective integration methods in heat exchanger systems to achieve sustainable operations in the industry. By mimicking indirect heat transfer approaches linking spray dryer exhaust with various heat sinks, significant reductions in hot utility consumption—21%—have been recorded. This proves the disruptive potential of heat exchangers to become the key driver of economic efficiency with environmental integrity in intricate cooling systems (Atkins et al., 2011). As a conclusive remark, the efficiency of heat exchangers in cooling systems cannot be overstated. Being key factors that facilitate successful heat transfer, they play a long way in determining the overall performance of industrial processes globally. Application of advanced design techniques and integration strategies can hopefully go a long way towards making such systems more sustainable. Such synergistic interaction between heat exchangers and cooling systems not only optimizes thermal interactions but also plays a long way towards saving energy in numerous industries. Therefore, understanding the basic principles of heat transfer and their application in designing effective heat exchangers is essential for engineers and technicians who seek to innovate and create existing cooling and heating systems in the context of current engineering (N/A, 2017) (Figure 2.12).

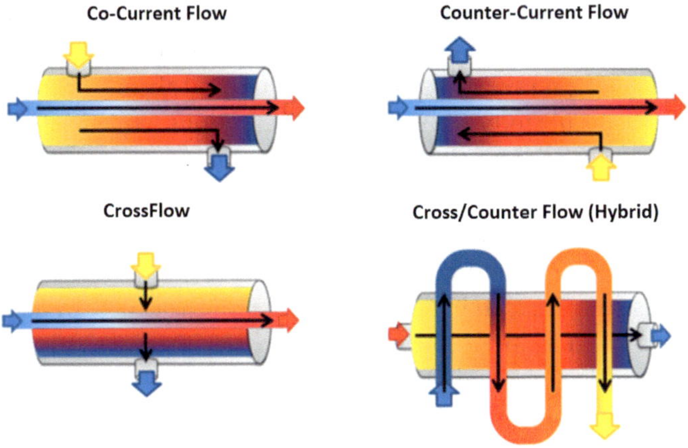

Figure 2.12. Heat exchanger flow configurations.

Source: https://www.thomasnet.com/articles/process-equipment/understanding-heat-exchangers/

2.4. RADIATION

Radiation is an important method of heat transfer that is the subject of thermodynamics, specifically defined as energy transfer through electromagnetic waves. Conduction and convection require a medium to transfer heat, while radiation can pass through a vacuum, meaning that the sun's energy can travel the distances of space and warm the Earth. This process is governed by the Stefan-Boltzmann law, which states that the total energy emitted depends significantly on temperature and places focus on the exponential relationship of heat and radiation. Understanding the mechanisms of radiation is fundamental, particularly in thermal applications where efficiency is a necessity. Moreover, radiation-matter interaction like transmission, reflection, and absorption controls thermal management in engineering systems and therefore is a fundamental design parameter of thermal insulation and energy systems, as well as green building technologies. One of the most important applications of radiation in medicine is therapeutic treatment using water-filtered infrared-A (wIRA), which exploits some wavelengths to deliver energy to tissue without excessive thermal load. The sunlight's warm sensation in temperate latitudes is caused by water vapor filtering specific infrared wavelengths, something that, when utilized in clinical therapy, raises tissue temperature and oxygenation very effectively. The wIRA mechanism facilitates extensive energy transfer with minimal pain and discomfort, reduced inflammation, and wound healing. Existing literature indicates that wIRA is beneficial for acute and chronic wounds and unveiled a remarkable reduction in pain and inflammatory reactions, enhancing recovery results. These new applications are a testament to the importance of studying radiation interactions with biological matter, which have the potential to significantly contribute to medical therapy while also being subject to heat transfer laws (Hoffmann et al., 2006). Additionally, advances in aeropropulsion systems prove the complex interaction between heat transfer, fluid flow, and combustion processes, attesting to the universal nature of radiation. Research emphasizes the significance of understanding radiative heat transfer in engines, particularly with more advanced engine designs, tighter efficiency, and performance demands. The complexity demands computational methods in order to adequately model interactions between combustion's various stages of combustion and thermal radiation. Examining flame stabilization, heat transfer, and turbulent reacting flows allows designers to optimize combustor geometries and minimize energy losses. Thus, radiation is not only a passive transfer process; it actually dictates the configurations and efficiencies of modern thermal systems, which points to its core significance in several fields of engineering, ultimately leading to the creation of sustainable technology (Mularz et al., 2025).

2.4.1. Stefan-Boltzmann Law

The Stefan-Boltzmann Law is a basis law in thermal radiation studies, outlining the relationship between an object and the radiant energy it emits. According to the law, the total energy radiated per unit surface area of a black body varies with the fourth power of the absolute temperature of the black body. Mathematically written as $E = \sigma T^4$, where E is the energy per unit area; T is the absolute temperature; and σ is the Stefan-Boltzmann constant, this equation emphasizes the role of temperature in describing radiative heat transfer. Thus, as the parameters such as temperature increase, the emission rate of heat shoots up considerably, influencing a broad spectrum of applications ranging from thermal engineering to environmental science, and gaining an unflappable grasp of this law becomes critical in predicting heat transfer conditions accurately. Practical application of the Stefan-Boltzmann Law is critical in cases such as the design and analysis of heat sinks, where heat dissipation efficiency may decide efficacy. In aluminum fin heat sink and foam experiments, the application of this law explains how radiative heat transfer a function of material properties and surface treatment is, such as in a comparison between different methods of producing aluminum foams. They demonstrate that, as heat sinks undergo thermal gradients, the interaction between surface treatment and heat transfer becomes more critical. Specifically, painted or anodized finishes realized up to a 50% increase in heat transfer rates, which emphasizes the key interaction between thermal radiation and material properties based on principles established by the Stefan-Boltzmann Law (Billiet et al., 2015) (Figure 2.13).

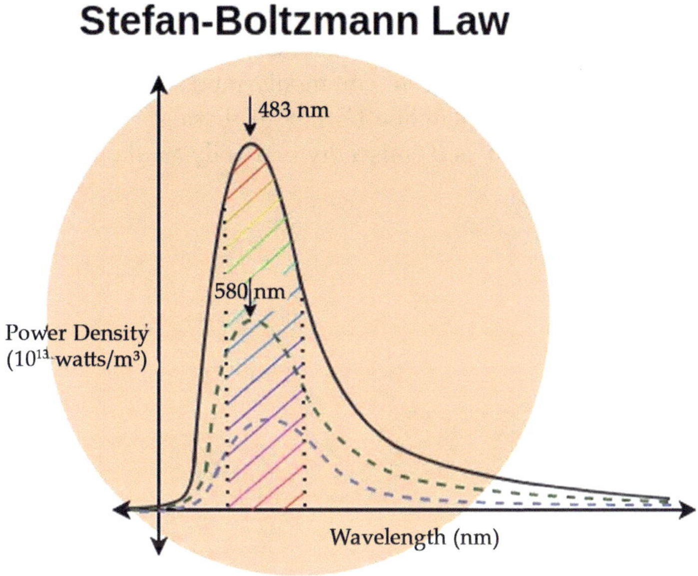

Figure 2.13. Stefan-Boltzmann law.

Source: https://www.geeksforgeeks.org/physics/stefan-boltzmann-law/

Furthermore, the applicability of the Stefan-Boltzmann Law extends beyond conventional thermal systems to innovative applications such as concentrating solar energy. Progress in thermal conductive focusing on recent years demonstrates the nature of harnessing direct and diffusive solar radiation in order to achieve substantial temperature rises at solar receivers. In contrast to traditional methods relying heavily on optical concepts, this method targets conductive heat transport in order to reach its maximum performance under various environment conditions. With the derivation of analytical solutions of temperature and concentration rises, scientists emphasize the applicability of the laws to new technologies, particularly in cases of significant diffuse radiation. The developments not only extend the applicability of the Stefan-Boltzmann Law but also necessitate further studies to enhance heat exchange processes in various mediums and circumstances (Montenegro et al., 2016).

2.4.2. Emissivity and Absorptivity Concepts

Both absorptivity and emissivity are key to the topic of radiative heat transfer within the broader discipline of thermal physics. The ability of a material compared to an ideal blackbody to emit thermal radiation is its emissivity, which for an ideal blackbody is equal to one. Compared between materials, emissivity is highly variable in relation to surface characteristics, such as temperature and the wavelength of emitted radiation. Conversely, absorptivity is the ability of a body to absorb incident radiation, commonly represented as a complement of reflectivity. According to Kirchhoff's law of thermal radiation, at thermal equilibrium, the absorptivity and emissivity of a body are the same.

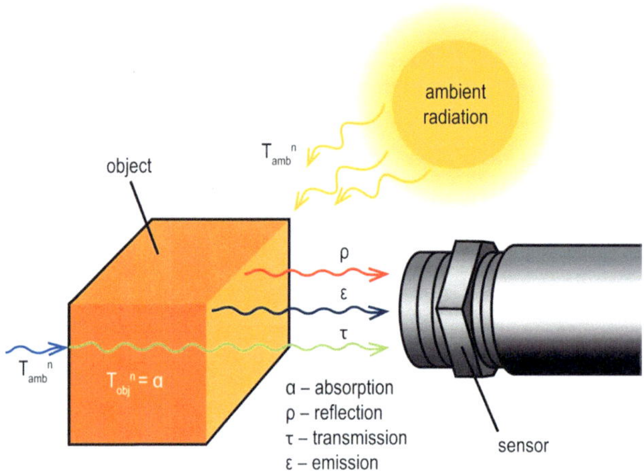

Figure 2.14. Kirchhoff's law of thermal radiation.

Source: https://optris.com/wp-content/uploads/2024/07/Kirchhoffs-law-of-thermal-radiation-1000x0-c-default.webp.

This connection goes a long way in describing how certain materials can efficiently absorb and emit thermal energy, which has applications as diverse as building materials to thermal insulation systems in engineering, particularly in heat exchange processes in surfaces and enclosures (Janna, 2018). The connection between emissivity and absorptivity is of extensive utility in practical applications, particularly optimizing thermal effectiveness in engineering systems. For instance, in building design, higher-emissivity materials are employed for heat-dissipation surfaces because they effectively dissipate thermal energy from the building. In contrast, low-emissivity materials have key roles to play in thermal-insulation requirements. Such knowledge also extends to photovoltaic systems, where selective materials can be created to provide maximum absorptivity of solar radiation while minimizing thermal losses. Radiative heat transfer not only depends upon the inherent properties of individual materials but also upon the interactions between them in a given environment. Ambient temperature, roughness of surfaces, and atmospheric conditions may alter the behavior of materials in terms of radiative heat exchange. The complexities of emissivity and absorptivity allow for the engineering of solutions to enhance energy efficiency across many industries. It is also necessary to address the frequency dependence of such properties because emissivity and absorptivity are also a function of wavelength, a factor that must be considered in high-performance applications. The study of these subjects is thus not merely academic but has engineering implications in the design and optimization of thermal systems in natural and engineered environments (Wong, 2003) (Figure 2.14).

2.4.3. Applications in Solar Energy and Thermal Imaging

The combination of solar energy uses and principles of heat transfer is increasingly important in addressing the demand for global energy and sustainability. Solar panels receive photovoltaic energy and convert it to electricity in terms of electric fields in semiconductor materials. The process demands basic understanding of heat transfer mechanisms, conduction, and convection, because excess heat can adversely affect the efficiency of solar cells. Thermal management is optimized to maximize the performance and lifespan of solar panels. Technologies such as passive cooling systems and heat sinks are employed to dissipate excess heat, maintaining energy conversion to be efficient and effective, thereby promoting mass adoption of solar technology by households, industries, and businesses. Thermal imaging is a state-of-the-art device in the overall branch of heat transfer technology, particularly in applications like urban climatology and building energy audits. Using infrared sensors, thermal imaging provides a non-destructive means of temperature difference mapping on surfaces, which can reveal critical information regarding energy loss in buildings. This becomes important since urban environments generate unique heat patterns that are influenced by human use and built-up materials. For instance, thermal imaging used in the detection of heat leaks in buildings has the potential to reduce energy consumption significantly and improve overall environmental performance. The integration of advanced thermal imaging techniques with modeling techniques provides improved thermal property examination, which contributes to better energy management planning during urban development and infrastructure construction, as indicated by recent studies and research on remote sensing technology (Pease et al., 2025). Further, the integration of thermal imaging with multisensory data helps to better understand thermal dynamics in various applications including environmental monitoring and disaster management. Phenomenological models analyzing large datasets facilitate accurate estimation of surface heat fluxes and thereby efficient resource management. Thermal and visual imagery recognizing thermal patterns focuses on physical properties of utmost significance to surface interactions and energy exchanges. This approach is drastically different from conventional methods in that it focuses on intrinsic property-based discriminatory features and not just signal intensities, thus improving classification processes in remote sensing application (Nandhakumar et al., 2025). All these advancements not only improve thermal science knowledge but also facilitate proactive measures in urban planning and renewable energy policy, giving a positive hue to sustainability goals.

2.5. HEAT TRANSFER IN FLUIDS

When learning about heat transfer in fluid, one has to know the mechanisms to understand different engineering applications. Fluids operate under diverse conditions that have a significant effect on their thermal properties and rate of transfer. The three principal modes of heat transfer—conduction, convection, and radiation—operate differently under fluid systems. For instance, convective heat transfer can be forced or buoyancy-induced, based on whether the fluid motion results from temperature gradients or applied forces. Such multifaceted dynamics need accurate modeling and simulation to predict the behavior of fluids in practical applications, especially systems in which thermal performance is crucial, e.g., heat exchangers and reactors. When industry demands shift, the optimization of these processes remains a significant priority for scientists and engineers as well. Research of fluids under harsh conditions gives further knowledge of their heat transfer capacity. The investigation of supercritical fluids, for example, has become interesting depending on their unique physical characteristics enhancing thermal transfer capability. These fluids enter a stage where specific liquid and gas properties blend, producing exceedingly high heat transfer coefficients compared to conventional phases. Heat transfer near solid boundaries and in compressible flow conditions becomes relevant, as documented in specific studies. Approximations of heat transfer performance from empirical data and theoretical foundations are central to the evolution of technology in heat exchange modes to less wasteful systems with supercritical fluids and exploring new operation regimes (Ito et al., 2017). One key area of research in fluid heat transfer is multiphase flow, e.g., in Taylor flows and complex liquids. These flows present unique challenges due to the interactions between the continuous phase and the dispersed phases, which play a significant role in influencing heat transfer rates. It has been discovered through research that study of gas-liquid hydrodynamics in these systems can lead to the development of innovative heat transfer methods that provide maximum performance in industry applications. In addition, the intersection of theory and practice in research at the university level also supports the necessity of exhaustive exploration, as emphasized by Alberini et al. (2010). Since academic research is more and more sophisticated, its discoveries will definitely refine the principles of heat transfer and stimulate technological development (Figure 2.15).

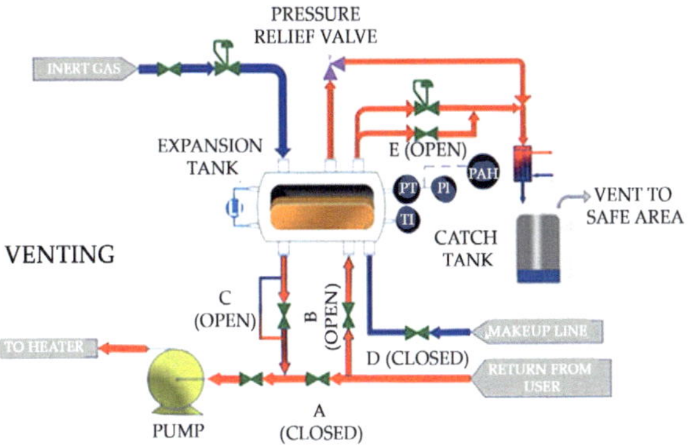

Figure 2.15. Heat-transfer fluid systems must be vented for residual water at start-up and for fluid degradation products during operation.

Source: https://external-content.duckduckgo.com/iu/?u=https%3A%2F%2Fwww.cradle-cfd.com%2Fdcms_media%2Fimage%2Fen_column_basic_fig4-7.jpg&f=1&nofb=1&ipt=5190e6e51a725a6f5dda9d1fffd4f90aa534f0ffe4829154579e2e5ca65628f9.

2.5.1. Properties of Fluids Affecting Heat Transfer

Fluid properties play the central role in determining heat transfer efficiency processes. Among properties, thermal conductivity is the most critical as it determines the ability of the fluid to conduct heat. Different fluids possess varying thermal conductivities based on their molecular structure and composition. For instance, water possesses the comparatively high thermal conductance to a variety of other fluids relative to several others, and as such, is the selection in many industries. Furthermore, the utilization of nanofluids, or specially designed colloids with nanoscale particles, can significantly enhance heat properties. Research indicates that the phenomenon of nanofluids is a function of flow regimes and even outperforms conventional fluids such as water under certain conditions, and hence the evaluation of heat transfer efficiency becomes more complicated (Asproulis et al., 2014). Other than the thermal conductivity, viscosity is another essential parameter in fluid heat transfer. Viscosity dictates the behavior of flow and consequently the convective heat transfer rates. High-viscosity fluids are more resistant to flowing, generating reduced convection and lower rates of heat transfer. Lower viscosity fluids, conversely, induce elevated flow dynamics and improved rates of heat transfer. The relevance of these properties is of even greater concern in turbulent flow uses, where shear forces and convective currents are the dominant factors. Most importantly, the fluid's dynamic characteristics in heat transfer can be unstable; turbulent flow, for example, can increase mixing and increase the surface area for the transfer of heat, thereby increasing overall heat transfer efficiency (Kenjereš et al., 2016). Lastly, particular heat capacity of a fluid is important in establishing the amount of heat energy required to cause a change in temperature. High specific heat capacity fluids can absorb and store more heat, thus being ideal for thermal energy applications. This characteristic becomes increasingly vital in applications such as heat exchangers where constant temperature gradients are present, and optimum thermal performance is required. The combination of specific heat capacity with other properties such as thermal conductivity and viscosity complicates analysis of heat transfer in fluids. These specific heat capacity measurements also differ widely as a function of phase (liquid or gas) and fluid composition, necessitating accurate assessments to project thermal performance in a wide range of situations (Asproulis et al., 2014).

2.5.2. Boundary Layer Theory

Theory of boundary layer lies at the heart of the explanation of heat transfer mechanics in fluid flows, particularly at solid boundaries. Based on the classical fluid dynamics, this theory elucidates how the motion of a fluid is significantly influenced by the presence of a solid boundary. When the fluid flows over a surface, the velocity undergoes a transition from zero at the wall, as per the no-slip condition, to nearly the free stream velocity far away from the surface. This gradient results in a velocity profile that is accountable for the determination of heat transfer rates. This behavior of this layer is capable of governing the effectiveness of heat processes in engineering and is observed particularly in microscale systems where fluid behavior varies from classical expectations, placing emphasis on the relevance of modern interpretations of boundary layer theory. At the micro and nanoscale regime, classical boundary layer approximations become invalid due to rarefied gas effects, when the mean free path of gas molecules is close to characteristic flow dimensions. This necessitates the incorporation of advanced models such as extended hydrodynamics that are an extension of Navier-Stokes-Fourier equations. These models adapt to fit non-equilibrium conditions and help to account for the accommodation of classical theories by empirical observations, mainly in micro-cavities where temperature gradients induce complex flow patterns. For instance, regularized moment equations like those discussed in more recent work are inclusive of the slip and temperature jump effects, hence enhanced boundary layer predictions in cases where conventional approaches might not be able to understand the intricate dynamics under consideration, for instance, gas dynamics in confined spaces under varied thermal conditions (Emerson et al., 2009) (Figure 2.16).

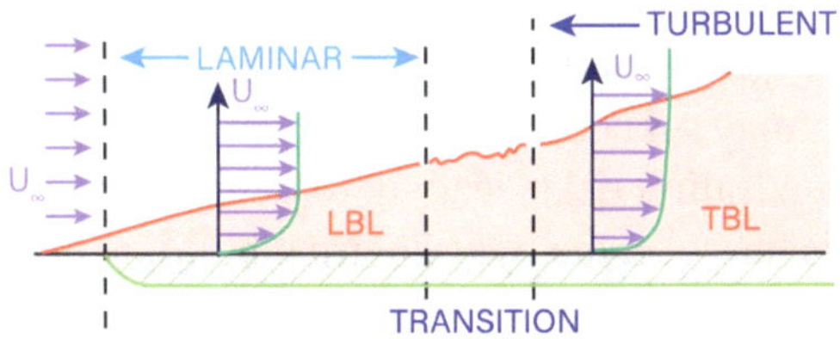

Figure 2.16. The boundary layer theory.

Source: https://i.ytimg.com/vi/k37vPSA3E1g/maxresdefault.jpg.

In addition, boundary layer theory is not just for theoretical calculation; it is vital in determining heat transfer efficiency in engineering systems, particularly microstructure-based systems. Research has indicated that gas flow and heat transfer characteristics may be significantly influenced by the aspect ratio of micro-cavities and thermal gradients via walls, thus influencing overall system performance. With the study of these interactions as boundary layers, the design of microelectronic and other new technologies' thermal management systems can be optimized. This interaction is not theoretical, for it also provides empirical thoughts to system design further exploiting heat transfer principles in small geometries, thereby showing the continued relevance of boundary layer theory to modern engineering challenges (Alaoui et al., 2019).

2.5.3. Applications in HVAC Systems

Application of heat transfer methods in operating HVAC systems plays a crucial role with regard to ensuring maximum energy efficiency in domestic and commercial buildings. The operational efficiency of the system relies on an understanding of conduction, convection, and radiation as far as the transmission and management of thermal energy is concerned. By applying advanced technologies, engineers and designers can make sure that HVAC systems function with zero wastage of energy while ensuring optimum indoor comfort standards. For instance, the application of strategic insulation and high thermal performance materials can be an excellent method for reducing heat loss or gain via building envelopes in order to enhance overall system efficiency. The thermal performance of various insulation materials and configurations should be investigated and compared, as presented in the evaluation of energy-efficient technologies for industrial building retrofits (Fisher et al., 2011). This assessment is important in providing assurance that HVAC systems are not only efficient but also energy-sustaining. In addition to material consumption, the operating conditions of HVAC systems play a vital role in establishing their general performance. Differential ventilation approaches, particularly in the tropical region, have been identified as both favorable and unfavorable due to their impact on indoor air quality and thermal stability. Sometimes, overcooling on air-conditioned sides may subconsciously augment non-air-conditioned building envelope sides' condensation risks, leading to potential damage and mold (Boji et al., 2016). For this purpose, the hygrothermal building element behavior gets important enough in order to avoid such scenarios. This comprehension allows for an improved overall plan when developing HVAC systems that try to balance energy efficiency with humidity control and ultimately achieve healthier interior spaces. The utilization of smart technologies in HVAC systems is also another major application of heat transfer principles. The creation of smart thermostats, smart sensors, and automatic ventilation controls has transformed the process of heating and cooling in modern buildings. Such technologies allow real-time monitoring and adjustment of HVAC operation according to occupancy levels and ambient conditions, thereby optimizing energy efficiency. Dynamic load control of the thermal load in controlling a desired indoor climate led to reduced reliance on traditional energy sources and thus lower operational costs as well as carbon footprint. As energy-efficient options become more in demand, further research, and development in the area of HVAC technology will be important so that these systems also keep evolving to meet the complexities of thermal management and energy preservation.

2.6. HEAT TRANSFER ANALYSIS TECHNIQUES

Heat transfer analysis techniques are at the core of the understanding and optimization of thermal performance in numerous applications within engineering. They encompass a wide range of techniques including conduction, convection, and radiation, all of which play an important role in heat transfer operations. Conduction, for instance, refers to the process of heat transfer through the use of solid materials, expressed mathematically as Fourier's law that represents the temperature gradient and heat flux relationship. Convection, nevertheless, combines thermal conduction and fluid flow, involving advanced modeling to incorporate parameters like flow behavior and heat transfer coefficients. Engineers can design systems that maximize energy efficiency and thermal comfort through an understanding of these underlying methods, which is becoming ever more critical in sustainability and environmental terms.

The comprehension of heat transfer processes also requires robust knowledge of a number of diagnostic and computational approaches. As an example, computational fluid dynamics (CFD) has been shown to be a potent approach to modelling heat transfer in complex geometries and flow configurations. It enables engineers to produce thermal profiles and identify likely hotspots in systems and design more efficient decisions. Furthermore, experimental methods such as thermography and heat flux sensors provide empirical information to cross-validate theoretical models. These sophisticated analysis techniques are particularly valuable for use in industrial applications where energy efficiency can lead to significant cost reductions. Thus, integrating these methods into the engineering design process is imperative to the development of complex thermal systems, as highlighted by the latest energy-saving trends on the topic of discussion within the industry (Fisher et al., 2011). In addition, incorporating heat transfer analysis techniques into project-based learning environments promotes real-world engineering students' expertise. Experimental design in the context of heat transfer allows students to bridge theoretical research to reality. By working with projects that entail in-depth examination, students are exposed to the design cycle, problem definition, and evaluation of solutions. Not only do they become ready to address modern-day engineering challenges, but through these experiences, they are also taught the significance of sustainability and economics in product development. By understanding the different facets of heat transfer and how it applies, students are better equipped to innovate and help move mechanical and manufacturing engineering forward, an idea mentioned again in problem-based learning project-oriented literature (N/A, 2017) (Figure 2.17).

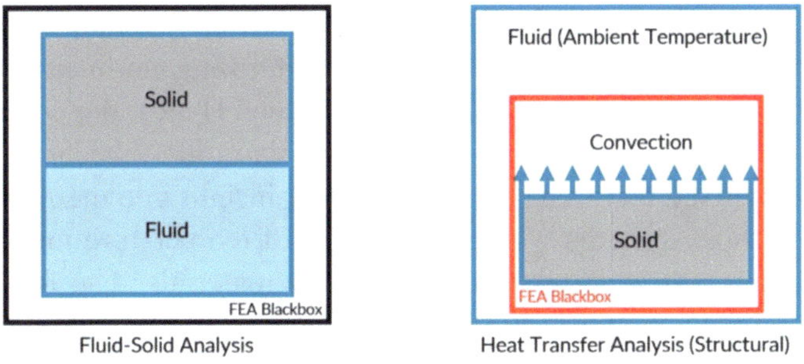

Figure 2.17. Structural heat transfer analysis compared to fluid–solid analysis, modeled within the conjugate heat transfer module.

Source: https://www.simscale.com/docs/simwiki/heat-transfer-thermal-analysis/what-is-heat-transfer/

2.6.1. Analytical Methods

The analytical methods form the core of the study and use of heat transfer principles, allowing scientists and engineers to address complex thermal issues. By means of mathematical modeling, the methods allow researchers to understand the behavior of heat within various materials and constructions. One leading method in this group involves the use of differential equations, which define temperature distribution within a system. For instance, in microfluidic devices such as micro-heat exchangers, lattice Boltzmann has been proved successful. Lattice Boltzmann, derived from the Boltzmann equation, possesses a unique application in micro-level problems where traditional fluid dynamics fail to apply, hence demonstrating the efficacy of analytical techniques in modern heat transfer issues (Asproulis et al., 2014). The theory of frameworks that address analysis methods has the tendency to involve simplifying assumptions that enable solutions of heat transfer equations. In cylindrical coordinates, the theory of heat conduction enables the solution of linear ordinary differential equations describing the thermal conditions of material. This is very important in the field of nuclear engineering, where the thermal performance of nuclear fuels composed of ceramics is very essential. By modeling the temperature profile and taking into account heat sources, researchers can gain valuable information concerning thermal properties of nuclear fuel elements. The investigation determines that thermal conductivity and the coefficient of heat transfer play a major role in influencing temperature profiles in these fuel elements, determining the necessity of analytical methods to provide accurate thermal estimates (Chibisov et al., 2008). Besides, increasing heat transfer scenarios' complexity necessitates a solution with analytical capacity to cater for varied conditions and material characteristics. Temperature gradient and thermal properties interaction has a tendency of forming nonlinear relationships that cannot be addressed by analytical tools solely. Hence, the convergence of numerical procedures and analytical solutions is quickly becoming a shared practice. This mixed research allows for greater insight into thermal dynamics, especially in those systems where the control of temperature must be strictly regulated, such as in energy systems and new materials. The formulation of analytical methods continues to be a vital component in the evolution of heat transfer research, allowing for theoretical development as well as practical applications.

2.6.2. Numerical Methods and Simulations

Numerical methods and simulation are powerful tools in the context of heat transfer, where the complex analysis of complex heat systems is made possible. Computer methods allow modeling various phenomena like conduction, convection, and radiation in scenarios that would be impossible to simulate with analytical solutions. Through the approximation of continuous domains in terms of discrete, manageable units, such as finite volumes or elements, engineers can utilize computational fluid dynamics (CFD) to precisely predict temperature distributions and fluid flow patterns. The application of these numerical methods is most relevant when handling phase-change processes, where traditional methods would not work since they are too complex in their nature. Thus, the coupling between numerical methods and heat transfer phenomena significantly helps in designing and optimizing thermal systems for a broad array of engineering applications. The enthalpy-porosity technique, which is widely used in numerical phase change simulations, illustrates the complexity and intricacy involved with modeling heat transfer. This technique employs a permeability coefficient, sometimes referred to as the mushy zone constant, to model material change from solid to liquid. Uncertainty in accurate determination of this coefficient, however, introduces significant variability in results obtained through simulation. Studies highlight how variations in mushy zone thickness and velocity fields can have significant influences on the morphology of the solid-liquid interface and therefore the overall rate of phase change. Therefore, sensitivity analyses are valuable, demonstrating that the numerical predictions can be insensitive to this coefficient in certain situations, particularly for finely discretized meshes. This indicates that robust computational platforms are required to achieve consistent results in difficult heat transfer problems (Ebrahimi et al., 2019). In the last few decades, there has been a huge surge in the application of CFD in thermal and energy engineering applications because of developments in computing capabilities and numerical simulation codes. This has enabled researchers and engineers to simulate power generation processes, renewable energy systems, and thermal management systems with unprecedented accuracy. The repeated character of CFD permits repeated design iterations and optimizations, hence speeding up the innovation process in engineering projects. Additionally, the all-encompassing character of CFD facilitates an integrative way, in which different physical phenomena are addressed at once. Therefore, the rising amount of literature centered on CFD applications serves to validate the revolutionizing impact of numerical methods and simulations on heat transfer research and their applications in energy systems (Paricio et al., 2019).

2.6.3. Experimental Techniques for Validation

Experimentation to confirm experimental methods is the most important facet of heat transfer in order to validate numerical models for reliability and accuracy. This experimentation involves purposefully designed experiments to derive results that will be used for the validation of theory predictions or simulations. Thermal conductivity tests, where samples are put through controlled heat or cold to calibrate thermal response, are the most commonly used method. The accuracy of such experiments depends mainly on experimental design, e.g., measurement instrument calibration and control of environmental conditions. In validating numerical models, comparison of simulated output to empirical data is of great importance as this results in finding differences and changing model parameters, ultimately yielding better predictions of heat transfer behavior in materials and systems. In guaranteeing the efficacy of simulation results, the choice of appropriate experimental techniques can significantly impact results of heat transfer investigations. For instance, the permeability coefficient influencing fluid speeds in porous media for solidification and melting processes is challenging when numerical simulations are employed. Different experimental techniques can produce dissimilar coefficients of this value, leading to ambiguity in phase-change process simulations. The wide range of uncertainty for this parameter necessitates rigorous experimental exploration to establish a reasonable range to verify. Research studying the sensitivity of numerical simulations to the permeability coefficient shows that there is a well-defined experimental scenario that can yield precious information on how mushy zone dynamics can affect phase-change accuracy, particularly with regard to solid-liquid interfaces during solidification and melting (Ebrahimi et al., 2019). Moreover, the use of experimental techniques in model validation adds to the introduction of hybrid approaches capable of addressing interactions among complex processes, such as welding and machining. Operations chaining in manufacturing demands a precise understanding of phenomena involved in heat transfer, and experiment data can be utilized not only to parameterize single-process simulation but even their cumulative effect. Experiments have shown that dedicated methods have been established and proved through collaborative efforts, e.g., round robin across the world, to replicate these processes repetitively (Bergheau et al., 2013). By focusing on residual stresses and surface integrity as a result of various stages of operation, researchers can create robust models that can predict the final state of materials after complex thermal cycling. This interconnectedness highlights the importance of the experimental techniques' role in confirming numerical models to ensure that they meet industrial standards and safety requirements.

SUMMARY

The investigation of heat transfer fundamentals brings to the fore the inherent importance of understanding the various mechanisms, such as conduction, convection, and radiation, responsible for energy transfer in various environments. Effective heat management is not only a fundamental component in technology and engineering but also crucial in sustainable solutions with the aim of minimizing energy consumption. With the increasing demand for efficient energy use across industries, returning to fundamentals in heat transfer is of paramount importance in developing innovative solutions to meet contemporary energy demands. The incorporation of new data, including that given by the Strathclyde University-based Energy Systems Research Unit, provides insightful information on material properties to aid engineers in developing improved structures and applications for thermal systems (Clarke et al., 2009). Furthermore, the research showcased during the 2nd Micro and Nano Flows Conference illustrates the complexity of the micro and nanoscale heat transfer process. The research showcases the importance of ongoing investigation and research on boiling heat transfer phenomena, which form the basis of a variety of applications such as cooling systems, heat exchangers, and thermal management devices. By addressing the inconsistencies identified in earlier experiments, and focusing on the multifactoring concepts presented, academia is able to advance knowledge of thermal dynamics, particularly in confined systems simulating micro-scale environments (Emerson et al., 2009). This foundation not only supports the development of existing technology but also paves the way for novel equipment based on these precise heat transfer principles. Lastly, the interconnectedness of heat transfer concepts with physics, materials science, and engineering necessitates cross-disciplinary interaction. As technology in heat transfer applications leads the overall field of energy optimization, the collection of detailed data and methods becomes indispensable. Continued research and application will continue to advance the predictive function in heat transfer, allowing for the development of more efficient systems that can respond appropriately to the evolving needs of world energy consumption. Here, the synergy on both professional and academic lines will bring about colossal amounts of efficiency in the field of energy, ultimately giving rise to a greener, cleaner, and sustainable future (Clarke et al., 2009; Emerson et al., 2009).

Fundamental understanding of the principles of heat transfer is essential in most engineering disciplines, including aerodynamics and thermodynamics. The foundations of these principles are rooted in the mechanisms of heat transfer: conduction, convection, and radiation. Conduction is the movement of thermal energy through materials and largely depends on material properties and temperature gradient. Convection is the process of heat transmission by fluids, where fluid movement is responsible for increasing rates of heat transfer. Lastly, radiation is the process of heat transfer through electromagnetic waves and does not require a medium. All of these processes play significant roles in several real-life applications, influencing design, efficiency, and use across a broad range of industries like aerospace and energy systems. In aeronautical engineering, it is highly important to study heat transfer for the design of advanced propulsion systems. Recent research has shown why it is important to study novel methods for optimizing heat handling in the systems, namely for combustion processes. For example, studies have shown that aerothermodynamic process-related problems have a decisive effect on engine efficiency and performance. As future propulsion systems are put under even stricter requirements, it is necessary to include cutting-edge heat transfer technologies that address the challenges directly. These areas of interest include flame stabilization, combustor aerodynamics, and turbulent reacting flows, which highlight where fluid dynamics converges with thermodynamics in modern engineering applications (Mularz et al., 2025). Besides, new studies are pioneering breakthroughs in heat transfer technologies that concentrate on innovative solutions for reducing skin friction in various systems. Techniques such as suction, gaseous injection, particle additives, and compliant wall technologies have all been encouraging in experimental conditions. Advances in these methods overcoming their challenges necessitate unified efforts to streamline their application in practical engineering systems. While considerable progress has been made, it is evident that additional research must be done to enhance their performance to the fullest, particularly in high-performance applications like those employed by aircraft. Designing long-lasting systems to optimize heat transfer while limiting frictional losses is at the core of innovation in heat transfer technology in the future (Ash et al., 2025).

REFERENCES

1. Alaoui, M., Baliti, J., & Hssikou, M. (2019). Continuum analysis of rarefaction effects on a thermally-induced gas flow. http://arxiv.org/abs/1904.08023 (accessed on 23rd October 2025).
2. Albdour, S. A., Addad, Y., Alyammahi, N., & Afgan, I. (2024). Steam condensation heat transfer in the presence of noncondensable gases (NCGs) in nuclear power plants (NPPs): A comprehensive review of fundamentals, current status, and prospects for future research. International Journal of Energy Research. https://www.semanticscholar.org/paper/b26acf9f841e150afd998cc579ead912ce66a1dc (accessed on 23rd October 2025).
3. Alberini, F. (2010). Gas-liquid hydrodynamics in Taylor flows with complex liquids. https://core.ac.uk/download/14699521.pdf (accessed on 23rd October 2025).
4. Albini, F., Albini, F., Albini, F., Albini, F., Anderson, H. E., Anderson, D. H., Sullivan, A. L., et al. (2007). A review of wildland fire spread modelling, 1990–present, 1: Physical and quasi-physical models. http://arxiv.org/abs/0706.3074 (accessed on 23rd October 2025).
5. Allegrini, E., Micangeli, A., Michelangeli, E., Ricci, R., et al. (2016). Heat transfer simulation of evacuated tube collectors (ETC): An application to a prototype. https://core.ac.uk/download/154947586.pdf (accessed on 23rd October 2025).
6. Alves-Filho, O., Bergmann, J., & Häussling Löwgren, B. (2020). A numerical implementation of the Soret effect in drying processes. https://core.ac.uk/download/322792227.pdf (accessed on 23rd October 2025).
7. Anderson, T. R., Hawkins, E., & Jones, P. D. (2016). CO_2, the greenhouse effect and global warming: From the pioneering work of Arrhenius and Callendar to today's Earth system models. Endeavour, 40, 178–187. https://doi.org/10.1016/j.endeavour.2016.07.002 (accessed on 23rd October 2025).
8. Arias Montenegro, F. J. (2016). On the use of thermal conductive focusing for solar concentration enhancement. https://core.ac.uk/download/81578522.pdf (accessed on 23rd October 2025).
9. Ash, R. L., & Fischer, M. C. (2025). A general review of concepts for reducing skin friction, including recommendations for future studies. https://core.ac.uk/download/pdf/42898092.pdf (accessed on 23rd October 2025).

10. Atkins, M. J., Hoffmann-Vocke, J., Neale, J. R., Walmsley, T. G., et al. (2011). Tube shape selection for heat recovery from particle-laden exhaust gas streams. https://core.ac.uk/download/29202520.pdf (accessed on 23rd October 2025).
11. Atkins, M. J., Neale, J. R., & Walmsley, M. R. W. (2011). Integrating heat recovery from milk powder spray dryer exhausts in the dairy industry. https://core.ac.uk/download/29199009.pdf (accessed on 23rd October 2025).
12. Bai, C., Danielewicz, K., El-Genk, M. S., Grubb, D., et al. (2015). Experimental and numerical investigation of an air-to-water heat pipe-based heat exchanger. https://core.ac.uk/download/29140162.pdf (accessed on 23rd October 2025).
13. Bergheau, J. M., Brosse, A., Gilles, P., Rech, J., et al. (2013). Chaining of welding and finish turning simulations for austenitic stainless-steel components. https://core.ac.uk/download/294831356.pdf (accessed on 23rd October 2025).
14. Billiet, M., De Paepe, M., De Schampheleire, S., Huisseune, H., et al. (2015). Influence of orientation and radiative heat transfer on aluminum foams in buoyancy-induced convection. https://core.ac.uk/download/55848277.pdf (accessed on 23rd October 2025).
15. Boji, E., Calautit, J. K., Cerolini, S., Galvin, R., Haw, L. C., Hens, N., Ibrahim, H., et al. (2016). Moisture condensation on building envelopes in differential ventilated spaces in the tropics: Quantitative assessment of influencing factors. https://core.ac.uk/download/286715648.pdf (accessed on 23rd October 2025).
16. Bothe, D., & Dreyer, W. (2013). Continuum thermodynamics of chemically reacting fluid mixtures. https://core.ac.uk/download/289298815.pdf (accessed on 23rd October 2025).
17. Broukal, J. (2014). Effervescent breakup and combustion of liquid fuels: Experiment and modelling. https://core.ac.uk/download/30311594.pdf (accessed on 23rd October 2025).
18. Bubbico, R., D'Annibale, F., Mariani, A., et al. (2014). Experimental evaluation of heat transfer coefficient for nanofluids. https://core.ac.uk/download/29139654.pdf (accessed on 23rd October 2025).
19. Clarke, J. A., & Yaneske, P. P. (2009). A rational approach to the harmonization of the thermal properties of building materials. https://core.ac.uk/download/9025303.pdf (accessed on 23rd October 2025).
20. Ding, Y., Song, P., & Witharana, S. (2011). Structure–property relationships for nanofluids. https://core.ac.uk/download/9633359.pdf (accessed on 23rd October 2025).

21. Dombrovsky, L. (2022). Specialty grand challenge for heat transfer and thermal power. 2. https://www.semanticscholar.org/paper/b237312f655f58ce9105624f8529a0f7d91b9a39 (accessed on 23rd October 2025).
22. Ebrahimi, A., Kleijn, C. R., & Richardson, I. M. (2019). Sensitivity of numerical predictions to the permeability coefficient in simulations of melting and solidification using the enthalpy-porosity method. http://arxiv.org/abs/2004.14168 (accessed on 23rd October 2025).
23. Emerson, D. R., Lockerby, D. A., & Reese, J. M. (2009). Gas dynamics at the micro-scale: A review of progress in hydrodynamic modelling. https://core.ac.uk/download/9633463.pdf (accessed on 23rd October 2025).
24. Engg, P. K. W. (2010). Pabrik Chromium Trioxide Dari Sodium Dichromate Dan Asam Sulfat. https://core.ac.uk/download/12217535.pdf (accessed on 23rd October 2025).
25. Fisher, D., Hand, J., Liesen, R., Wilson, E., et al. (2011). Screening of energy efficient technologies for industrial buildings' retrofit. https://core.ac.uk/download/9033399.pdf (accessed on 23rd October 2025).
26. Gorban, A. N., Barenblatt, G. I., Barenblatt, M., Bertini, L., Blesgen, T., Boudart, M., Briggs, M., et al. (2011). Quasichemical models of multicomponent nonlinear diffusion. http://arxiv.org/abs/1012.2908 (accessed on 23rd October 2025).
27. Goupil, C., Müller, E., Seifert, W., Snyder, G. J., et al. (2011). Thermodynamics of thermoelectric phenomena and applications. https://core.ac.uk/download/4890584.pdf (accessed on 23rd October 2025).
28. Guo, N., Zhao, Z., Yan, H., & Chen, M. (2023). Dynamic thermal radiation regulation for thermal management. Next Energy, 1, 100072–100072. https://doi.org/10.1016/j.nxener.2023.100072 (accessed on 23rd October 2025).
29. Ha, J. (1991). Modelling lingered engine blocks. https://core.ac.uk/download/pdf/1568946.pdf (accessed on 23rd October 2025).
30. Hachemi, M., & Zarita, R. (2014). Microchannel fluid flow and heat transfer by lattice Boltzmann method. https://core.ac.uk/download/29139712.pdf (accessed on 23rd October 2025).
31. Hoffmann, G. (2006). Grundlagen und Wirkprinzipien von wassergefiltertem Infrarot A (wIRA) in Bezug zur Wundheilung. https://core.ac.uk/download/14505659.pdf (accessed on 23rd October 2025).
32. Iranzo Paricio, J. A. (2019). CFD applications in energy engineering research and simulation: An introduction to published reviews. https://core.ac.uk/download/288003442.pdf (accessed on 23rd October 2025).

33. Ito, Y. (2017). Heat transfer of supercritical fluid flows and compressible flows. https://core.ac.uk/download/322428815.pdf (accessed on 23rd October 2025).
34. Janna, W. (2018). Engineering Heat Transfer. https://www.semanticscholar.org/paper/f6712aa82f2b0e6f699af39feeb2b8cc0e43518a (accessed on 23rd October 2025).
35. Kenjereš, S., Kidess, A., Kleijn, C. R., Righolt, B. W., et al. (2016). Marangoni driven turbulence in high energy surface melting processes. http://arxiv.org/abs/1603.07124 (accessed on 23rd October 2025).
36. Kubakh, S., & Tsapar, V. (2023). Simulation of a shell-tube heat exchanger in the oil and gas industry. Proceedings of the NTUU "Igor Sikorsky KPI." Series: Chemical Engineering, Ecology and Resource Saving. https://www.semanticscholar.org/paper/07d43e96a1feb146829a4db5b364c5b11ab403f3 (accessed on 23rd October 2025).
37. Madhlopa, A., & Ngwalo, G. (2007). Solar dryer with thermal storage and biomass-backup heater. https://core.ac.uk/download/9023892.pdf (accessed on 23rd October 2025).
38. Mularz, E. J. (2025). Future fundamental combustion research for aero propulsion systems. https://core.ac.uk/download/pdf/42845245.pdf (accessed on 23rd October 2025).
39. Nandhakumar, N. (2025). A phenomenological approach to multisource data integration: Analysing infrared and visible data. https://core.ac.uk/download/pdf/42820282.pdf (accessed on 23rd October 2025).
40. Novoselov, K. S., Mishchenko, A., Carvalho, A., & Castro Neto, A. H. (2016). 2D materials and van der Waals heterostructures. Science, 353. https://doi.org/10.1126/science.aac9439 (accessed on 23rd October 2025).
41. Pease, R. W. (2025). Climatology of urban-regional systems. https://core.ac.uk/download/pdf/10302970.pdf (accessed on 23rd October 2025).
42. Schillaci, E. (2017). DNS of multiphase flows: Study of atomization and free-surface phenomena. https://www.semanticscholar.org/paper/06f1497ce6ab26c5ad66f8ac71b96c381dc0aeb8 (accessed on 23rd October 2025).
43. Shekriladze, I. G. (2009). Multifactoring concept–A key to investigation of forced-boiling in microsystems. https://core.ac.uk/download/9633382.pdf (accessed on 23rd October 2025).
44. Wang, H., Zhang, L., Chen, Z., Hu, J., Li, S., Wang, Z., Liu, J., et al. (2014). Semiconductor heterojunction photocatalysts: Design, construction, and photocatalytic performances. Chemical Society Reviews, 43, 5234–5234. https://doi.org/10.1039/c4cs00126e (accessed on 23rd October 2025).

CHAPTER 3

THERMAL PROPERTIES OF MATERIALS

Contents

INTRODUCTION

Study of thermal material characteristics is crucial for a complete grasp of how different substances respond to heat transfer. As emerging building technologies and the growing frequency of climate change, efficient materials to help improve thermal performance are more essential than ever. Above all, the insulation role of the wall system is a critical component, especially in regions beset by harsh temperatures. Historically, research has established that thermal performance of the building envelope strongly impacts indoor thermal comfort and energy consumption. This aligns with the demand for newer materials that ensure maximal thermal efficiency with minimized environmental impacts. Sustainable material manufacturing practices go beyond energy conservation to support greater environmental stewardship in the construction industry, aligning with a heightened global emphasis on sustainability. In the context of thermal performance, new avenues that entail the utilization of waste materials in the construction of building components have emerged. Research indicates a twofold interest in overcoming sustainability while improving the thermal characteristics of common building materials. For instance, the incorporation of locally available waste materials, such as palm oil production wastes and glass powder, can result in significant improved thermal control without compromising structure. Thermally enhanced sustainable hybrid (TESH) brick design is one of the best examples of this concept, demonstrating improved thermal efficiency and minimized energy consumption compared to traditional materials. Through the employment of a Design Science methodology, scientists are looking for material alternatives that not only meet building codes but also enable reduced energy consumption during the life of a building, even bridging the gap between innovation and sustainability in building construction. Moreover, advanced materials are not limited to their traditional applications; they can be turned into advanced materials by imparting them with current innovations, such as nanotechnology. Adding nanoparticles, e.g., TiO_2, in bulk materials has been promising in changing thermal and electrical properties. Studies indicate that these additions have led to significant decreases in thermal conductivity, which can lead to improved heat management. The nanoparticle interactions with the matrix substance can create additional phonon scattering centers and modify the overall thermal property. All these developments reflect the dynamic endeavors in increasing the performance and efficiency of the material through traditional as well as advanced techniques. All in all, these developments in the thermophysical properties of materials emphasize the necessity of interdisciplinary studies to address contemporary construction and material science problems.

Definition of Thermal Properties

Thermal properties are inherent properties that determine the response of materials under changing temperature and heat flux conditions. Thermal conductivity, specific heat capacity, and thermal expansion are some of these properties, and each of them plays an important role in determining the thermal performance of a material. Thermal conductivity, for instance, is a measure of how efficiently a material can conduct heat, a very significant factor in uses like insulation and heat exchangers. Specific heat capacity, on the other hand, measures the amount of energy required to change the temperature of a material and influence its ability to absorb and store thermal energy. It is crucial for engineers and scientists to understand these properties when selecting materials for specific functions, especially in energy-efficient and heat-management applications. A more in-depth analysis of thermal properties reveals their multifaceted relationship with microstructure and design parameters. For example, new methods have been employed in recent studies to calculate thermal properties in complex microstructures. Tiling technique, as applied in the numerical homogenization study, is applicable for creating statistically representative realizations of microstructures with better understanding of their apparent thermal properties. Applications of these methodologies can lead to improved designs in static materials and active systems and to improved predictions of thermal performance under various conditions. As highlighted in the findings, the partition theorem and statistical sampling techniques are fundamental in the construction of coherent confidence intervals of thermal properties and thus contribute significantly to valuable insight into material choice and design optimization for engineering purposes (Doškář et al., 2018) (Figure 3.1).

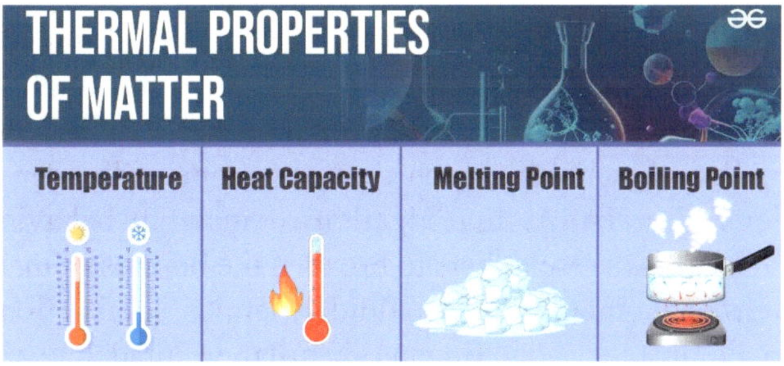

Figure 3.1. Thermal properties of matter.

Source: https://external-content.duckduckgo.com/iu/?u=https%3A%2F%2Fmedia.geeksforgeeks.org%2Fwp-content%2Fuploads%2F20240130162401%2FThermal-Properties-of-matter-.webp&f=1&nofb=1&ipt=e29b7b2bae5465d431c86895aa5528cb9dc0c4d6eb525211c6b770fa45e19422.

The influence of design parameters on thermal performance becomes more pronounced in specialized applications, i.e., solar thermal collectors-based active façades. Through a diligent case study of an unglazed collector within sandwich panels, principal design parameters, such as hydraulic circuit configurations, were assessed for their impacts on thermal efficiency. This research emphasizes the importance of understanding how different factors affect the overall efficiency and energy capture of thermal systems, which is important for sustainable architecture and energy-efficient designs. The efficiency curves developed provide a comprehensive picture of energy performance under different conditions and emphasize the actual advantages of a thorough thermal property analysis in design processes (Blanco et al., 2020). The findings developed through such research allow for the creation of novel solutions for the utilization of solar power efficiently and the promotion of advancements in material science and engineering.

Importance of Studying Thermal Properties

The study of thermal properties is at the heart of a wide variety of scientific and industrial processes, particularly those concerned with optimizing processes in which heat transfer is critical. Understanding the thermal properties of materials in food processing industries can also lead to significant gains in terms of energy efficiency. When processes like frying are investigated through systematic analysis with energy balance models, it is understood that detailed analysis of heat and mass transfer becomes important in identifying inefficiencies (Jouhara et al., 2010). Not only does it enhance energy conservation but also maintains the quality of the final product along with minimizing operating cost. Therefore, thermal properties knowledge is not of purely theoretical importance; it is a practical need with immediate economic and environmental sustainability implications in many sectors. Material thermal properties also have significant relevance in high-tech activities, such as those conducted under microgravity by organizations like NASA. Investigation of material behavior in various conditions of heat allows researchers to broaden the borders of material science and discover new phenomena related to fluid dynamics, combustion science, and crystallization. For instance, experiments aboard the United States Microgravity Laboratory have given information on convection driven by surface tension and imply prospects for producing high-end materials that perform more effectively in space (Accardi et al., 2025). The ability to manipulate and understand thermal properties under such environments not only facilitates innovation but also ensures novel material design in terms of how they transfer heat under extreme conditions. Besides utility, applicability in researching thermal properties is also observed in the case of environmental conditions. As global temperatures are rising and energy efficiency gains increased significance, there is a greater than ever demand for materials that will withstand severe thermal fluctuations without wasteful consumption of energy. Investigations into the thermal response of new materials can provide us with solutions that not only maximize system performance but also assist in reducing carbon footprints across industry. Through interaction with such thermal properties, scientists can create materials that are more sustainable in nature and capable of meeting future energy needs. Thus, the study of thermal properties caters to both immediate industrial needs and long-term global sustainability requirements, making it an indispensable role in advancing science and technology.

3.1. TYPES OF THERMAL PROPERTIES

Thermal properties of materials include a wide variety of properties that define how materials respond to heat and changes in temperature. Thermal conductivity, specific heat capacity, and thermal expansion are but a few such characteristics, each making a material highly suitable for a variety of applications. Thermal conductivity measures the ability of a material to conduct heat; high conductivity is especially important in cases of good heat dissipation, e.g., in electronic components. Conversely, low thermal conductivity materials are used for insulation, which lower heat transfer. Engineers and scientists should be aware of these properties when they select materials for use in energy systems, building materials, and consumer goods that eventually find application in resulting innovations toward thermal efficiency and performance enhancement.

Specific heat capacity is another important thermal property that is used to quantify the quantity of energy in terms of heat required to raise the temperature of a unit mass of a material by one degree Celsius. The property is of key use in thermal management, particularly in phase-change applications, e.g., refrigeration systems and heat exchangers. Specific heat capacity also assists in estimating temperature change on being subjected to heat, allowing for better performance design. For example, substances with high specific heat capacities can level off temperature fluctuations, which is beneficial in processes such as food production and storage. With each new material that scientists discover, knowledge of specific heat capacity enables more efficient thermal systems, enabling greater energy efficiency in various applications. Besides, thermal expansion is also an essential thermal characteristic that determines how materials change in size or volume with temperature variations. Materials expand when heated, and this must be carefully accounted for in applications where dimensional stability is crucial, e.g., aerospace and construction. Incompatible thermal expansion coefficients in the joined materials can lead to structure failure, implying the necessity for compatible material and design selection. Thermal expansion calculations are increasingly becoming sophisticated with better machine learning architectures, which provide predictive data that informs material selection processes. For instance, AFLOW-ML software makes it easier to extract thermal property predictions, the incorporation of such critical information into engineering workflows easier, and more effective thermal management concepts within new material design (Asenjo et al., 2012; Carrete et al., 2017) (Figure 3.2).

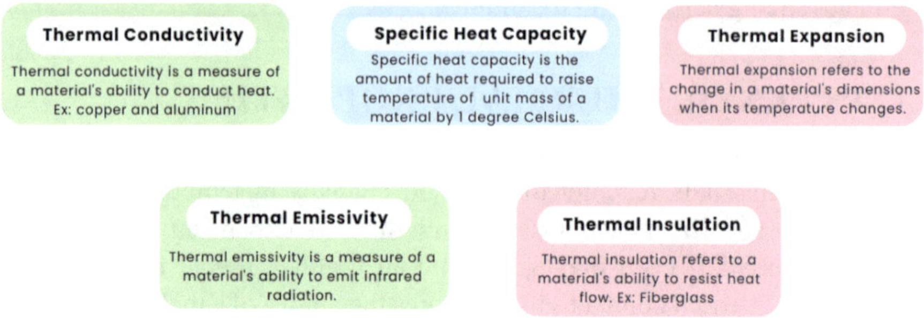

Figure 3.2. Types of thermal properties of matter.

Source: https://images.examples.com/wp-content/uploads/2024/07/Types-of-Thermal-Properties-of-Matter-.png.

3.1.1. Thermal Conductivity

Thermal conductivity is a significant thermal property which governs the manner in which materials conduct heat transfer. The property is most significant in numerous applications, from electronic devices requiring efficient heat dissipation to insulators designed to minimize thermal loss. Thermal conduction primarily involves the process of phonon transport—quantized vibration modes of a material. The interaction of the phonons with the materials structure and atomic lattice is at the center of determining its thermal conductivity. Understanding such dynamics allows scientists and engineers to tailor the material to respond to particular thermal performance requirements, rendering them more suitable for a broad range of technological applications.

Simulation of thermal conductivity has advanced considerably with advances in high-throughput approaches. Recent implementations, such as the quasi-harmonic Debye approximation in the AFLOW and Materials Project suites, demonstrate a revolutionary improvement in our ability to foresee material properties at an affordable cost. The technique provides an ordinal consistent ranking of thermal conductivities across a wide set of semiconductor materials and demonstrates that the Debye temperature, a crucial descriptor of lattice dynamics, can perform better predictions than calculated thermal conductivities. As testified in a study of 182 materials, this method emphasizes the need to take advantage of computational techniques in order to advance our understanding of thermal properties and, thus, pave the way to material innovation across all engineering domains (Asta et al., 2014) (Figure 3.3).

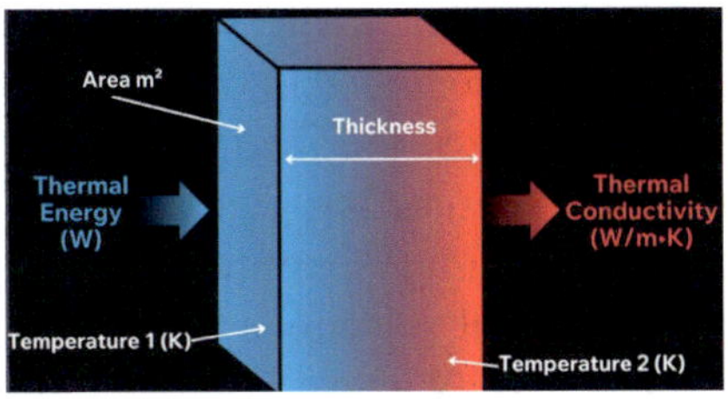

Figure 3.3. Illustration of thermal conductivity.

Source: https://thermtest.com/what-is-thermal-conductivity.

Specific material classes, i.e., monolayer transition metal dichalcogenides, display the intricacies of phonon interactions to show factors influencing thermal conductivity. Material analyses of MoS_2, $MoSe_2$, and WS_2 have established with certainty how the average atomic weight and interatomic bonding characteristics significantly impact thermal performance. MoS_2, for instance, is the most thermally conducting among the group because its lower average atomic weight enables better phonon transportation. Anharmonic vibration and covalent bond contribution is also tremendous; materials with stiff interatomic bonds, like WS_2, have high thermal conductivities despite their more complex lattice dynamics. Such advanced analysis of thermal conductivity provides us with important insights into not just the underlying physics but also ongoing real-world applications in development for material science (Peng et al., 2015).

3.1.2. Specific Heat Capacity

Specific heat capacity is a root thermal material property that quantifies the amount of heat energy it would require to raise the temperature of a unit mass of a substance by one degree Celsius. This property is vital in application for predicting how materials would respond to variations in thermal energy under varying circumstances. For example, high specific heat capacity materials can handle high quantities of heat without a noticeable temperature change and thus are well suited for uses where thermal stability is desired. Materials of low specific heat capacity, on the other hand, heat rapidly, and this can bear implications upon how useful they are in thermal management systems. It is therefore crucial to know specific heat capacity in engineering applications where temperature is important. Measurement of specific heat capacity requires careful experimental techniques, which commonly utilize methods such as differential scanning calorimetry. The accuracy of the measurements in this case is vital, as any deviation would lead to considerable inconsistencies in thermal performance assessment. For instance, research has proven the effectiveness of an integral process where experimental values of thermal conductivity and diffusivity are integrated to obtain specific heat capacity with precision. This process emphasizes the interplay of various thermophysical properties, as suggested by recent studies that describe how such parameters dictate material selection in high-temperature processes. Engineers are thus able to make appropriate material selections for specific thermal conditions. In addition, specific heat capacity plays a crucial role in the operational efficiency of materials in industrial processes. For example, in a factory, the control of temperature is essential to ensure quality output and process stability. The inherent heat retention capability of materials impacts not just their safe operation temperatures but also processing behavior and energy efficiency (Harničárová et al., 2014) (Figure 3.4).

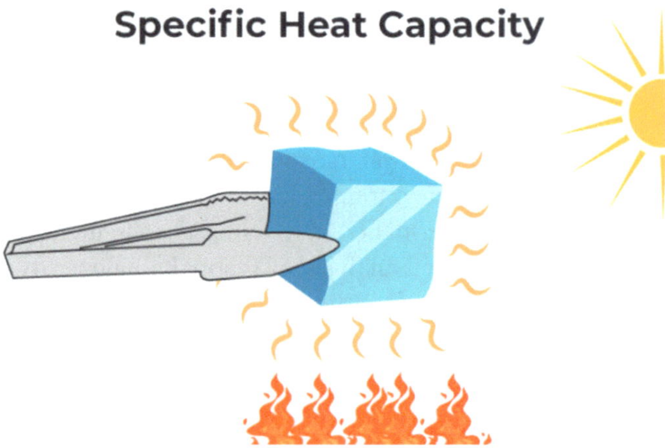

Figure 3.4. Specific heat capacity.

Source: https://media.geeksforgeeks.org/wp-content/uploads/20220916121147/SpecificheatCapacity.png.

Since companies increasingly strive for high performance at minimal energy consumption, the importance of specific heat capacity cannot be overstressed. Analytical methods such as the lumped capacitance model allow one to establish this parameter with good accuracy from cooling curve analysis, which provides an efficient way of designing thermal management systems (Harničárová et al., 2014). By understanding and application of the principles of heat capacity, the engineers can enhance significantly both functionality and reliability of heat systems.

3.1.3. Thermal Expansion

Thermal expansion refers to the physical property of materials that describes how their sizes react to changes in temperature. Most materials increase in size as temperature increases, and this is an aspect that is generally very significant in engineering and construction. This is mathematically described by the use of the coefficient of thermal expansion (CTE), which describes to what extent a material expands in terms of the change in temperature. An understanding of the CTE is essential to ensuring structural integrity since it allows one to predict how materials will respond under the influence of varying temperatures. For instance, when composite materials are utilized, thermal expansion coefficient mismatches can lead to stress concentrations which result in failure if not addressed accordingly. Designers must employ precise values of CTE to design long-lasting systems for a specific thermal environment. Individual materials possess unique thermal expansion characteristics that play a role in defining their application. Ceramics and certain oxides, for example those with a magnetoplumbite structure, are typically characterized by extremely high thermal stability and significant thermal expansion. Recent studies have placed into perspective the thermal expansion coefficients of various oxide materials, such as $LaMgAl_{11}O_{19}$ and $GdMgAl_{11}O_{19}$, with an average value of approximately $9.6 \times 10^{-6}/°C$ upon heating from room temperature to 1500°C. The material is typically used in high-temperature thermal barrier coatings due to its ability to survive harsh conditions without compromising its structural integrity. Also, the thermal expansion coefficient's lack of sensitivity to oxide composition reflects the importance of crystal structure, showing how material design can couple thermal properties to applications (Bansal et al., 2007) (Figure 3.5).

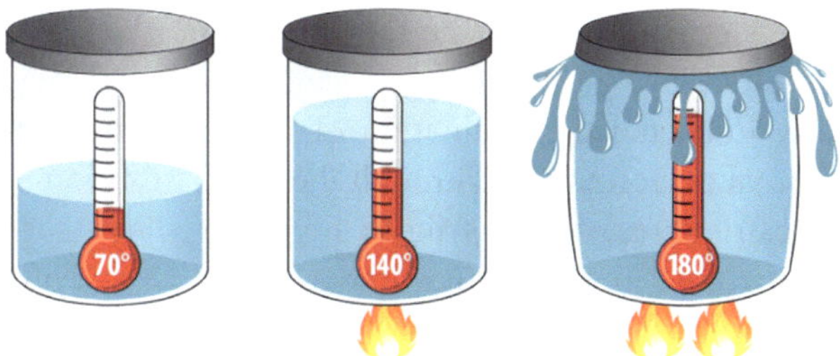

Figure 3.5. Thermal expansion.

Source: https://www.plumbingsupply.com/thermal-expansion.html.

Research continues to evolve in the field of thermal expansion and its use for thermal barrier coatings (TBC). Ionic oxide materials, particularly doped compounds such as Yttrium and Dysprosium, continue to be more prominent due to their enhanced thermal capabilities. For instance, research has indicated that doping ZrO_2 with Dy^{3+} ions significantly enhances the coefficient of thermal expansion compared to undoped counterparts. This connection highlights the extent to which even minimal variation in chemical composition can have dramatically different impacts on thermal properties of materials at elevated temperatures. Beyond this, thermochemical compatibility of the doped material, particularly with Al_2O_3, is crucial for thermal stress resistance such that the TBC systems can be run in temperatures up to 1500°C without drastic degradation (Choy et al., 2016). As research continues to evolve, these technologies become imperative in shaping technology and maximizing energy efficiency.

3.2. MEASUREMENT TECHNIQUES

Measurement techniques are key to understanding the thermal properties of materials, particularly on the micro and nanoscale. A new technique is the suspended thermoreflectance (STR) technique, which employs an optoelectric device that is optimized for greater spatial and temporal resolution of thermal analysis. In contrast to conventional techniques limited by either surface morphology or heat capacity, the STR system generates unconstrained thermal conductivity measurements by studying the change in reflectivity of light while varying the temperature of the material. This is indicative of the sensitivity of the STR method to materials' properties, enabling accurate thermal mapping necessary for application in advanced areas of engineering where thin films are a crucial parameter. These updates point to the necessity of using advanced measurement techniques in ensuring accuracy while evaluating thermal property values, particularly in view of the progression of material science towards more developed structures and operation (Sarkar et al., 2025). In addition to optical measuring techniques, advances in digital image processing have significantly impacted the evaluation of microstructural properties that contribute to thermal performance. Through the mechanization of the analysis of intricate microstructures, such as those found in alloys such as Ti_6Al_4V, researchers are now able to utilize algorithms that are especially designed to identify and quantify features such as grain size and composition. This method not only enhances the precision of microstructural analysis but also accelerates the speed and reproducibility of findings, which are otherwise handicapped by traditional approaches. As a result, the integration of automated measurement techniques provides a more consistent understanding of materials' thermal performance at varied conditions. Thus, such technological improvements present an important change in the measurement techniques that provide the foundation for the analysis of thermal properties in varied materials (Campbell et al., 2018). The demand for precise measurement techniques also extends to verification and calibration of new techniques. For instance, the evolution of STR techniques demonstrates the critical interaction among model theory and experimental confirmation. By comparing results of the STR technique against finite element models and the literature, scientists are able to confirm the accuracy and validity of their measurements of thermal conductivity. The two-pronged mechanism guarantees that the findings are not new alone but scientifically sound as well. Moreover, as new materials are being investigated, flexible measurement techniques will become imperative to characterize their thermal response effectively. As a result, continued refinement of measuring devices and rigorous verification procedures are the focal point for development in material science and for enhancing the performance of heat application in technology (Sarkar et al., 2025).

3.2.1. Steady-State Methods

The concept of steady-state methods plays an essential role in the evaluation of material thermal properties, particularly building and construction. The procedures assume that a material has reached a state of steady-state temperature distribution to facilitate more convenient as well as more effective measurement methods. This is especially useful where transient effects may complicate analysis of thermal behavior. For example, steady-state techniques can be used to determine thermal conductivity by measuring the flow of heat through a material over a period of time with constant temperatures. Such techniques enable comparisons among materials, conformity to regulations and standards, and assistance in developing energy-efficient building design. The analytical approach of steady-state methods enhances our understanding of the thermal response of different materials in real-world applications, ultimately guiding more environmentally friendly building design. Steady-state methods are particularly well-suited when calculating the thermal transmittance, or U-value, of insulating construction materials. A study accounted for the thermal conductivity of materials such as un-fired clay bricks and straw bales by employing a steady-state air-to-air thermal transmittance method. These products showed rates of thermal transmittance below 0.35 W/m^2 K, such that they could meet the stringent UK Building Regulations for energy efficiency. Using steady-state methods, researchers and engineers can determine and optimize the performance of insulation so that buildings are still habitable while using less energy. This measurement of sustainable building material is critical in order to enable advances in green architectural methods. With increasing focus on energy efficiency, steady-state analysis is still the way for introducing new materials in building construction. Apart from thermal transmittance measurement, steady-state methods may also provide information on material classification and performance reliability critical for sustainable design of a building. Thermal property classification with such methods facilitates efficient screening of building materials, their suitability for various applications. For instance, in a study by the Energy Systems Research Unit, it was emphasized how important it is to carefully curate existing thermal property data to allow guidelines for energy-efficient construction to be formed (Clarke et al., 2009). By upgrading the classification systems, scientists can ensure that materials are selected based on their defined thermal performance, enhancing building efficiency in general. Moreover, with more innovations in green design on the rise, steady-state analysis is a benchmark against which compliance with current codes is not only tested but also for paving the way in environmentally friendly materials answering adapting energy requirements (Allen et al., 2004) (Figure 3.6).

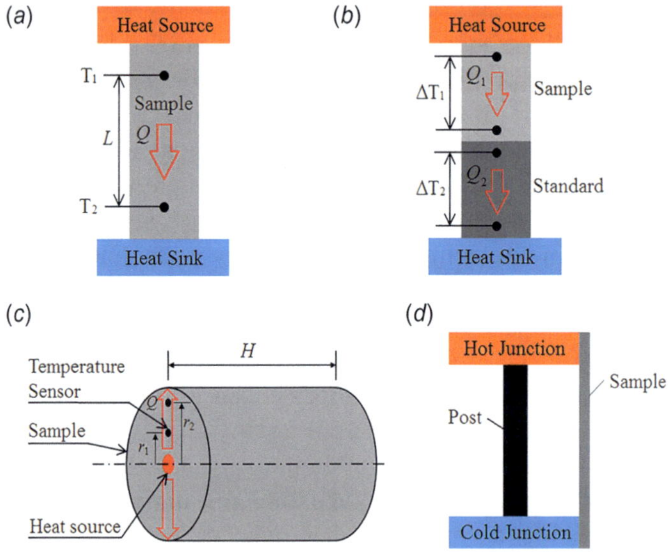

Figure 3.6. Schematic of steady-state methods commonly used to measure the thermal conductivity of bulk materials.

Note: (a) Absolute technique: the sample is placed between a heat source and a heat sink, with known power output, resulting in a temperature drop across a given length of the sample. (b) Comparative cut-bar technique: a standard material with known thermal conductivity is mounted in series with the test sample. (c) Radial heat-flow method: a cylindrical sample is heated from its axis, and as heat flows radially outward, a steady-state temperature gradient is established in the radial direction. And (d) parallel thermal conductance technique: used for measuring small needlelike samples that cannot support heaters and thermocouples. A sample holder or stage is placed between the heat source and heat sink, and a thermal conductance measurement for the sample holder is carried out first to quantify associated heat loss.

Source: https://www.researchgate.net/figure/Schematic-of-steady-state-methods-commonly-used-for-measuring-the-thermal-conductivity-of_fig10_303655808.

3.2.2. Transient Methods

The study of transient methods in the frame of thermal properties of materials is highly demanded due to their ability to provide information related to thermal conductivity under natural conditions. Traditional laboratory techniques are not equipped with the ability to register the complexity of natural conditions, i.e., varying moisture content and the building material's heterogeneous composition. Transient methods, including thermal probe technology, allow for *in situ* measurements that capture the dynamic properties of material responses to heat loading. Utilizing portable measurement kits, researchers are able to acquire data representative of actual thermal performance within buildings, thereby addressing fundamental issues about the validity of thermal properties used in engineering calculations. The findings indicate a notable difference between in-place measurements and laboratory values and highlight the urgent necessity for approaches that are closer to what it is to measure the behavior of material in position (De et al., 2008). In addition to thermal probe technology, transient grating spectroscopy is one of the newer techniques to examine thermal transport within materials. This technique is particularly valuable for the study of thermal phonon transport within materials at size scales comparable to mean free paths of phonons. Unlike previous techniques, transient grating spectroscopy has the capability to examine heat conduction in a two-dimensional system, which has the potential to better elucidate the manner in which heat conducts through various materials in-plane and between planes. The greatest achievement of this method lies in its ability to eliminate phenomenological fitting parameters, which have a tendency to complicate the analysis and interpretation of experimental data. With an analytic Greens function derived from the Boltzmann equation, scientists can infer phonon mean free paths through transient grating measurements, thereby expanding the scope of this method to encompass dense and opaque bulk materials (Minnich et al., 2015). Aside from simple measurement, the implications of utilizing transient techniques extend and essentially redefine the foundation of thermal property knowledge through many materials. As researchers investigate further and create such technologies, the potential for material design and engineering practice creativity widens. Through accurate modeling of material response under transient conditions, engineers can create buildings and components that are more energy efficient and responsive to temperature fluctuations. The ability to combine actual data in the field with thermal models raises predictive analysis accuracy, eventually giving way to wiser decisions on construction and material selection. In general, development of transient approaches significantly contributes to the global Thermal Properties discourse in that more actual connection between theoretical understanding and actual application can be made in material science (Figure 3.7).

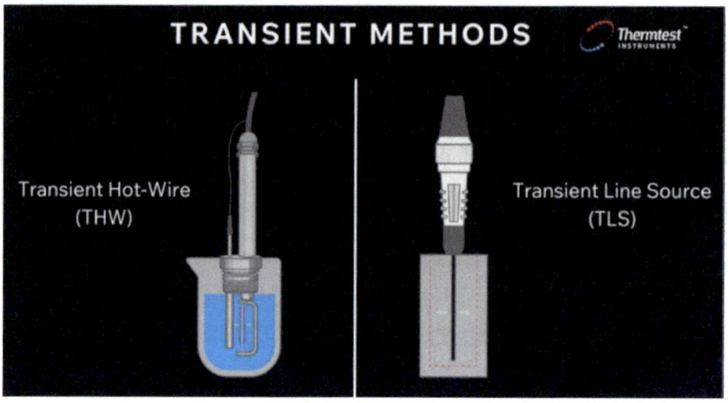

Figure 3.7. Illustrating transient methods for thermal conductivity measurement.

Source: https://thermtest.com/what-is-thermal-conductivity.

3.2.3. Calorimetry

Calorimetry is significant in establishing the thermal characteristics of materials as it enables one to ascertain the heat transfer associated with chemical reactions or phase transitions. This establishment is crucial in evaluating the response of materials to temperature changes, particularly in materials science and engineering. Using various calorimetric techniques, scientists can establish specific heats, phase transition enthalpies, and thermal stability profiles. Engineers and designers needed to select appropriate materials for the construction of buildings when the specifications entail energy efficiency and fire resistance need these values. Calorimetry serves to optimize the performance of materials under various thermal conditions to ensure that materials used meet safety and durability requirements. One major application of calorimetry is in the assessment of wood as a building material, specifically its fire characteristics. Wood, being an organic substance, thermally degrades when ignited through intricate mechanisms, producing gases and byproducts with relevance to its integrity, as well as response to fire. Techniques such as cone calorimetry and thermogravimetric analysis provide close observation of wood combustion and heat release rates when exposed to heat. These reactions are significant, given that wood fire resistance contributes significantly to the safety of buildings in residential and commercial establishments. Besides, since wood is sustainable, continued studies helped by calorimetric methods can lead to new treatments and coatings that enhance fire resistance without diminishing environmental benefits, striving for safer use in construction (Iglauer et al., 2013). For the purpose of creating fire retardant technologies, calorimetry enlightens safer and more effective materials. The calorimetric techniques can measure the impact of silicon-based products and nanocomposite paints on wood thermal characteristics. Such research is important as it determines the efficiency of these materials in inhibiting fire and increasing fire resistance. Further, with the evolution of codes of construction to address sustainability and safety, the usage of calorimetry to examine emerging fire-retardant chemistries increases its significance. The learning gained through this research not only ensures compliance with safety measures but also results in new techniques that promote the use of eco-friendly materials in architecture. Calorimetry, ultimately, transcends measurement; it sets the course of material sciences and determines how we approach safety in thermal use, particularly thermal use involving wood (Hull et al., 2013) (Figure 3.8).

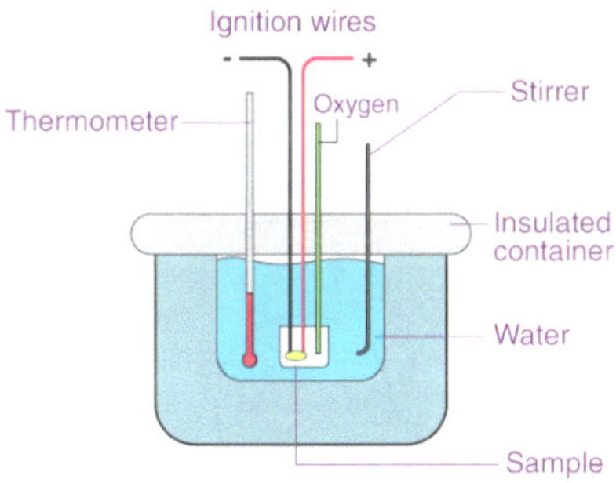

Figure 3.8. Calorimetry measurement of heat transfer.

Source: https://chemistrytalk.org/calorimetry/

3.3. FACTORS AFFECTING THERMAL PROPERTIES

Material thermal properties are highly dependent upon their chemical structure, which determines inherent properties such as thermal conductivity, specific heat capacity, and thermal expansion. Metals, for instance, are prone to having high thermal conductivity due to the presence of free electrons that facilitate heat flow. Insulating materials such as polymers, in contrast, are less conductive thermally because their molecular structure inhibits the flow of heat. This fundamental difference explains the reason various materials are utilized for specific purposes, from construction to advanced electronics. The thermal properties of tungsten carbide, as emphasized by recent studies, demonstrate how material properties are composed, such as the addition of cobalt as part of the production process, to create materials with increased thermal resistance, which is most beneficial in high-stress applications (Agote et al., 2019). The processing methods employed in material manufacture play an important role in determining their thermal properties. Such processes as sintering and field-assisted sintering possess the ability to create microstructure variations, thereby impacting thermal behavior. For instance, the field-assisted sintering technique (FAST) has been able to decrease processing time while facilitating detailed control of the end properties of the material.

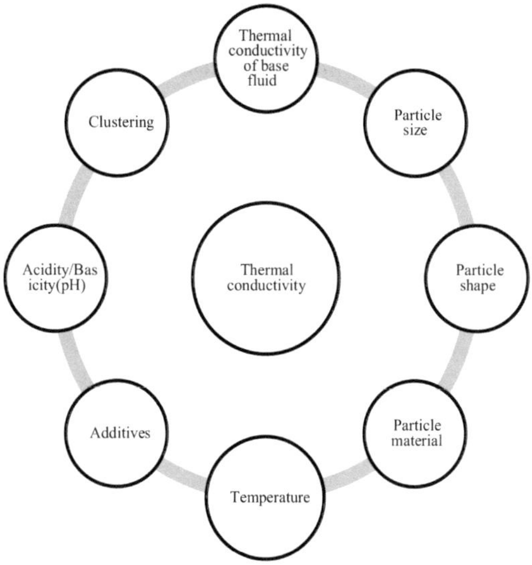

Figure 3.9. Factors affecting thermal conductivity.

Source: https://www.researchgate.net/figure/Factors-affecting-thermal-conductivity_fig2_370537089.

It matters for processes that require rapid thermal stabilization and reliability, especially in metal cutting and wear part production. Studies investigating the correspondence between processing parameters, including compaction pressure and current density, have revealed how processing parameters contribute to the optimally desired thermal resistance in materials like tungsten carbide (Agote et al., 2019). Therefore, processing conditions need to be optimized to tailor the thermal properties of engineered materials. Another major factor that may influence thermal properties is the addition of additives, such as metal ions, during synthesis of organic material. The humic substance example demonstrates how adding metal ions modifies thermal stability through the creation of binding sites that alter the interaction with heat. For instance, the thermosetting features of lignitic humic acid combined with sodium and copper ions show a complex interaction under which stability to thermo-oxidation can be improved with structured ionic modification. Thermogravimetric analysis is one approach utilized for the characterization of such modifications, illustrating the formation of porosity with increased water retention capacity, an important feature in numerous environmental and industrial applications. Therefore, knowledge of interactions between additives and base materials is required for improving the thermal properties of composite substances in real-world applications (Figure 3.9).

3.3.1. Material Composition

Material composition and thermal characteristics are of utmost significance in material engineering and science. At the atomic level, the composition and kind of atoms within a material inevitably influence its heat response. Specifically, materials composed of heavy atoms possess higher thermal conductivity due to the increased mass facilitating energy transfer by atomic vibration. Conversely, materials with less dense or more complex atomic configurations can block heat flow and create lower thermal conductivity. Hence, the understanding of the atomic composition allows one to strategically design materials capable of properly regulating the conduction of thermal energy, the importance of which is at its peak in applications such as thermal barrier coatings and insulation technology. The studies of some compositions, such as those based on magnetoplumbite oxides, determine certain thermal properties due to their ordered structures. Studies have shown that materials such as $LaMgAl_{11}O_{19}$ and its doped counterparts exhibit improved thermal stability and varying thermal conductivities upon exposure to elevated temperatures. For instance, thermal expansion coefficients from room temperature to 1500°C suggest an average value of approximately 9.6×10^{-6}/°C for such oxides, supporting the fact that the composition has a significant influence on thermal response. Quite interestingly, doping some compositions may alter their thermal properties, as represented by reduced thermal conductivity in the case of $Gd_{0.7}Yb_{0.3}MgAl_{11}O_{19}$ compared to its undoped form. This accounts for the subtle interplay between material composition, dopant elements, and thermal performance, requiring material scientists to consider these elements in their high-temperature designs (Bansal et al., 2007). In addition, a study of material composition also reveals its effect on variable thermal properties beyond thermal conductivity, such as melting and boiling points and heat of atomization. Studies have shown that these properties are linked by a common set of parameters that include the number of atoms per unit volume and bonding properties of the atoms. Such findings imply that the material density and underlying atomic structure can be used to quantify a variety of thermal properties quantitatively. As illustrated with the case of studies on solid parts and alkali halides, such comprehension of these connections is paramount to establishing the impact of variations in material constitution on thermal behavior in various substances, ultimately guiding the development of materials of specific thermal functions (N/A, 2025).

3.3.2. Temperature Variations

Temperature changes play an essential role in the regulation of material thermophysical properties, their regulation under diverse environmental conditions. Upon temperature changes in the materials, their thermal expansion coefficients are critical, particularly in construction and manufacturing technology. For instance, a difference in thermal expansion coefficients of pore fluid and solid matrix can lead to thermal pressurization effects. A good example of this is in the hardened cement paste, where abrupt temperature changes can induce pore pressure variations, thus affecting structural integrity of structures, for instance, oil well cement sheaths. Understanding such interactions is of significant importance to the production of materials that are insensitive to thermal stresses and thus remain stable and dependable in the long term for various applications (Ghabezloo et al., 2008). Temperature differences also offer insights into the phenomenon of thermal rectification in radiative heat transfer between materials. The effect becomes significant when two materials such as silicon carbide (SiC) and silicon dioxide (SiO_2) are subjected to different thermal environments. Research indicates that increasing the temperature difference between two materials also amplifies the thermal rectification effect, a characteristic critical to the creation of new thermal control systems. Upcoming technologies will likely entail thermal diodes that capitalize on this process of rectification to increase energy efficiency in electronic devices. The near-field thermal coupling demonstrates how drastically the radiative heat flux is modified with the temperature difference and distance of separation, shedding light on the thermal interaction of materials going beyond their bulk nature (Drevillon et al., 2015) (Figure 3.10).

Besides, the implications of temperature variations extend beyond mere theoretical models to practical use in material science. For instance, compositing material that effectively responds to thermal stresses due to temperature variations becomes important in aerospace and automobile engineering. Material composition could be optimized through research on thermal properties of their constituents under varying conditions. Researchers continue to experiment with the addition of chemicals that offer thermal stability or hold back adverse reactions to extreme thermal fluctuations, ultimately creating materials that can endure extreme conditions and maintain structural stability. The pursuit of testing temperature fluctuations not only advances the boundaries of material science as a discipline but also initiates technological and engineering innovations, reflecting the interrelation between thermal properties and material performance (Ghabezloo et al., 2008).

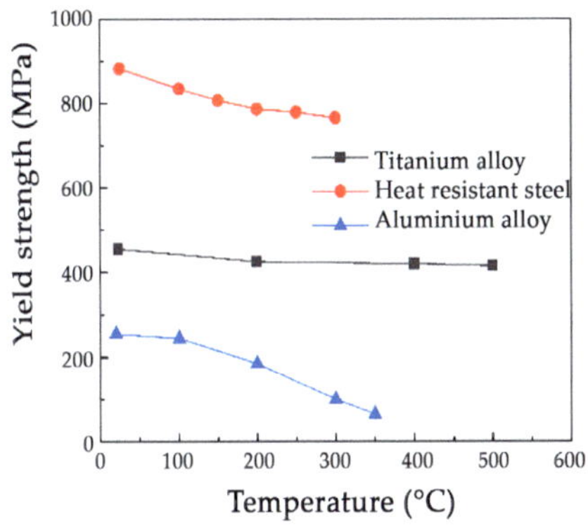

Figure 3.10. Variation of thermal conductivity with temperature for different materials.

Source: https://www.researchgate.net/figure/ariation-of-thermal-conductivity-with-temperature-for-different-materials_fig1_338292949.

3.3.3. Structural Changes

The thermal material properties are intimately connected with their structural changes, particularly with temperature changes. In heating, atomic reorganizations may occur, leading to expansion, phase transitions, or even structural damage. Such changes are of crucial relevance in applications ranging from aerospace engineering to biomedical devices where the material response must be regulated with high accuracy. For instance, while planning for surgery, it is important to have a clear understanding of bone structural changes. A haptic interface will be advantageous to surgeons as it provides a virtual model that mimics such changes in real time, enhancing bone manipulation procedures during surgery, thus avoiding potential complications (Arnez et al., 2012). Through this interacting set of thermal properties and structural integrity, the materials can be made to withstand harsh environments, having a direct influence on their operational and reliability performance in actual use. In addition, the longevity and resistance of materials could be significantly influenced by thermal-induced structural alterations, manifesting as microcracking or other forms of degradations. In application in high-performance fields like aerospace and automotive, employment of high-temperature-resistant advanced materials with the capability to maintain structural integrity at high temperatures is required. Research highlights the importance of high-temperature-resistant materials and composite structures with high resistance against thermal stress and degradation (Davies et al., 2025). This innovation not only increases functionality, but also performance and safety of components during cyclic thermal loading. Although scientists and engineers are working to create new materials, it is simple enough to design the systems that will work under harsh environments with consistent performance if one understands how thermal conditions influence structural changes. The range of possible applications is expanded. It is also critical to forecast the structural changes in materials as thermals fluctuate for predictive modeling and to design future materials. Nondestructive techniques can be employed to monitor these changes effectively, giving instantaneous material integrity measurements without diminishing their utility. As materials progress with greater design considerations and life predictive methods, precise thermal behavior simulation is necessary (Davies et al., 2025). Not only does such simulation assist in creating longer-lasting materials, but it also assists in planning for usage based on real conditions within a controlled environment. Lastly, the interlinking between the thermal properties and the structural transformation facilitates innovation, setting new pathways for engineered materials to achieve sought-after performance characteristics in different fields, thus contributing to material science and technology advancement.

3.4. APPLICATIONS IN ENGINEERING

The finding of thermal properties in engineering is of increasing relevance, particularly with the necessity of developing materials with the ability to efficiently regulate heat. Nanotrusses have been a groundbreaking innovation under such a scenario. Made up of hollow nanoscale beams arranged in a periodic truss structure, nanotrusses are able to overcome the common limitations of low density, low thermal conductivity, and high stiffness all at once, all of which are properties that are traditionally incompatible with traditional materials. As observed from recent studies, the thermal conductivity of such materials can be quantified using two parameters: wall thickness and solid fraction, which identify their unique thermal and mechanical characteristics. One is able to design thermal performance and expand new engineering applications using this new design, particularly in fields where engineering components with very good thermal characteristics and light weight are required, such as aerospace and automobile engineering (Dou et al., 2015). Carbon-based material applications in engineering demonstrate the rich variety of possibilities obtained through thermal property manipulation. Interestingly, carbon allotropes such as graphene and carbon nanotubes show high thermal conductivity with a value range extending over five orders of magnitude. This range is crucial for engineers to provide ideal materials for applications involving thermal management. New disclosures of the thermal and thermoelectric properties of these carbon materials open doors in numerous industries, from electronics to energy systems. Graphene's ability to enable efficient heat conduction while remaining light in weight, for instance, positions it well as a potential solution to enhance the thermal performance of electronic components. Besides, the unique size-dependent thermal conduction in two-dimensional materials like graphene provides new solutions for engineering improved thermal dissipation in high-performance devices (Balandin et al., 2011). New materials with engineered thermal performance are the drivers of the future in sustainable engineering. With the increased demand for energy efficiency and sustainable engineering, engineers must not only consider the material's mechanical properties but also its thermal control capabilities. The development and utilization of novel structures, such as nanotrusses and carbon allotropes, provide the potential to transcend boundaries in traditional engineering practices. Enhanced thermal properties make these materials essential in applications from building thermal insulation to solar energy system heat exchangers. Not only do such applications improve efficiency but also contribute to energy and emission saving, aligning engineering practice with global sustainability targets. With emphasis on materials that provide improved thermal performance combined with mechanical strength, engineers can move ahead to a greener future by taking advantage of the thermal characteristics of cutting-edge materials to address the challenges of tomorrow (Balandin et al., 2011; Dou et al., 2015).

3.4.1. Insulation Materials

From the thermal properties point of view, insulation materials play a critical role of enhancing energy efficiency in buildings. They are designed to limit the transfer of heat, thereby controlling indoor temperatures and reducing heating and cooling demands. From among a variety of insulation materials, natural alternatives, such as hemp-lime, have been framed as green alternatives to traditional ones. Specifically, hemp-lime insulation can be applied to the outside face of monolithic earth walls like cob to improve their thermal performance significantly. The literature has confirmed that conventional earth walls typically have poor thermal performance, with U-values of over 1 W/m^2K, which is not of modern building standard. While the retrofitting of the walls with 250 mm hemp-lime insulation can reduce the U-value to the current target of 0.3 W/m^2K, demonstrating the potential that can be tapped from eco-friendly materials in offering more efficient conventional building processes (Goodhew et al., 2017) (Figure 3.11).

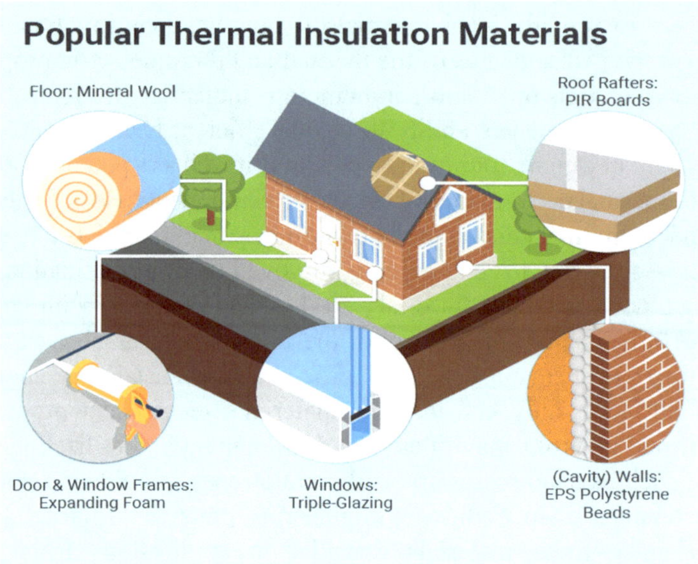

Figure 3.11. Common thermal insulation materials.

Source: https://cdn.greenmatch.co.uk/cdn-cgi/image/format=auto/2/2024/08/image-2.png.

As the construction industry seeks to reduce its carbon emissions, the design of cementitious binders has been of interest. Replacement of conventional Ordinary Portland Cement (OPC) with lower-carbon alternatives is one of the main strategies to reduce construction material emissions. Some of the new binders composed of metakaolin, silica fume, and nano-silica have been spotted by researchers as having significantly improved heat performance compared to OPC. These new materials have thermal conductivities 58–90% lower than traditional cements and thus possess good insulation qualities with less mass. Even a similarly thin application of these new binders can put household building energy emissions into compliance with current building standards. These innovations highlight the dual benefit of saving operational energy needs along with solving the urgent problem of carbon emissions in buildings (Hamilton et al., 2018). While using high-performance insulation materials has several advantages, one has to consider the bigger picture of how they would be brought into mainstream building practice. The performance, sustainability, and building life trade-offs must carefully be balanced. Notably, the use of high-performance insulation materials should never cause problems with moisture, particularly in heritage buildings that rely on breathability for durability. Products like hemp-lime provide a permeable solution, preventing condensation and moisture retention problems from non-permeable ones. Additionally, next-generation cementitious binders not only improve on thermal performance but also bring recyclability to the table, further enhancing their green reputation as sustainable solutions. Thus, the decision-makers in the building construction industry need to take into account these key considerations while selecting insulating materials with the aim of balancing building structural integrity, environmental considerations, and thermal efficiency (Goodhew et al., 2017; Hamilton et al., 2018).

3.4.2. Heat Exchangers

Heat exchangers are at the heart of numerous engineering applications, as they accomplish effective heat transfer of thermal energy between mediums without actual contact. Their operation is based on thermal characteristics such as thermal conductivity, heat capacity, and flow arrangement, which identify effectiveness and efficiency. The design must accommodate an array of operation parameters including temperature gradients, fluid velocities, and pressure drops. A recent study highlighted that heat recovery in fuel cell power plants can significantly enhance energy availability by utilizing more appropriate heat transfer designs to replace traditional air-cooled heat exchangers to transfer heat to integrated utility systems. In a 40-kW power generation plant, this approach indicated that as much as 84 percent of the fuels lower heating value could be harnessed, emphasizing the critical position optimized heat exchanger design occupies in improving energy utilization in industrial applications (Clausi et al., 2025). Materials development used in heat exchangers is at the core of how their performance can be improved and their applications expanded. High-performance materials such as metal foams and metal matrix composites (MMCs) have emerged as possible alternatives due to their thermally and structurally engineered features. The materials contain tailored properties that maximize heat transfer with better efficiencies and thus suit many forms of heat exchangers from liquid-liquid to gas-gas heat exchangers. Recent research has reviewed these materials properties and examined their suitability in other heat exchanger systems and found that accurate material selection is essential to ensure optimum operation efficiency. Despite the suitability challenges, there are still manufacturing issues to be dealt with, requiring further investigation and optimization to facilitate wide-scale usage of these materials in heat exchange operations (Han et al., 2012). The efficiency of heat exchangers is intrinsically related to their design parameters, and these have to be optimally optimized to accommodate unique operating conditions. Typical design configurations include counterflow, parallel flow, and crossflow configurations, each with characteristic strengths and weaknesses with respect to thermal performance and pressure loss. Counterflow heat exchangers, for instance, are prone to offering superior efficiency in the form of uniform temperature gradient along the direction of flow. In the realm of emerging technologies, the integration of high-performance materials into heat exchanger design holds the potential to break current constraints, thereby enhancing thermal conductivity while reducing weight and spatial demands. This marriage of innovative design and integrated materials analysis highlights the vital relationship between heat exchanger design and materials' inherent thermal response, and as such, it is a core area of research for future energy systems and environmental studies programs (Clausi et al., 2025; Han et al., 2012) (Figure 3.12).

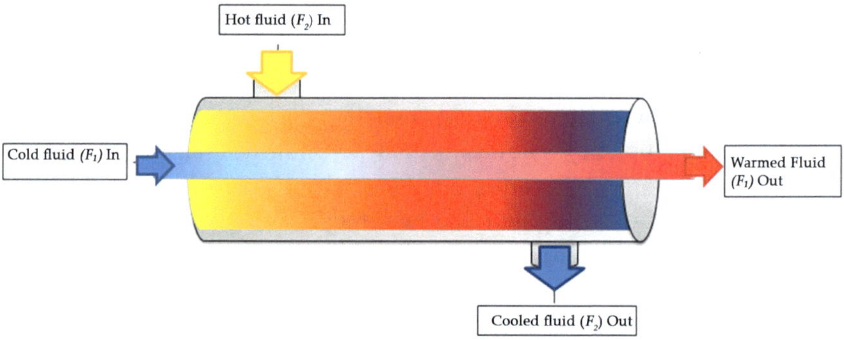

Figure 3.12. Heat transfer in a double pipe heat exchanger.

Source: https://heatexchan.blogspot.com/2019/12/types-of-heat-exchangers.html.

3.4.3. Thermal Barrier Coatings (TBCs)

Thermal barrier coatings (TBCs) are an important materials science development, particularly for applications that call for increased thermal resistance and toughness, including aerospace and power plants. TBCs are used to form an effective thermal insulation layer that protects substrates from aggressive temperatures and oxidative wear. By effecting the significant reduction of heat conduction to the substrate materials, TBCs enhance high-temperature component life and efficiency. Their performance relies on several factors, e.g., thermal conductivity, thermal expansion coefficients, and microstructural integrity under service conditions. As the synthesis technology of materials has developed, e.g., sol-gel processing and plasma spraying, TBCs can be tailored at the microstructural level to have specified thermal properties without sacrificing mechanical strength. Recent research has unveiled possible candidate materials for TBCs, particularly those of magnetoplumbite structures, which exhibit high thermal stability and low thermal conductivity. For instance, oxide compounds like $LaMgAl_{11}O_{19}$ and $GdMgAl_{11}O_{19}$ were found to possess an average thermal expansion coefficient of approximately 9.6×10^{-6}/°C over a temperature range of up to 1500°C. Doping such compounds with elements like Yb has the effect of reducing thermal conductivity, thus improving the insulating characteristics of TBCs. Besides, the coatings have high sintering resistance, with their protective properties still intact even when subjected to temperatures almost reaching 1600°C. These properties render magnetoplumbite-based TBCs extremely good materials for high-stress thermal environments, showing how composition-performance interaction plays a crucial role in thermal barrier technology (Bansal et al., 2007). The mechanical integrity of thermal barrier coatings is also as valuable as their performance effectiveness, especially in applications such as rocket thrust chambers, where conditions suddenly change. The effective thermal conductivity of metal–ceramic composite coatings was found to be between 0.7472 and 4.483 W/(m·K), with excellent thermal insulating capability during rocket flight. The coatings are successful because the controlled particle-to-particle interaction achieved in the plasma spraying process plays a more important role in reducing contact resistance than the intrinsic thermal conductivities of the materials involved. This finding brings to focus the importance of processing techniques in ensuring maximum thermal characteristics of TBCs, and it suggests that future advances in deposition technologies are likely to bring even superior thermal insulators to enhance the overall system performance under harsh thermal conditions (Price et al., 2025).

3.5. ROLE IN ENERGY EFFICIENCY

The importance of thermally material properties in enhancing energy efficiency cannot be emphasized sufficiently, given that the world is currently grappling with the challenges of sustainable production and utilization of energy. Materials with good thermal properties allow for effective heat transfer and control, which plays an important role in numerous applications, varying from building insulation systems to components in renewable energy devices. In perovskite solar cells, for example, the molecular packing and charge transport properties have been directly linked with energy efficiency outcomes. Shorter alkyl chain substituents have been found to allow tight structural assembly, leading to increased hole mobilities and reduced charge recombination. This relationship requires the selection of materials with improved thermal properties in order to enhance the photovoltaic efficiencies, as evidenced by the high-performance values assigned to some derivatives, where methoxy-shaped molecule achieved a power conversion efficiency of as high as 5.63% in operating solar cells (Agarwala et al., 2018). Furthermore, the design and use of thermoelectric materials play a significant role in the production of energy efficiency technology. Thermoelectric converters convert heat energy to electrical energy and, on the other hand, electrical energy to heat through the thermal and transport properties of materials used in their manufacture. While up to now the efficiency of thermoelectric converters has been low, recent studies attempt to develop materials with high thermoelectric properties to operate under the conditions of modern energy systems. The focus on electron crystal-phonon glass type materials seeks to advance thermal management, which in return can drive improved efficiencies in conversion. By advancing the working fluid's mechanics and the thermodynamic characteristics of the systems, researchers seek to bridge the gaps that have been identified and maximize the energy output of thermoelectric devices, thus taking a central role in energy efficiency for various applications (Benenti et al., 2016). Lastly, the intersection of energy efficiency and thermal properties is an extremely critical point in modern material engineering and science. With the synthesis of materials getting better by the day, the thermal conductivities and thermodynamic properties have to be maximized in most energy systems. Understanding how specific materials respond to thermal stress, which can radically affect energy performance, will lead to more efficient systems and less energy loss in applications from industrial processes to consumer electronics. The synergy of leading materials with enhanced thermal capabilities has the prospect to reshape energy efficiency across a range of industries, making way for sustainable activity that is increasingly essential to combating climate change and enabling resource conservation. With advancing research across disciplines, it will try to make innovations that continue to increase energy efficiency provided by thermal properties.

3.5.1. Impact on Building Design

The impact of material thermal characteristics on the design of buildings becomes increasingly critical with the rising global environmental concerns. Building, as a cause of nearly 30% of carbon dioxide emitted globally, is under scrutiny regarding its energy and material usage efficiency. Emerging designs, often rooted in dynamic simulation tools, demonstrate that there is a large performance gap wherein real energy efficiency differs from theoretical calculations on the basis of material characteristics. This performance gap, as it has come to be known, is a call for the balancing of theoretical and practical aspects of building material performance, with specific emphasis being placed on designing with an explicit understanding of the thermal performance of materials. Through this attention to ensuring physical building material properties are carefully thought through, energy efficiency can be significantly enhanced, as well as reduce the impact of the built environment on the environment (Jankovic et al., 2019). In the quest for improved building design, sustainable design matters a lot, particularly in the thermal performance of the wall systems. Studies on the use of waste materials as components of the building are a very promising field, particularly in economically deprived regions, such as Malaysia. Development of affordable housing must take into consideration the thermal performance, where conventional materials are not enough. With material mix optimizations, such as waste glass and palm oil processing by-products, scientists are creating wall systems with better thermal efficiency at no additional cost.

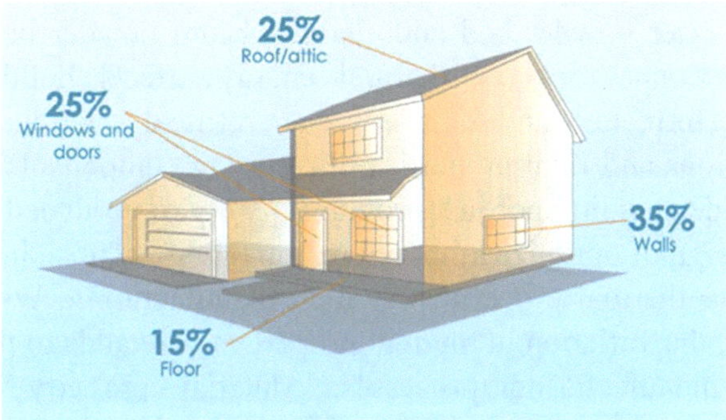

Figure 3.13. Mapping out heat loss on a standard house design.

Source: https://thermtest.com/thermal-resistance-thermal-conductance-of-building-materials.

These innovations not only talk about environmental sustainability but also present practical answers to increase energy efficiency in low-income housing conditions, having a vision for new building works (Narendra et al., 2017). Further, the application of advanced material science to building construction comes with employing analytical technology such as simulation software to accurately predict thermal performance. Application of software like ANSYS WORKBENCH allows architects and engineers to analyze temperature distribution and rates of heat transfer through sophisticated wall systems. This level of analysis can give huge differences in energy consumption; for instance, thermally enhanced sustainable hybrid (TESH) wall systems have been shown to reduce energy consumption by over 10% compared to conventional systems. Lower operating energy costs reaffirm the economic benefits of making an investment in innovative materials, solidifying their place not only in sustainable design but also in cost-effective building solutions. Hence, embracing the thermal properties of materials makes way for a more robust and more efficient built environment (Narendra et al., 2017) (Figure 3.13).

3.5.2. Influence on Manufacturing Processes

The influence of thermal properties in manufacturing is substantial, especially in advanced processes such as additive manufacturing (AM). As industries increasingly depend on AM for the manufacture of complex components, an understanding of thermal behavior of materials is imperative. For instance, NASA Glenn Research Center is investigating the titanium, copper, and nickel alloy thermal profiles during rocket propulsion component manufacturing. These studies utilize sophisticated characterization techniques that monitor the microstructural changes induced by thermal treatments. By investigating processes like laser powder bed and electron beam powder bed, researchers are able to account for how thermal energy affects build density and mechanical performance and thus enlighten critical relationships between thermal conditions and final properties of resulting components (Carter et al., 2025). This understanding not just provides quality to produced parts but also informs the design of optimized thermal treatment procedures, highlighting the strong influence thermal properties have on manufacturing. With composites manufacturing, the variation in thermal properties only adds to the complexity of affecting manufacturing processes. Material property changes and boundary conditions can greatly influence the outcome of each manufacturing process. This statistical fluctuation can cause defects which compromise the structural integrity of composites. The interplay among process parameters and material thermal properties demands a cautious method of manufacturing design. Advanced stochastic simulation techniques are good at quantifying the variabilities and provide the insights necessary for engineers to develop optimum process designs for composite parts. Understanding the mechanisms through which thermal properties influence process-induced defects, therefore, becomes fundamental to quality assurance and manufacturing consistency. The understanding can contribute significantly to reducing variability risks throughout composite manufacturing processes, guaranteeing the supply of more consistent end products (Long et al., 2014). Furthermore, integration of the results from thermal property research into the industrial manufacturing setting can yield significant benefits in efficiency and performance. With the emerging industries bringing with them new frontiers of material performance, thermal property manipulation through processing becomes a competitive advantage. Materials' thermal history—pre-production to post-processing—does not only affect their mechanical property but also their suitability for specific functions.

This complete process allows manufacturers to tailor processes to the demanding performance requirements of industries such as aerospace, automotive, and medical devices. Recent research indicates that when thermal properties are tweaked and analyzed in detail, companies are able to reduce wastage, enhance product quality, and push up production timelines. With this perspective of the pivotal position of thermal properties in dictating manufacturing processes, industries can continue to build on their innovation and advance technological frontiers (Figure 3.14).

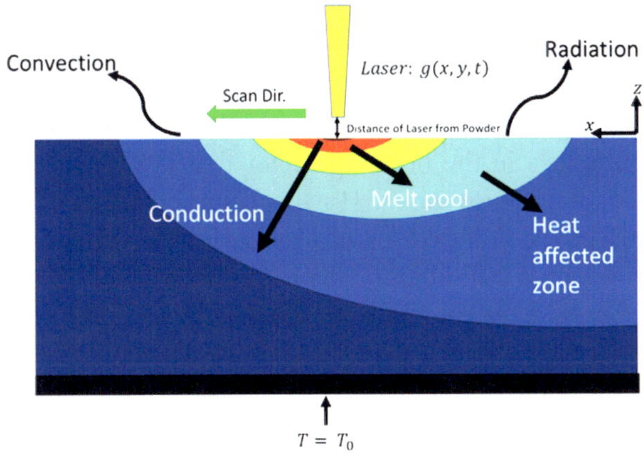

Figure 3.14. Heat transfer in laser-based metal additive manufacturing.

Source: https://www.mdpi.com/2504-4494/2/3/63.

3.5.3. Contribution to Sustainable Practices

The drive towards sustainable building practice is increasingly known as a key driver to curb climate change. The thermal properties of building material contribute significantly to energy efficiency and overall environmental impact. Enhanced thermal conductivity and insulating properties are capable of reducing the necessity for artificial heating and cooling systems, in turn, reducing the greenhouse gas contribution. Employing materials of superior thermal performance not only conserves energy but is also inherent in creating comfortable and cost-effective environments. Sustainable practices often involve material usage and choice innovations, enhancing the shift towards greener building practices in conjunction with global efforts to reduce carbon profiles. When considering pervious concrete pavements (PCPs), their thermally distinctive properties bring to light their contribution towards sustainability, particularly in cold climates where winter poses a challenge. Studies have shown that the efficiency of ice and snow control on such surfaces is directly related to their thermal performance. Through the establishment of theoretical models such as the Enhanced Integrated Climatic Model, practitioners can forecast the surface temperature of PCPs and maximize the application of deicing agents. Such practices offer not only protection but also minimize the negative impact on the environment that was heretofore associated with the application of deicing chemicals. The integration of effective thermal management into construction processes directly furthers the general goal of enhancing urban infrastructure sustainability, thereby demonstrating a practical approach to environmental stewardship (Chen et al., 2017). At the broader level of building practices, policy initiatives aimed at stimulating green building practice to a great extent shape the use of green building methods. The UK legislations such as the Code for Sustainable Homes encourage the utilization of environmental materials and practices in the building sector that is well known for its high emission rate of greenhouse gases. However, the critiques of green building advocates highlight the deficiencies of current law in driving innovative and contextually appropriate building practices. Through the questioning of current standards and the pushing for more inclusive green building definitions, the stakeholders can drive the industry toward means not only in compliance with sustainability but also actually cutting carbon emissions on a large scale. Thus, the ongoing dialogue in the construction sector is required to continue development of sustainable processes that align with future global standards (Gibbs et al., 2015).

3.6. RECENT ADVANCES IN THERMAL MATERIALS

The development of thermal materials has gained immense pace in recent times, particularly in the context of Nuclear Thermal Propulsion (NTP) systems. Cermet materials, which are composite materials consisting of ceramic particles dispersed in a metal matrix, have been identified as being highly applicable for high-temperature applications. The materials exhibit an extremely high ability to trap fission products and fuels and also show high thermal shock resistance as well as high strength. NASA's recent activities included net shape production of ceramic and refractory metal components, for example, tungsten-rhenium-hafnium carbide (W-Re-HfC), to facilitate the development of efficient cermet fuels. This is critical to the enhancement of thermal stability and performance of propulsion systems to accommodate the extreme conditions of operation. Further, the reduction of low-cost production processes through means like Vacuum Plasma Spray (VPS) is important for utilization and can completely revolutionize the material science associated with NTP (Hickman et al., 2025). At the same time, two-dimensional materials and their heterostructures have evoked increased interest in the research associated with Electrochemical

Energy Storage Systems (EESS) to achieve a thorough understanding of the thermal transport behavior of these materials. Such materials, such as graphene and transition metal dichalcogenides, are of interest due to their high surface area and electrical conductivity. Successful incorporation in supercapacitors and batteries requires that there be a deep understanding of the thermal transport mechanisms in the materials. Experimental characterization of the thermal dynamics in these advanced structures is still a challenge, and this results in the use of advanced theoretical and simulation tools. Techniques such as first-principles calculations and molecular dynamics simulations are highly informative in thermal transport but computationally expensive. The availability of machine learning (ML) methods has thus emerged as an effective means of enhancing predictions of thermal properties and addressing the demands of rapid material characterization (Datta et al., 2023). The synergy of the most recent advancements in thermal materials and computational approaches provides promising directions for future research.

Figure 3.15. Nuclear thermal propulsion (NTP) systems.

Source: https://singularityhub.com/uploads/2024/10/nuclear_thermal_propulsion-concept_NASA.jpeg?auto=webp&auto=webp&optimize=high&quality=70&width=1440.

Thermally transport performance matching with mechanical stability in low-dimensional heterostructures derived from materials remains an ongoing challenge. Optimization of interlayer distance in bi/multilayer structures holds significant implications for both thermal conductivity and structural stability. Moreover, research into new composite structures has the capability to develop materials that outperform current standards for performance in thermal properties. Future research needs to concentrate on integrating ML techniques into material design to allow for simultaneous optimization of several properties. The multi-parameter approach offers a systematic way of finding materials with better performance in various applications, particularly in energy storage and propulsion technologies. Finally, the future detection of thermal materials, combined with advanced computational approaches, will be essential to meeting the increasing needs of contemporary technology (Datta et al., 2023) (Figure 3.15).

3.6.1. Nanomaterials

The finding of nanomaterials has revolutionized our understanding of thermal properties in various materials. In contrast to their bulk counterparts, nanomaterials exhibit unique behaviors due to their smaller dimensions, which significantly affect thermal conductivity, strength, and stability. For instance, nanomaterials can possess enhanced thermal conductivity depending on their high surface area to volume ratio. This particular characteristic enables improved heat transfer capabilities in applications ranging from electronics to aerospace engineering. Nanoscale size effects lead to unmatched performance, and this has encouraged scientists to investigate the potential of enhancing thermal management systems. The developments indicate the significance of understanding thermal properties at the nanoscale, as they provide avenues for the development of more efficient and longer-lasting materials across various industries. Interaction with the environment by nanomaterials also defines their thermal properties. Under various external conditions like temperature and loading rates, nanomaterial mechanical stability can also switch violently. For instance, in recent graphene monolayer experiments, researchers discovered that the failure strength distribution is sensitive to such parameters and exhibits complex behaviors governed by thermal and rate-dependent mechanisms. This trend, also known as a crossover between regimes, indicates that consideration must be given to the studying of nanomaterials in both the traditional material science and new theories such as extreme value theory. These interdependencies are not only broadening the knowledge horizon surrounding thermal properties but also aid in material design for a particular thermal performance requirement (Sellerio et al., 2015). In addition, the nanomaterials that have been developed by integrating nanomaterials with traditional polymers have displayed significant shifts in thermal properties. Experiments on the development of polyamide 6 (PA6) composites with exfoliated graphite have demonstrated that the incorporation of nanomaterials has significantly enhanced the thermal and mechanical properties of polymers. Atomic force microscopy (AFM) shows that the nanoscale interactions have the ability to increase dispersion and compatibility between the matrix and reinforcing phase. Further, transmission and scanning electron microscopy provide insights into the morphologies of the composite materials themselves yet again highlighting the implications for thermal conductivity and mechanical strength. These advances thus point to the potential for nanomaterials not only to revolutionize the very thermal behavior of known materials but also to create new applications across industries requiring thermal management solutions of high performance (Bocchini et al., 2012).

3.6.2. Phase Change Materials (PCMs)

Phase change materials (PCMs) have attracted significant attention from researchers working in the field of material thermal properties, primarily for their ability to absorb and relinquish heat energy during phase change. The materials, normally solid, liquid, or gases at normal conditions, are exceptionally useful where increased thermal control is required. Through the phase transition—normally from solid to liquid to solid—PCMs are able to store appreciable latent heat capacities.

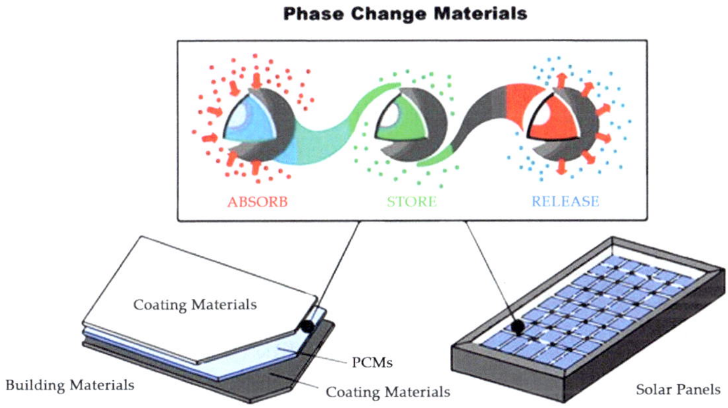

Figure 3.16. Phase change material (PCM).

Source: https://imgur.com/obs3lKL.

This property makes them useful in applications where temperature control is crucial, such as thermal energy storage in buildings. The use of PCMs can significantly improve efficiency in terms of energy use, particularly in places with big differences in temperature. This functionality not only aids in energy demand and supply balancing but also facilitates sustainability in energy utilization by reducing the requirements for active heating and cooling systems (Hale et al., 1972). In building construction, the use of Phase Change Material Wallboards (PCMWs) has emerged as an excellent solution to aid thermal performance. PCMWs offer superior energy management through effective moderation of indoor temperatures to mitigate the usual thermal changes that would necessitate additional heating or cooling. Numerical modeling of PCMWs shows that melting temperature, latent heat, and thermal conductivity are key factors to consider in the performance of thermal performance. For instance, walls with embedded PCMs are capable of providing thermal buffering, allowing for greater energy storage at temperatures at their peak and releasing the stored energy when needed. All these properties warrant application of PCMs in modern buildings, contributing to energy efficiency and sustainability in building design and building rehabilitation processes (Shire et al., 2013) (Figure 3.16).

Phase change material study is not confined to architectural application but is extended to other areas like electronics and textiles. In electronics, PCMs are applied to maintain equipment at their optimal temperatures to avoid overheating and ensure maximum lifespan. In textiles, they allow for wear that adapts to the ambient temperature, providing users with added comfort. Continued research assures optimization of the thermal capacity of PCMs is still a priority. Identifying and learning the impact of melting ranges and heat transfer coefficients are essential to optimizing the performance of these materials. Ultimately, continued research into phase change materials addresses an important integration between innovative designs and practical applications in the overall energy conservation and thermal management practices (Hale et al., 1972).

3.6.3. Smart Materials

Smart materials are increasingly well recognized for their unique thermal characteristics, enabling them to respond dynamically to environmental signals. These materials have abilities such as thermal expansion, memory of shape, and other adaptive capabilities necessary in a variety of applications. For instance, piezoelectric materials like lead zirconate titanate are exploited in advanced technologies for thermal management and energy harvesting in aerospace technology, such as cubic reconnaissance satellites. The materials modeled in order to assess these materials demonstrate the significant influence of temperature, frequency, and resistance variations on their functionality, establishing the connection between thermal properties and material behavior. It is a crucial relationship to understand because not only does it enhance the efficiency of the materials but also facilitates effective integration of such materials in complex systems, indicating an obvious direction for future engineering development in smart material science (Butt et al., 2017). The integration of smart materials has groundbreaking impacts on maximizing building thermal efficiency, including the use of novel composites on buildings such as Trombe walls. The uses include lightweight polycarbonate panels with nanoporous aerogel insulation to maximize thermal control in energy-efficient housing. The studies indicate that the systems can realize high energy savings while maintaining little risk of overheating, particularly in diverse UK constructions. With parametric modeling, studies have estimated that retrofitting traditional housing with such smart materials allows for greater energy savings compared to other forms of traditional insulation. This indicates not just a technological change but also a mindset and usage change towards building materials, where smart materials are the spotlight in becoming major actors to sustainable construction and energy-saving projects (Ashby et al., 2014). Besides, with growing need for durable infrastructure, research on smart materials for applications in structures goes on increasing. The ability of the materials to react and readjust themselves under changing thermal conditions offers possible application in many fields of engineering. For instance, using piezoelectric materials in structural components can lead to on-site monitoring and response to thermal conditions, significantly enhancing security and longevity. On top of that, integration of smart materials into standard building systems can potentially optimize energy efficiency while providing opportunities for passive and active temperature regulation. With ongoing research investigating the thermodynamic behavior of such emerging materials, it will enable their more extensive application across many industries and solidify the crucial role of smart materials in addressing contemporary thermal management concerns (Ashby et al., 2014; Butt et al., 2017).

SUMMARY

The study of thermal material properties showcases the critical role these properties play in determining the performance and durability of materials across various applications. It is evident from the findings that materials experience drastic changes in their thermal properties under varied environmental conditions, with implications for application. For instance, as presented in the case of the solar array blanket assembly used on the Space Station Freedom, the elastic-plastic stress due to thermal fluctuations creates the risk of fatigue failure and suggests that materials are to be effectively selected to balance the thermal demands with the mechanical stability (Armand et al., 2025). This figure emphasizes the necessity of high-intensity testing and thermal property analysis to ensure material reliability, particularly in high-reward situations, where material failure could have important consequences. Material degradation related to thermal properties are also critical in ascertaining long-term performance of implants and other devices. The characterization results of the polypropylene-PTFE composite mesh indicate that thermal properties are dynamic and evolve with material aging as a result of oxidative and hydrolytic degradation. Both carbonyl group detection and weight loss exhibit extreme changes in thermal stability throughout their *in vivo* lifespan, which indicates that thermal properties are linked with chemical stability and overall material performance (Grant et al., 2008). Such information goes into the development of stronger materials that can endure thermal and chemical stresses, thereby expanding their lifespan and feasibility in medical applications. Thus, an understanding of the implications of thermal properties goes into innovative durable materials that meet contemporary demands. Ultimately, integration of knowledge on thermal properties of materials not only dispels their elementary importance in different disciplines but also drives future advancement towards better material performance. Understanding of the underlying mechanisms that govern thermal behaviors allows researchers and engineers to develop strategies for avoiding the detrimental effects of thermal stress and degradation. Also, as we progress toward more complex applications, such as aerospace or biomedicine, the impacts of such discoveries become more tangible, requiring an interdisciplinary approach in material science. As can be observed, by the inclusion of thermal property analysis in materials development, we are able to enhance safety and effectiveness to a very large extent, setting a stage for future innovations that will propel material strength in most applications (Armand et al., 2025; Grant et al., 2008).

In studying the thermal characteristics of materials, one is to consider the intrinsic nature of the various materials under varying thermal conditions. Thermal conductivity, a key factor, determines the capacity of a material to transfer heat. Various materials can have widely divergent thermal conductivities; for instance, metals like copper possess high thermal conductivity, whereas insulators, such as polystyrene, possess much smaller values. A knowledge of these properties is required for industrial applications and scientific research purposes alike, where efficient heat management can make or break the operation and longevity of materials. Additionally, material thermal expansion plays a critical role in nature, as different materials respond to temperatures differently, affecting their form and performance in actual applications. Another salient aspect in the field of thermal properties is the effects of background noise in experimental systems. The production of high-energy particles during certain measurements can significantly affect data taking, especially in neutron scattering facilities. Instrumental performance numbers, as defined mainly by signal-to-noise ratio (S/N), can be impaired by such background effects, as detailed by Ansell et al. (2014). Therefore, getting rid of this background noise issues through a systematic process can yield measurement quality and reliability improvements. Increased instrument performance not just improves scientific research but also translates into economic gains by reducing the price of inefficient data accumulation. A clear understanding of both thermal properties and the impact of experimental conditions is therefore essential for material science advancement. Last but not least, thermal properties are discovered to play an important role in the calculation of multi-material structures under cyclic loading conditions. The newer numerical analysis methods, such as the use of Melan's theorem with ratchet analysis procedures, have enabled one to determine the nature of complex structures both under thermal and mechanical loads. Using such sophisticated analytical models, one can obtain lower bounds to structural performance that are relevant to practical applications on pipe intersections and other such structures that are likely to undergo variable temperature environments, as demonstrated by Chen et al. (2013). The implications of how changing thermal material properties behave under repetitive loading states are crucial to designing resilient systems that can support rigorous operational stresses over a period. Thus, critical examination of such key points makes the knowledge base in the field of thermal properties of materials more robust.

REFERENCES

1. Accardi, D., McMahan, T., Mikatarian, J., Powers, T., et al. (2025). The first United States microgravity laboratory. https://core.ac.uk/download/pdf/42810667.pdf (accessed on 23rd October 2025).
2. Agarwala, A., Ameen, A., Anthony, T., Bach, P., Back, M., Bi, J., Bi, C., et al. (2018). Effect of alkyl chain length on the properties of triphenylamine-based hole transport materials and their performance in perovskite solar cells. https://core.ac.uk/download/145240978.pdf (accessed on 23rd October 2025).
3. Agote, I., Astacio, R., Cintas, J., Cuevas, F., et al. (2019). Hard metal production by ERS: Processing parameter roles in final properties. https://core.ac.uk/download/186637583.pdf (accessed on 23rd October 2025).
4. Aivazis, M., Goddard, B., Meiron, D., Ortiz, M., et al. (1999). The 1999 Center for Simulation of Dynamic Response in Materials Annual Technical Report. https://core.ac.uk/download/4890872.pdf (accessed on 23rd October 2025).
5. Allen, R., Goodhew, S., Goodhew, J., Incropera, F., Griffiths, R., Goodhew, R., Szokolay, S., et al. (2004). Sustainable earth walls to meet the building regulations. https://core.ac.uk/download/30647390.pdf (accessed on 23rd October 2025).
6. Ansell, S., Bentley, P. M., Cherkashyna, N., et al. (2014). Overcoming high energy backgrounds at pulsed spallation sources. http://arxiv.org/abs/1501.02364 (accessed on 23rd October 2025).
7. Armand, S. C., Liao, M. H., & Morris, R. W. (2025). A transient plasticity study and low cycle fatigue analysis of the Space Station Freedom photovoltaic solar array blanket. https://core.ac.uk/download/pdf/42824524.pdf (accessed on 23rd October 2025).
8. Arnez, V., Corney, J., Govea-Valladares, E., Hernandez, J., et al. (2012). A novel haptic model and environment for maxillofacial surgical operation planning and manipulation. https://core.ac.uk/download/9559140.pdf (accessed on 23rd October 2025).
9. Asenjo, I., Lacaze, J., Larrañaga, P., Sertucha, J. A., et al. (2012). Influence of 1 wt-% addition of Ni on structural and mechanical properties of ferritic ductile irons. https://core.ac.uk/download/16754656.pdf (accessed on 23rd October 2025).
10. Ashby, M., BERR, B., Bynum, R. T., Dowson, M., Duffie, J. A., et al. (2014). Trombe walls with nanoporous aerogel insulation applied to UK housing refurbishments. https://core.ac.uk/download/29203889.pdf (accessed on 23rd October 2025).

11. Asta, M., Curtarolo, S., de Jong, M., Levy, O., et al. (2014). High-throughput computational screening of thermal conductivity, Debye temperature and Grüneisen parameter using a quasi-harmonic Debye model. http://arxiv.org/abs/1407.7789 (accessed on 23rd October 2025).

12. Baker, D. J. (2011). Thermal characterization of carbon composites. https://core.ac.uk/download/322523403.pdf (accessed on 23rd October 2025).

13. Balandin, A., Balandin, A., Casher, A., Dhar, A., Khitun, A., Lippi, A., Reina, A., et al. (2011). Thermal properties of graphene, carbon nanotubes and nanostructured carbon materials. http://arxiv.org/abs/1106.3789 (accessed on 23rd October 2025).

14. Bansal, N. P., Eslamloo-Grami, M., & Zhu, D. (2007). Thermal properties of oxides with magnetoplumbite structure for advanced thermal barrier coatings. https://core.ac.uk/download/pdf/10537612.pdf (accessed on 23rd October 2025).

15. Benenti, G., Goupil, C., & Ouerdane, H. (2016). The thermoelectric working fluid: Thermodynamics and transport. https://core.ac.uk/download/580036380.pdf (accessed on 23rd October 2025).

16. Blanco, J. M., Chica, J. A., Elguezabal, P., Lopez, L., et al. (2020). CFD model-based analysis and experimental assessment of key design parameters for an integrated unglazed metallic thermal collector façade. https://core.ac.uk/download/228298458.pdf (accessed on 23rd October 2025).

17. Bocchini, S., Cavalleri, O., Cozza, E. S., Frache, A., et al. (2012). Simple method for the preparation of composites based on PA6 and partially exfoliated graphite. https://core.ac.uk/download/pdf/11425034.pdf (accessed on 23rd October 2025).

18. Butt, Z., Elahi, H., Eugeni, M., Gaudenzi, P., et al. (2017). Effects of variable resistance on smart structures of cubic reconnaissance satellites in various thermal and frequency shocking conditions. https://core.ac.uk/download/132157099.pdf (accessed on 23rd October 2025).

19. Campbell, A., Ion, W., Marshall, S., Murray, M., et al. (2018). New methods for automatic quantification of microstructural features using digital image processing. https://core.ac.uk/download/145241394.pdf (accessed on 23rd October 2025).

20. Carrete, J., Curtarolo, S., Gossett, E., Isayev, O., et al. (2017). AFLOW-ML: A RESTful API for machine-learning predictions of materials properties. http://arxiv.org/abs/1711.10744 (accessed on 23rd October 2025).

21. Carter, R., Cooper, K., Draper, S., Ellis, M., et al. (2025). Materials characterization of additively manufactured components for rocket propulsion. https://core.ac.uk/download/pdf/42702478.pdf (accessed on 23rd October 2025).
22. Cederkrantz, D., Heijl, R., Nygren, M., Palmqvist, A. E. C., et al. (2012). Thermoelectric properties of $Ba_8Ga_{16}Ge_{30}$ with TiO_2 nano inclusions. https://core.ac.uk/download/pdf/70596722.pdf (accessed on 23rd October 2025).
23. Chen, H., Tipping, D., & Ure, J. M. (2013). Calculation of a lower bound ratchet limit part 2: Application to a pipe intersection and dissimilar material join. https://core.ac.uk/download/9039518.pdf (accessed on 23rd October 2025).
24. Chen, Z., Haselbach, L., Nantasai, B., Nassiri, S., et al. (2017). Prediction of thermal behavior of pervious concrete pavements in winter. https://core.ac.uk/download/286999887.pdf (accessed on 23rd October 2025).
25. Choy, K. L., Qu, L., & Wheatley, R. (2016). Theoretical and experimental studies of doping effects on thermodynamic properties of (Dy, Y)-ZrO_2. https://core.ac.uk/download/79549968.pdf (accessed on 23rd October 2025).
26. Clarke, J. A., & Yaneske, P. P. (2009). A rational approach to the harmonization of the thermal properties of building materials. https://core.ac.uk/download/9025303.pdf (accessed on 23rd October 2025).
27. Clausi, J. V., Grasso, A. P., & King, J. M. (2025). Study of fuel cell powerplant with heat recovery. https://core.ac.uk/download/pdf/42888481.pdf (accessed on 23rd October 2025).
28. Datta, D., & Lee, E. S. (2023). Exploring thermal transport in electrochemical energy storage systems utilizing two-dimensional materials: Prospects and hurdles. http://arxiv.org/abs/2310.08592 (accessed on 23rd October 2025).
29. Davies, R. L., Glasgow, T. K., Halford, G. R., Lauver, M. R., et al. (2025). Progress in materials and structures at Lewis Research Center. https://core.ac.uk/download/pdf/42862524.pdf (accessed on 23rd October 2025).
30. De Wilde, P., Goodhew, S., Griffiths, R., Pilkington, B., et al. (2008). Thermal probe technology for buildings: The transition from laboratory to field measurements. https://core.ac.uk/download/30634692.pdf (accessed on 23rd October 2025).
31. Doškář, M., Jarušková, D., Novák, J., Zeman, J., et al. (2018). Wang tiling aided statistical determination of the representative volume element

size of random heterogeneous materials. http://arxiv.org/abs/1709.05926 (accessed on 23rd October 2025).

32. Dou, N. G., & Minnich, A. J. (2015). Heat conduction in multifunctional nanotrusses studied using Boltzmann transport equation. https://core.ac.uk/download/216229066.pdf (accessed on 23rd October 2025).
33. Drevillon, J., Ezzahri, Y., Joulain, K., Meneses, D., et al. (2015). Radiative thermal rectification between SiC and SiO_2. https://core.ac.uk/download/52994860.pdf (accessed on 23rd October 2025).
34. Ghabezloo, S., Saint-Marc, J., & Sulem, J. (2008). The effect of undrained heating on a fluid-saturated hardened cement paste. https://core.ac.uk/download/48355055.pdf (accessed on 23rd October 2025).
35. Gibbs, D., O'Neill, K. (2015). Building a green economy? Sustainability transitions in the UK building sector. https://core.ac.uk/download/151156937.pdf (accessed on 23rd October 2025).
36. Goodhew, S., & Griffiths, R. (2017). Thermal properties of cob retrofitted with external hemp–lime. https://core.ac.uk/download/82971100.pdf (accessed on 23rd October 2025).
37. Grant, S. A., & Schmitt, C. (2008). Materials characterization of an explanted polypropylene–PTFE hernia mesh [abstract]. https://core.ac.uk/download/62756395.pdf (accessed on 23rd October 2025).
38. Hale, D. V., Hoover, M. J., & Oneill, M. J. (1972). Phase-change materials handbook. https://core.ac.uk/download/pdf/10296671.pdf (accessed on 23rd October 2025).
39. Hamilton, A., Maddalena, R., & Roberts, J. J. (2018). Can Portland cement be replaced by low-carbon alternative materials? A study on thermal properties and carbon emissions of innovative cements. https://core.ac.uk/download/151397423.pdf (accessed on 23rd October 2025).
40. Han, X. H., Jacobi, A. M., Park, Y. G., et al. (2012). A review of metal foam and metal matrix composites for heat exchangers and heat sinks. https://core.ac.uk/download/55763987.pdf (accessed on 23rd October 2025).
41. Harničárová, M., Hlinka, J., Jančíková, Z., et al. (2014). On experimental thermal analysis of solid materials. https://core.ac.uk/download/94761745.pdf (accessed on 23rd October 2025).
42. Hickman, R., Panda, B., & Shah, S. (2025). Fabrication of high temperature cermet materials for nuclear thermal propulsion. https://core.ac.uk/download/pdf/10516413.pdf (accessed on 23rd October 2025).

43. Hull, T. R., & Lowden, L. (2013). Flammability behaviour of wood and a review of the methods for its reduction. https://core.ac.uk/download/19576180.pdf (accessed on 23rd October 2025).
44. Iglauer, S., Lebedev, M., Zhang, Y., et al. (2013). Flammability behaviour of wood and a review of the methods for its reduction. https://core.ac.uk/download/19576180.pdf (accessed on 23rd October 2025).
45. Jankovic, L. (2019). Improving building energy efficiency through measurement of building physics properties using dynamic heating tests. https://core.ac.uk/download/287581666.pdf (accessed on 23rd October 2025).
46. Jouhara, H., Karayiannis, T. G., Tassou, S. A., et al. (2010). Analysis of energy use in crisp frying processes. https://core.ac.uk/download/2733850.pdf (accessed on 23rd October 2025).
47. Long, A. C., Mesogitis, T., & Skordos, A. A. (2014). Uncertainty in the manufacturing of fibrous thermosetting composites: A review. https://core.ac.uk/download/42143883.pdf (accessed on 23rd October 2025).
48. Minnich, A. J. (2015). Multidimensional quasiballistic thermal transport in transient grating spectroscopy. https://core.ac.uk/download/33122503.pdf (accessed on 23rd October 2025).
49. N/A. (2025). Correlation of materials properties with the atomic density concept. https://core.ac.uk/download/pdf/42888857.pdf (accessed on 23rd October 2025).
50. Nováčková, K. (2009). Stability of salts and complexes of humic acids. https://core.ac.uk/download/30278457.pdf (accessed on 23rd October 2025).
51. Peng, B., Shao, H., Xu, Y., et al. (2015). Thermal conductivity of monolayer MoS_2, $MoSe_2$, and WS_2: Interplay of mass effect, interatomic bonding and anharmonicity. http://arxiv.org/abs/1509.01391 (accessed on 23rd October 2025).
52. Price, H. G., Jr., Quentmeyer, R. J., & Schacht, R. L. (2025). Reliability and effective thermal conductivity of three metallic-ceramic composite insulating coatings on cooled hydrogen-oxygen rockets. https://core.ac.uk/download/pdf/42899686.pdf (accessed on 23rd October 2025).
53. Raut, A. N. (2017). Development of sustainable material for hybrid wall system to improve indoor thermal performance. https://core.ac.uk/download/12005656.pdf (accessed on 23rd October 2025).
54. Sarkar, D. (2025). Development of suspended thermoreflectance technique and its application in thermal property measurement of semiconductor

materials. https://core.ac.uk/download/322365077.pdf (accessed on 23rd October 2025).

55. Sellerio, A. L., Taloni, A., & Zapperi, S. (2015). Fracture size effects in nanoscale materials: The case of graphene. http://arxiv.org/abs/1508.05720 (accessed on 23rd October 2025).
56. Sercel, D. (2012). Materials for spring fabrication. https://core.ac.uk/download/30292752.pdf (accessed on 23rd October 2025).
57. Shire, G. S. F., Tian, Y. (Y.), & Zhou, D. (2013). Parametric analysis of influencing factors in phase change material wallboard (PCMW). https://core.ac.uk/download/19553603.pdf (accessed on 23rd October 2025).

CHAPTER 4

HEAT GENERATION IN ELECTRONIC DEVICES

Contents

INTRODUCTION

In the current context of technology, heat production in electronic systems is a prime area of research and development. With more powerful and compact devices, the issue of heat management efficiently has become more critical. Notably, overproduction of heat can lead to performance degradation, reliability loss, and ultimately failure of electronic components. This is especially seen in high-performance computing applications such as telecommunications, where efficient thermal management becomes critical to supporting longevity and functionality. Researchers are therefore confronted with the challenge of exploring novel materials and thermal cooling technologies with the potential to address the thermodynamic hurdles created by heat in such devices. One of the exciting fronts of research is centered on thermoelectric materials and their use in heat management. Thermoelectric converters, for instance, can utilize waste heat and transform it into usable electrical energy, thus preventing the undesirable effects of excess heat buildup. Furthermore, advances in nanostructured thermoelectric devices show the promise of enhanced heat conversion efficiency. Research conducted in the area shows that energy filtering techniques have the potential to significantly improve the performance of thermoelectric generation, particularly in systems where incoherent scattering prevails. By refining and knowing the properties of the material, researchers aim to devise solutions that not just minimize heat production but harness it as well, ultimately achieving more efficient electronic systems (Muralidharan et al., 2017). The advent of solid oxide fuel cells (SOFCs) as energy-conversion devices with high efficiency reflects another facet of managing heat. The cells are operated at increased temperatures, which brings with it certain issues regarding heat production and material compatibility. Although SOFCs are noted for high efficiency and minimal emissions, the necessity of controlling operating temperatures has given rise to the development of intermediate temperature SOFC. Research indicates that optimization of cathode fabrication process, as well as infiltration techniques, can result in enhanced performance and stability and heat-related issues relief (Lee et al., 2020). Ongoing research into such technologies not only continues to shed more light on the generation of heat in electronic devices but also paves the way for innovations with the potential to significantly enhance energy efficiency and operational efficacy across a variety of applications.

Definition of Heat Generation in Electronic Devices

Heat in electronic devices is a side effect of energy conversion processes occurring when they are in use. Whenever electrical current flows through semiconductor materials, it encounters resistance, and energy is lost in the form of heat. The phenomenon is often referred to as Joule heating and is a result of the resistive properties of materials used in electronic devices. The amount of heat generated is based on several factors, including the type of material, device structure, and operating conditions. With time, as the devices became more complex and miniaturized, the problems of handling heat have also developed with a corresponding rise in proportion, necessitating an elaborate understanding of heat generation and its impact on device performance and longevity. Moreover, the heat generation description in electronic devices also plays a critical role for enhancing efficiency and reliability. Electronic devices that generate a large amount of heat could face various issues from decreased performance to lifespan, and even complete destruction in the worst case. The interplay between heat production and operational efficiency is pertinent; as elucidated in the latest research, maximizing the thermoelectric performance can reduce entropy production, which is in conformity with the concepts of heat control in electronics.

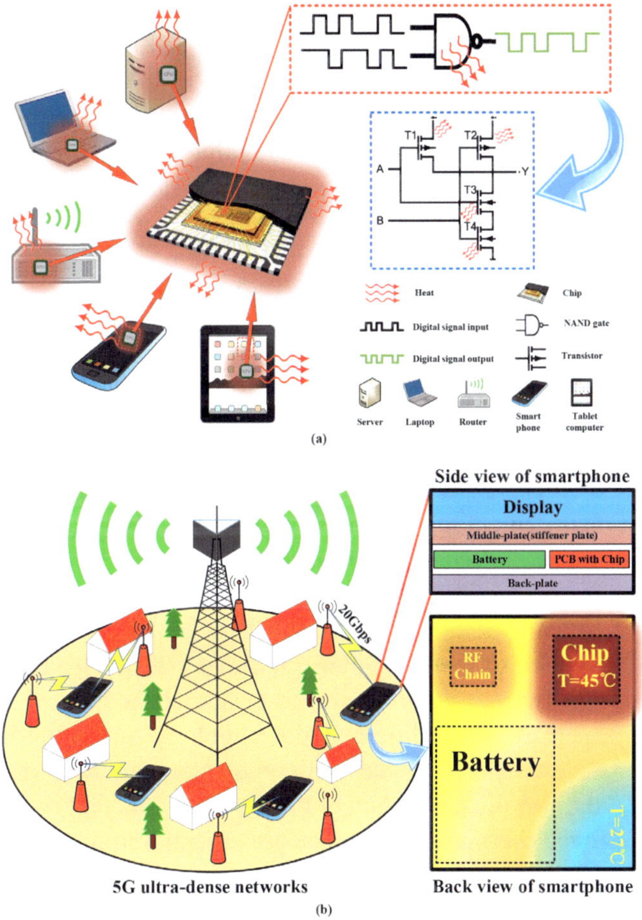

Figure 4.1. (a) Computation and heat generation in electronic products; (b) structure and temperature distribution of smartphones.

Source: https://static.scientificamerican.com/sciam/cache/file/4A4F47F4-D0F7-467A-85BF15A5AFF00F7C_source.jpg?w=1200.

By implementing effective thermal management techniques, which may include innovations in material science and refrigeration technologies, it becomes feasible to keep devices at temperatures that are best for operation, hence protecting sensitive parts and ensuring overall operational integrity (Goupil et al., 2011). In addition, technology advancements have made solutions in managing the impact of heat generation in electronic devices more imperative. The study of thermoelectric materials and their efficiencies provide a window not only for the recovery of waste heat but for performance improvement through effective dissipation of heat. They would be able to provide the dual service of reducing heat output during operation while, simultaneously, harnessing excess heat as a power source. This dual role is important in designing environmentally friendly electronic solutions. With technology continuing to advance, understanding the mechanisms of heat generation—and implementing good thermal management principles—will be crucial to shaping the destiny of electronic device operation and design (N/A, 1976) (Figure 4.1).

Importance of Studying Heat Generation

Understanding heat generation is essential in electronic device design and usage since excessive heat can lead to a drop in performance and ultimately result in component failure. Electronic devices, particularly those used in high-density applications such as computers and cell phones, generate large amounts of thermal energy due to the high integrated circuits in small spaces. Effective heat management methods must be developed to prevent overheating that can compromise the reliability of such devices. Heat generation research allows engineers to develop cooling systems, either as heat sinks, thermal pads, or even active coolers, to maintain temperatures within operational parameters. Prevention of thermal failures increases the lifespan and operation of devices, ultimately resulting in improved customer satisfaction and reduced maintenance expenses. In addition, nanotechnology and materials science innovations are directly associated with understanding heat generation in electronics. Studies on the development of new materials that can sustain heat or possess improved thermal conductance are essential in modern electronics. For instance, nanoscale-engineered materials can be utilized to advance thermal management methods so that components become more efficient and compact with no risk of overheating. The study of surface plasmon polaritons, for instance, indicates how evolving hybrid multilayer architecture can influence heating and cooling in nanooptics technology, as established by Temnov et al. (2012). By incorporating these new materials into device design, engineers can minimize heat accumulation and enhance thermal efficiency. Finally, the environmental impacts of heating produced inside electronic devices cannot be overlooked, especially with ongoing electronic usage. An understanding of how heat is produced and managed allows one to optimize energy consumption, something that directly leads to sustainability. In hybrid vehicle systems, as discussed by Capata et al. (2018), the design of the propulsion system must consider heat production for maximum energy use. Based on the study of such dynamics, engineers can design systems that minimize wastage of energy as well as emissions, something towards cleaner technology efforts across the world. Thus, the importance of heat generation studies extends beyond performance per se; it entails bigger environmental and social issues that need innovative solutions in the design of electronic equipment.

4.1. SOURCES OF HEAT GENERATION

The principal sources of thermal generation in electronic devices can be traced back to various factors that take place in the process of their operating cycles. Of significant concern is the inherent electrical resistance in elements of the circuit, such as resistors, capacitors, and wire, which produces heat as electrical current flows through them according to Joules law. This takes place in all types of electronic devices, from mobile phones to complex computer systems. The resulting heat not only affects the performance of the device but also leads to long-term degradation if not properly controlled. As devices become smaller and operate at higher frequencies, the proportion of generated heat rises, so it is important to learn effective thermal management techniques in order to combat potential dangers from overheating. Another prominent source of heat generation comes from power management components, particularly in mobile devices where battery life and functionality take center stage. Mobile electronics have witnessed an increase in energy requirements due to feature-rich capabilities like high-resolution displays and high-end processors. Power dissipation in application processors is directly related to their functionalities, thus generating significant thermal energy. Studies have shown that malfunctioning battery management systems can cause overcurrent situations, whereby batteries might overheat and introduce safety hazards, including thermal runaways or explosions. This serves to underscore that all of the devices design parameters, particularly the coupling between the application processor and the battery, are critical to delivering optimal thermal performance and safety as outlined in related analyses such as mobile electronics (Chen et al., 2019). In InP Gunn oscillators, which are high-frequency devices, thermal generation is a significant aspect that influences their efficiency and overall performance. InP Gunn oscillators, which are an integral part in the majority of applications within telecommunications and the generation of signals, have experienced tremendous advancements in design that allow for more levels of output power at ever more elevated frequencies. Optimization of doping profiles in InP Gunn devices has proven to manage heat generation, enabling operation at frequencies up to 500 GHz. The process presents the fine balancing act between performance and thermal management, where inefficiency in dissipating heat can lead to component degradation and compromised device reliability. The advancements within this area herald a rosy future with the establishment of high-tech heat management technologies that enhance the performance of high-frequency devices while offering safety and reliability (Eisele et al., 2004).

4.1.1. Electrical Resistance in Components

The electrical resistance effect in electronic devices is central to the investigation of heat generation. Resistance is felt when an electric current passes through a conductor, leading to energy dissipation as heat. This is very applicable in semiconductor devices, where resistance plays dual functions of device operation and heat management. When current is carried by a resistor, the atomic crystal structure of the material resists the movement of electrons and induces a temperature increase. Not only does the heat induced hinder the efficiency of the device, but longevity of operation is also impacted. An understanding of the phenomenon of resistance in the components is crucial because excess heat induces thermal runaway, whereby increasing temperatures continuously increase the resistance, creating a potentially deadly cycle that deteriorates electronic performance. When seeking effective electronic devices, minimizing electrical resistance is of utmost significance. The resistance value varies in materials; hence, the application of appropriate conductors and semiconductors is of utmost significance for optimal thermal performance. As highlighted by recent studies, the application of innovative materials can lead to considerable resistance reductions, thus facilitating improved heat management. For instance, research has established that fewer resistive materials reduce excessive heating, which improves overall device efficiency. The relationship between resistance and heating is even more critical in complex circuits where components engage each other in several ways. The total heat produced can raise ambient temperatures, leading to higher energy consumption and potential erosion of electronic devices if not properly controlled (Adam et al., 2018) (Figure 4.2).

Consequently, the developments in material science have a direct bearing towards the production of more efficient electronic devices. There is ongoing research aimed at enhancing the effective use of heat in electronic devices, as evidenced by the research into emerging technologies. The enhancement of the efficiency of thermodynamic systems that effectively convert low-grade heat to work is one area among many. This is particularly relevant in the case of portable devices, where energy source management with limited resources is essential to prolonged operation. The systems' operating behavior depends on their resistivity and the thermal relations they undergo with the surroundings. Electrical resistance, with high sensitivity and creative design methods, has the potential to enhance device efficiency and heat performance immensely (Markides et al., 2011). By focusing on the thermal-fluid dynamics involved, scientists are attempting to break through the challenges of heat generation, paving the way to the next generation of electronic devices that will not only be faster but also capable of sustaining more extreme operating conditions.

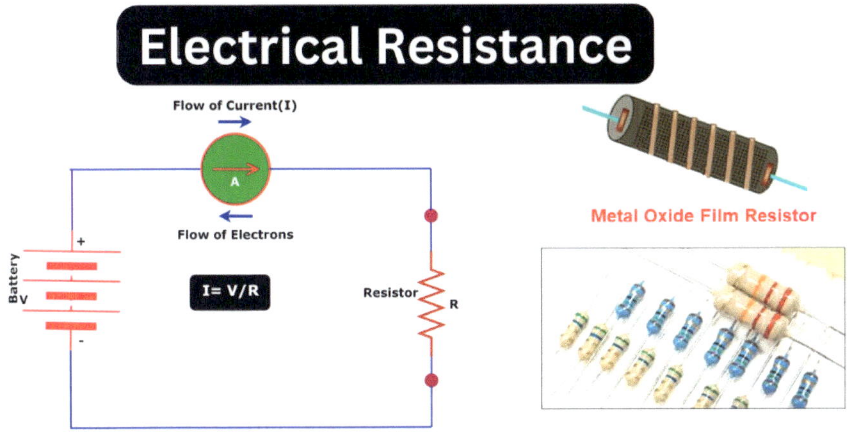

Figure 4.2. Electrical resistance.

Source: https://external-content.duckduckgo.com/iu/?u=https%3A%2F%2Felectricalampere.com%2Fwp-content%2Fuploads%2F2023%2F09%2FElectrical-Resistance.png&f=1&nofb=1&ipt=86ebc3bb24b9f8f2fed9f39179ba5f48a8858312f4daabed7efe681af1b6fa4c.

4.1.2. Power Consumption of Devices

Electronics' energy usage is a critical aspect that determines the overall heat generation in the systems. With the advancement in the complexity of the devices, the power needed by them will naturally rise with an accompanying rise in energy dissipation in the form of heat. For instance, mobile electronics have seen sweeping leaps in functionality where rising functionality is comfortably linked with rising power usage. This trend is particularly evident in smartphones, where processors must run increasingly complex applications. Consequently, the thermal energy generated under such demands necessitates adequate management methods for device functionality and safety assurance. The ignoring of such thermal complications has led to catastrophic occurrences, including device explosions due to inadequate thermal control. Hence, it is important to understand how power consumption and heat generation are interconnected so that electronic devices can be made safer and more efficient. Power consumption trends also significantly influence device function as well as design. For example, new low-noise InP Gunn devices have shown an excellent enhancement of RF output power while their direct current power requirements are comparable with traditional sources. These developments not only improve the efficiency of oscillators but also highlight a critical shift in the way manufacturers design for energy efficiency (Figure 4.3).

Electronics	Required Power
GPS receiver chip	15 mW
Cell phone (standby)	8.1 mW
PPG sensor	1.47 mW
Humidity	1 mW
Pressure	0.5 mW
3D accelerometer	0.32 mW
Temperature	27 μW
A/D conversion	1 μW
RF transmission	sub μW

Figure 4.3. Average power consumption of typical electronic devices and sensors.

Source: https://www.researchgate.net/figure/Average-power-consumption-of-typical-electronic-devices-and-sensors_tbl3_232742342.

The application of optimized doping concentrations in these devices reflects a heightened emphasis on balance between functionality and the need for power. Although energy use remains at the forefront of technological development, having the capability to identify the correspondence between innovative design and energy efficiency can provide great reward in many applications within electronic devices. Identification of this relationship is crucial in addressing both the performance and heat issues that accompany greater power demands. Finally, power consumption implications extend from individual devices to broader environmental conditions. With greater usage of electronic gadgets, their combined energy consumption aggravates, leading to tremendous heat generation and, consequently, undesirable environmental impacts. Without adequate thermal management, energy-intensive devices promote waste of energy, threatening device longevity as well as sustainability values. Cities, above all, are suffering from an expanse known as the urban heat island effect fueled by intense emissions of heat from extensive electronic usage. Efforts to rectify these power consumption problems are going to require concerted efforts in the direction of developing energy-efficient technologies, enhancing thermal management strategies, and altering consumer habits in relation to device usage. With the growing demand for electronic appliances, it is essential to practice conserving power to create a sustainable future for technology.

4.1.3. Environmental Factors Affecting Heat Generation

Environmental conditions play a critical role in the heating of electronic devices and also have effects on their performance and efficiency. Ambient temperature is among the most important factors of heat generation, which might have a powerful effect on electronic circuits' thermal behavior. For instance, electronic devices operating in high temperatures might experience increased resistance and thus cause more power dissipation with surplus heat emission. Conversely, lower temperatures can alleviate overheating issues but, at the same time, affect performance if the device must work with special thermal conditions to perform optimally. Further, improper ventilation can also increase heat buildup, resulting in potential thermal throttling or failure. Understanding ambient indoor temperature coupling with cooling mechanisms of the device is thus crucial to enhance longevity and operational dependability in electronic devices. The other important environmental parameter is humidity, which can adversely affect electronic equipment by increasing corrosion and thermal conductivity. In humid environments, water can deposit within electronic devices and produce short circuits and damage insulation characteristics. Not only does the moisture contribute to heat generation due to increased internal resistance, but it can also compromise the efficiency of cooling systems, particularly in systems designed for a specific level of humidity. Aside from that, humid conditions can interrupt the air flow necessary for effective thermal dissipation. It is essential, thus, to take into account the impact of humidity when designing electronic equipment, with materials and cooling mechanisms adequate enough for different environmental conditions, the ultimate goal being reliability and efficiency for different uses. The materials used in the construction of electronic devices also contribute to the production of heat and retention. Different materials possess distinct thermal properties, such as thermal conductivity and heat capacity, which can affect the way heat is absorbed, dissipated, or generated within components. For instance, high-thermal-conductivity metals like copper and aluminum are usually used in heat sinks to spread heat away from critical components efficiently. However, the selection of materials also carries problems with thermal expansion and stress with mechanical forces when devices are subjected to varying temperatures. Moreover, the application of recyclable materials in electronic devices, as per waste management research, can enhance the sustainability of device production as well as carry impacts on thermal dynamics due to the fact that various compositions may alter the pattern of heat generation upon usage

(Lysenko et al., 2018). Knowledge of these factors is critical in the formulation of thermal management strategies that ensure operational effectiveness while reducing environmental impact (Faiz et al., 2025).

4.2. TYPES OF ELECTRONIC DEVICES

The electronic devices world is complex and diverse, with an array of categories that have varied characteristics and applications. Among these devices, communication devices like smartphones and tablets have transformed the way we interact with technology. With increasing computing power in portable devices, there has been increased demand for increasingly sophisticated components capable of supporting sophisticated applications. These items rely on integrated circuits, batteries, and display technology that all generate heat when in operation. Management of this heat output is crucial not just to maximize performance but also to the safety of the user, particularly in light of the risks of overheating. With better miniaturization, the integration and design of effective thermal management mechanisms are the principal concern of manufacturers, providing long life and functionality in consumer products. Another vital class of electronic devices is personal computing systems, for example, desktops and laptops. Personal computing systems consist of some interdependent components, for example, processors, graphics cards, and storage devices, which together produce heat. CPU and GPU are big users of power, and their power-consumption operations produce a tremendous amount of heat. As noted in thermal management research, the cooling functionalities inside such devices need to be optimized so that they do not throttle performance or fail altogether. Additionally, since the components are being made more efficient and smaller in size, the heat dissipation techniques employed, like heat sinks and liquid cooling modules, are all the more significant. Therefore, knowing the heat behavior of personal computing devices is important in order to develop advanced designs that are able to handle growing power demands in a safe and heat-reduction fashion. Other than these, electronic cigarettes (ECs) form a niche but still applicable category in electronic devices. EC technology has revolutionized the design of atomizers with deep implications for their operation. As is displayed in studies that examined atomizer structures, differences in design have a huge influence on the generated heat when used. For instance, the initial versions had intricate geometries, while the newer versions have simplified the structures to suit the demands of consumers and the evolution of technology. Understanding how thermal dissipation occurs within these atomizers is crucial since improper thermal management can influence device operation as well as safety. The dialog between design and heat generation thus testifies to the pervasive need for EC technology innovation, mirroring progress in broader classes of electronics devices where heat management remains a dominant consideration in device design.

4.2.1. Consumer electronics (e.g., smartphones, laptops)

The spread of consumer electronics, including smartphones and laptops, has raised the concern about heat generation in electronic devices considerably. The devices that play a crucial role in our daily life are most often made compact and lightweight, often rendering these devices inefficient in heat dissipation. Heat generation usually comes as a byproduct of electronic operations, particularly during data processing and wireless communication activities. Overheating not only adversely affects the performance of the device but can even impact the hardware's durability. Furthermore, with growing device capabilities, with enhanced processors and graphics units, heat output is a crucial parameter to handle.

Figure 4.4. Consumer electronics.

Source: https://www.boydcorp.com/wp-content/uploads/Industries/Consumer-Electronics/Consumer-Electronics-1920x656-1.jpg.

Manufacturers are pushed towards designing innovative thermal management systems to counter such effects effectively while keeping devices user-friendly and efficient. This constant drive for balance indicates the need for extensive research on ways of minimizing heat production in consumer electronics. Heat management issues extend beyond individual devices, particularly where there are networked systems and the Internet of Things (IoT). Energy consumption phenomenon known as cross-factor has another considerable energy depleted as data packets travel through the operating systems network stack. This interdependency is particularly significant in wireless communication, in which numerous connections must be processed and transmitted at the same time. The impact of the cross-factor illustrates a key aspect of energy efficiency often overlooked in conventional energy models based solely on individual components. Optimization of these factors is essential to attaining device performance and reducing overall energy consumption to a maximum, especially as user performance expectations continue to rise. Efficient methods for energy debugging in mobiles and laptops, therefore, must account for such broader systemic interdependencies in a bid to make sustainable heat production solutions (Figure 4.4).

Additionally, research on new ways of measuring consumer appliance energy consumption is of great importance in making observations about heat management. Using a comprehensive measurement model, researchers can analyze energy usage patterns by device type and provide important information to manufacturers and developers for designing more effective products. These studies encourage improved comprehension not just of typical heat generation but also of the causes and mechanisms that are associated with high thermal performance. What these analyses provide works to enhance the application of innovative cooling systems and design innovations, ultimately contributing to enhanced reliability and performance in notebooks and mobile phones. The dynamic nature of technology demands collaboration between academics and industry in order to find such energy and heat management methods to make consumer electronics able to withstand the demands and needs of today's users without having a vast ecological impact (Azcorra et al., 2015; N/A, 2010).

4.2.2. Industrial Equipment (e.g., Servers, Machinery)

In industrial environments today, computation is addressed by servers and equipment operating around the clock, producing an enormous amount of heat. Servers are tasked with data processing and storage, but the tremendous workload in cloud computing and big data processing produces excessive heat. This heat energy not only poses problems to efficiency in operations but also requires extensive cooling systems to maintain optimum performance. Also, as explained by KEA (2013), advancements in server performance over time have ironically increased the need for better solutions to deal with heat, emphasizing the fact that technological advancements tend to exacerbate thermal challenges.

The relationship between machines in manufacturing businesses and heat generation must also be taken with utmost seriousness, especially regarding its efficiency when performing functions. Machines such as production robots and CNC machines are heavily loaded, generating much heat that, if not properly regulated, will lead to performance loss and premature machine failure. Integration of sophisticated cooling systems is essential in order to maximize service life and maintain accuracy in operations (Figure 4.5).

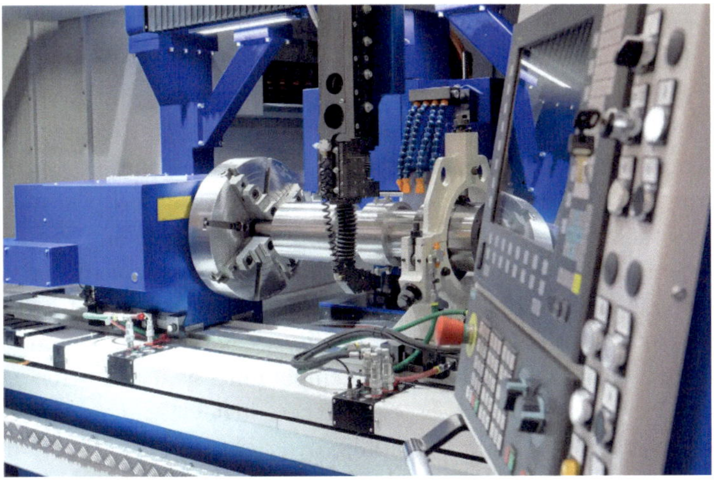

Figure 4.5. Industrial machinery and equipment.

Source: https://www.time-matters.com/wp-content/uploads/2023/07/industrial-machinery-components-spare-parts-logistics.webp.

The necessity for effective heat management is then added upon by shifting industries; as noted by Copeland et al. (2019), the Internet of Things and smart grids are transforming operating conditions, requiring smarter and more efficient equipment that will aid in diminishing heat generation while optimizing the handling of resources. In addressing heat generation issues in electronic devices, particularly within an industrial environment, the importance of considering the energy consumption implications as well as sustainability lies crucially. With industries operating to bring about greater operational efficiency, technology used—starting right from servers to heavy machinery—must be scrutinized for its heat generation as well as energy consumption. Emerging strategies such as the use of IoT-based monitoring systems can provide real-time thermal performance information, which can be utilized in predictive maintenance and resource optimization. The necessity to deploy such technologies not only addresses the short-term needs of heat generation problems but intersects with greater horizons of sustainability and energy conservation in industrial processes. Thus, integration of advanced cooling technologies and smart monitoring systems becomes a strategic necessity for industries looking to combat the pervasive issue of heat generation in electrical equipment.

4.2.3. Automotive Electronics (e.g., Electric Vehicles [EVs])

The evolution of car electronics, particularly in electric cars (EVs), has transformed the car industry dramatically. As demand for environmentally friendly transportation options increases, such vehicles rely heavily on complex electronic systems monitoring everything from battery efficiency to car diagnostics in real time. Advanced electronic control units (ECUs) play a crucial role in EVs in optimizing performance and efficiency. However, such systems also generate considerable heat, and thus the issues of thermal management. Adequate dissipation of heat becomes a priority in maintaining the components in the best working condition otherwise leading to catastrophic failure or degradation in efficiency. Hence, the development of materials and cooling systems capable of managing increased amounts of heat generation is essential for the sustainability and reliability of automotive electronics used in electric vehicles. In examining the temperature generation tendencies of automotive electronics, one can observe that certain devices are more susceptible to heat issues than others. For example, power electronics utilized in electric drives experience high current and frequency operation, resulting in high temperatures. This is particularly true for applications with switched reluctance machines (SRMs) in hybrid electric vehicles (HEVs), where power and torque density requirements call for the use of complex electronic components that undergo tremendous thermal stress. These harsh operating conditions call for extensive thermal management methods to prevent overheating and ensure optimal performance. With advanced cooling systems and thermal interfaces, designers can offset the impact of heating, thereby enhancing the reliability and performance of car electronics in electrical systems of such vehicles (Fraser et al., 2018) (Figure 4.6).

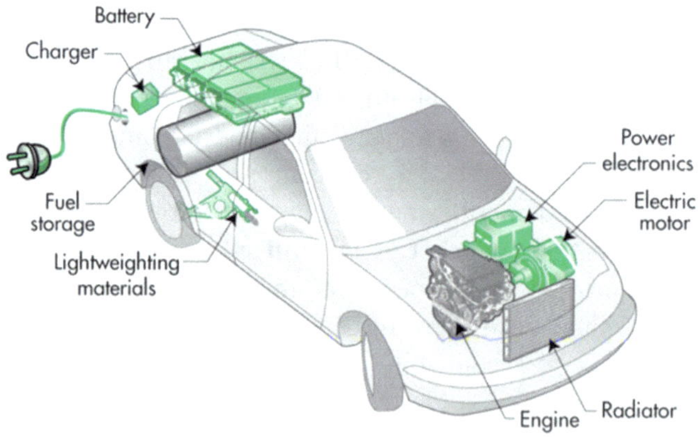

Figure 4.6. Automotive electronics.

Source: https://www.elprocus.com/wp-content/uploads/2014/09/major-power-electronic-blocks-in-automotive-electronics.jpg.

Also, the feasibility of electric vehicles is indelibly connected with their application of lithium batteries, the latter ones contributing to heat generation under operations as well. As the automobile sector and power sector invest more capital in lithium-ion technology, understanding how much lithium resources are present takes on monumental significance, most importantly in performance and operational efficiencies. Long-term prognostications of lithium resource availability are critical for long-term planning in EV manufacturing and technology development. As lithium supply fluctuations directly relate to battery cost and, ultimately, electric vehicle cost, an in-depth understanding of such is part of the overall discussion about automotive electronics. Improving the effectiveness of battery technologies is thus not just a matter of thermal management but is also part of the larger tale of resource sustainability and technology advancement in the automotive sector (Vargas et al., 2018).

4.3. MECHANISMS OF HEAT TRANSFER

The mechanisms of heat transfer—conduction, convection, and radiation—are significant in the course of electronic device heat generation. Conduction entails direct heat transfer through materials, an aspect that is very significant in the case of solid-state components. Materials' thermal conductivity is a significant factor in devices' heat dissipation. Effective heat conduction minimizes hotspots that lead to malfunction or reduced electronics lifespan. On top of that, material science progress has produced better thermal interface materials (TIMs) that allow for more efficient heat transfer with lowered thermal resistance. Hence, understanding conduction helps engineers design electronic circuits to withstand heat produced during use, ultimately resulting in better reliability and performance. Convection, the second heat transfer mechanism, is most relevant for the scenario of handling thermal impact in larger electronic systems. Heat is transferred by fluid movement in convective heat transfer, which can be natural or forced using mechanical devices such as fans. The design of convection-cooled systems constitutes a critical aspect of thermal management of high-speed computing equipment or power electronics. Heat sinks and fans, for instance, effectively increase air movement above warm components for the purpose of extracting heat from the surface to the ambient. This highlights the significant interplay between thermal performance and air flow dynamics, underscoring the necessity of optimized cooling in electronic device engineering. Effective convection methods can significantly enhance electronic equipment life by sustaining stable operating temperatures. Lastly, radiation is a less significant but nonetheless crucial input into heat transfer, particularly at high-temperature applications in electronic devices. Unlike conduction and convection, radiation does not require a medium of transmission and can occur even in vacuo. The method becomes a necessity in high-thermal-load systems where the surplus heat must be efficiently dissipated to prevent overheating. Solid oxide fuel cells, for instance, a standard feature of next-generation energy applications, are plagued by high operating temperatures (Lee et al., 2020) that result in enormous thermal management issues. By integrating material that encourages radiative heat transfer, it becomes possible for engineers to combat overheating issues, thereby improving efficiency as well as performance. Knowledge of the three conduction, convection, and radiation mechanisms provides a solid basis for the formulation of good thermal management schemes for ensuring reliability of electronic equipment (Figure 4.7).

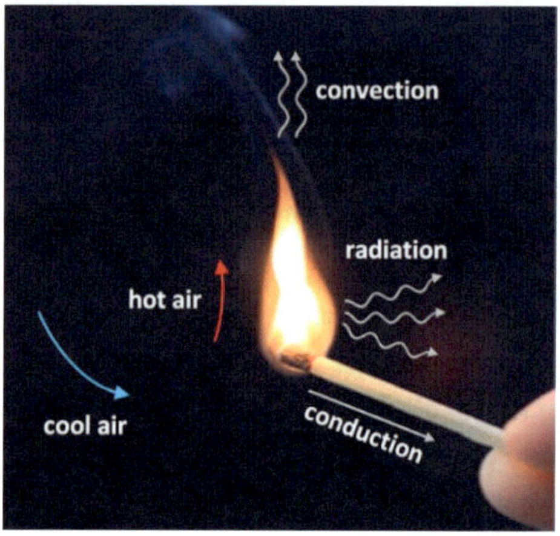

Figure 4.7. Basic mechanisms of heat transfer in a match flame.

Source: https://www.researchgate.net/figure/Basic-mechanisms-of-heat-transfer-in-a-match-flame-convection-allowed-by-hot-gases_fig1_337101866.

4.3.1. Conduction in Solid Materials

Examining conduction in solid materials is crucial while learning about heat generation in electronic devices. Thermal conduction effectiveness relies on the structure of atoms and bonding within materials. In metals, for instance, free electrons play an important role in conductive characteristics to enable quick dissipation of heat. Conversely, non-metals such as ceramics or polymers tend to possess lower thermal conductivities due to their tightly bound electrons that will serve as impediments to heat transfer. The disparity in conduction capacities becomes relevant in electronic components' thermal management, particularly as they miniaturize and functionality-intensify. Increased surface area and thermal interfaces can transport heat away from areas of high generation, thereby becoming less likely to induce thermal overload and device failure, a critical issue in modern electronics. New materials such as carbon nanotubes (CNTs) have been developed in recent breakthroughs in material science with immense potentials for increasing thermal conduction in electronic devices. Researchers discovered that thermal conductance in CNTs is length dependent, primarily due to the enhancement of low-frequency vibrational modes that enable energy transfer across a material. The findings, which utilized equilibrium molecular dynamics calculations and the Green-Kubo relation, show the role of solid structural constituents to thermal behaviors under working stress (Figure 4.8).

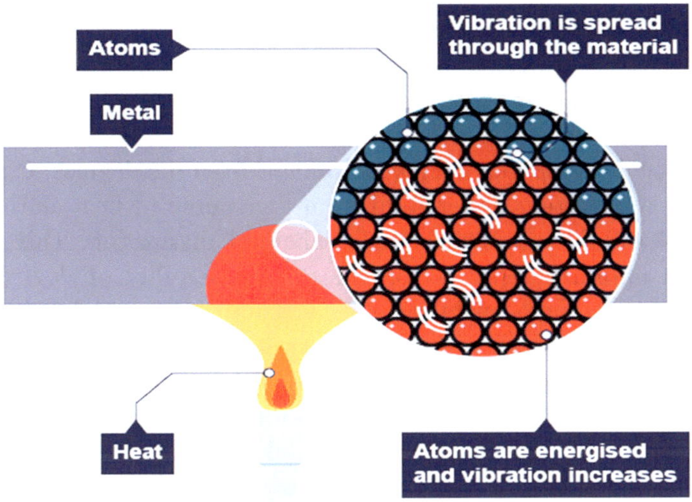

Figure 4.8. Conduction in a solid.

Source: https://bam.files.bbci.co.uk/bam/live/content/zbw2cqt/medium.

The uses of CNTs in microprocessors and high-performance electronics represent significant steps toward optimizing heat management methods. Through clever material integration with advanced conductive properties, device engineers can not only extend the lifetimes of operation but also enhance overall system efficiency and reliability in an increasingly dynamic technological landscape. The propensity of electronic devices to heat up requires a sophisticated understanding of the thermal conduction challenges facing solid materials. For example, the latest studies have involved researchers identifying crucial parameters like battery overcurrent and component positioning in creating components leading to cataclysmic failure, such as explosions in smartphones. As highlighted by recent research, a failure to effectively deal with the heat generated by batteries, often overlooked in traditional thermal studies, can result in major contributions to thermal runaway behavior. The use of effective cooling devices like heat pipes can address these challenges;

however, their effectiveness is dependent on specific operating conditions and physical configurations. With so intricate an interaction of these factors, an integrated strategy is required to design electronic devices that not only generate but also dissipate heat effectively, guaranteeing functionality and safety in more compact devices (Chen et al., 2019; Kelly et al., 2004).

4.3.2. Convection in Fluids

Fluid convection plays a pivotal role in the effective regulation of heat generation in electronic components, particularly in conditions where traditional cooling measures are not sufficient. Electronic devices such as processors generate tremendous amounts of heat when they are in operation, and this can lead to system overheating and breakdown. Natural and forced convection enables thermal energy movement from the devices. Convection operates by transferring hotter fluid from the source of heat and replacing it with cooler fluid to enhance higher thermal dissipation. This is a key principle in optimizing the performance and reliability of devices where the highest processing capability translates into increased heating, which necessitates advanced cooling methods that rely on fluid dynamics to maintain operations within a stable state. Recent research has demonstrated remarkable enhancements in rates of heat extraction for liquid cooling systems with improved convective processes. New heat sink structures based on concentric channel configurations, for instance, have exhibited significantly greater cooling performance compared to traditional heat sinks.

Combined with nanofluids—liquids containing suspended nanoparticles—such systems exhibited a significant enhancement in the efficiency of heat rejection, achieving enhancements of up to 17% in higher concentrations of nanoparticles. The interface between the fluid dynamics and geometries concerned underscores maximizing convection processes in electronic cooling from evidence in research studies into various volume fractions of nanoparticles and heat transfer performance effects on liquid-cooled systems (Ahmadi et al., 2020). Thermal management implications also extend beyond conventional coolants to the present relevant safety concerns in modern devices. The increasing functional sophistication in cell phones has introduced severe thermal analysis to ensure trustworthy operation and prevent thermal runaway situations. Thermal management failures have, on several occasions, resulted in catastrophic failure such as explosion of devices. These failures inform the requirement for a thorough knowledge of heat generation mechanisms in devices, including contributions from batteries. Because overheating also may come from other sources such as batteries, effective convection strategies must consider the intrinsic design of thermal management systems that integrate different types of cooling solutions such as heat pipes with conventional techniques to facilitate intensified heat extraction (Chen et al., 2019).

4.3.3. Radiation from Surfaces

Surface radiation is intimately connected with electronic equipment heat generation, as it is the passing of thermal energy in the form of electromagnetic waves. Electronic equipment cannot but generate thermal radiation during operation, which occurs as per blackbody principles of radiation. The effectiveness of heat dissipation largely relies on the surface material properties. For instance, more emissive surfaces radiate heat more effectively, and such is crucial in the regulation of optimum operating temperatures of parts like transistors and capacitors. Furthermore, understanding the wavelengths distribution of the emitted radiation can also determine the design of the thermal management systems that entail radiative heat transfer, ultimately leading to the improvement of the performance and reliability of electronic components across various working environments. Recent advancements in nanotechnology have also enhanced our understanding of radiation at the microscopic level. Notably, research on surface plasmon polaritons (SPPs) demonstrates how electromagnetic waves can be manipulated at metal-dielectric interfaces to enhance thermal radiation characteristics. SPPs enable sub-wavelength confinement and ultrasensitivity measurements on the nanometer level, playing a major role in the development of high-efficiency thermal emitters and absorbers in electronics (Temnov et al., 2012). By hybridizing multilayer structures, researchers are designing novel active plasmonic devices that optimize radiation characteristics. Apart from enabling enhanced heat dissipation, the process also defines new applications in the field of nanophotonics, thereby connecting the field of optics and thermal engineering in designing electronic devices. Furthermore, novel interfaces between magneto-plasmonics and acousto-plasmonics also hold promising capabilities for designing novel heat generation technologies. The ability to control heat radiation through electromagnetic interaction provides new possibilities for manufacturing thermal management solutions that are both compact and highly efficient. Investigating these interactions allows one to generate high-energy photoelectrons through ultrashort pulses of surface plasmons, which makes electronic devices highly efficient (Temnov et al., 2012). The greater the researchers investigate these complex processes, the more crucial the role of thermal design in future electronics. Radiation knowledge from surfaces takes on new meanings, shaping the direction innovations can head towards more effective and more potent thermal solutions for the electronics industry with maximum performance without breaching the hazard of overheating.

4.4. EFFECTS OF HEAT ON DEVICE PERFORMANCE

Heat impact on system performance is a significant problem with today's electronic systems, particularly with the demand for more functional and compact devices. Electronic components, e.g., processors and batteries, which are increasingly being shrunk in size, generate a lot of heat due to inefficiency in operation. Thermal generation is proportional to higher functionality in these devices and hence more thermal failure hazards.

Figure 4.9. Heating affects the performance of the phone.

Source: https://www.screenfixed.com.au/wp-content/uploads/2023/10/How-Heat-Affects-Your-Phones-Performance.jpg.

This has been especially true in mobile phones, where lack of effective thermal management has resulted in severe malfunctions like overheating and even catastrophic failures like explosions. These occurrences emphasize the necessity of advanced thermal management methods to maintain device integrity and function under different working conditions, proving that optimal thermal control is a prerequisite for today's electronics' reliability (Chen et al., 2019). Various studies have investigated alternative cooling technologies and heat dissipation systems to mitigate the detrimental effects of heat on electronic devices. Researchers have also been engaged in using thermally more conductive materials as well as integrating heat pipes to enable enhanced heat dissemination as well as dissipation. There are still some current limitations, especially the size as well as the design of such devices, which is limiting in performance when being applied to integrated device structures. Notably, the packaging of significant components such as application processors and battery packs plays an important role in influencing the efficiency of thermal management, with general device performance being impacted. Without the consideration of thermal dynamics across these components, heat generation can be further accelerated, with the reliability of the device being impacted (Chen et al., 2019) (Figure 4.9).

Such an endeavor is thus warranted to enhance the thermal analysis methods for electronic design advancement. Emerging technologies like thermoelectric generators are good examples of new means of heat control and enhancement of device performance. The application of nanostructured components in the generators presents a path towards the optimization of power generation while minimizing heat buildup. Through mechanisms such as energy filtering and inter sub-band scattering, researchers have postulated that one can greatly enhance thermoelectric generation in even incoherently scattered systems. Such results have applications beyond mere power generation; they give direction for efficient cooling processes pertinent to a wide range of electronic devices. The research explains the way in which some configurations and features of barriers within such generators are capable of affecting performance, requiring careful design considerations so that overheating vulnerabilities in electronic components are avoided.

4.4.1. Impact on Efficiency and Functionality

In electronic hardware, the generation of heat primarily impacts functionality and efficiency. Thermodynamic principles state that extraneous heat can lead to thermal runaway, a phenomenon where increasing temperature raises energy dissipation, thus lowering device performance and potentially causing it to fail. This is particularly pertinent in high-performance computing environments, where processors generate massive amounts of heat through computation. The combination of thermal management with energy efficiency is indicative of the reality that developments in heat dissipation ensure not only device integrity but also optimal functional capability. For instance, devices that maximize heat dissipation can be operated at higher clock speeds without thermal shutdown, maximizing productivity and efficiency of processing capacity. Therefore, heat generation becomes a critical factor in designing electronic devices for optimal performance. In order to combat the inefficiencies resulting from heat generation, there have been many advances in the field of material science, with the discovery of nanostructured materials. The interesting optoelectronic properties of nanostructured materials can improve thermal conductivity while minimizing resistive heating of electronic devices. For example, the integration of semiconductor nanostructures into photovoltaic cells has been shown to alter optical transitions and enhance electron transport, veering away from the limits of traditional bulk material. This new method not only allows for enhanced energy efficiency in products but also enhances product performance by enabling new applications as well as extending component lifespan. The creation and application of such high-tech materials have the capability to transform electronic device functionality by solving the issues of heat production and resulting inefficiencies of functioning. Further, combining energy management systems with intelligence in electronic devices offers a feasible way to increase overall efficiency and performance. Through the utilization of real-time communications data in intelligent grids and meters, appliances can dynamically shift energy load based on thermal state analysis. Optimization is emphasized in research pointing to the necessity of harmonized communications protocols to facilitate interoperability among disparate energy grid entities. Improved efficiency is realized when the devices can inform the surrounding hardware of their thermal status such that adaptive cooling measures and load balancing can be achieved, which collectively offset heat generation. Given emphasis upon the employment of integrated communication technologies, electronic devices not only become able to improve internal efficiency but also evolve in line with fluctuating functional demands, determining the future situation for energy management on the global scale of electronics (Aeberhard et al., 2013; Chin et al., 2011).

4.4.2. Risk of Thermal Throttling

The danger of thermal throttling in computing devices has now become a severe concern with the advancement of technology and with more powerful devices. Thermal throttling is when a device throttles its speed so it does not overheat, which could be because of long periods of heavy loading or inadequate thermal control. This is particularly a problem in high-power computing pieces of equipment such as CPUs and GPUs, where excessive heat generation leads to inefficiencies in function. A recurring condition of thermal throttling not only impairs the performance of devices, which run at lower processing rates, but also lowers user satisfaction. Such alignment of heat generation and throttling address the importance of effective thermal management systems, which must be designed to balance performance needs against the thermal constraints of discrete components. In addition, the effectiveness of thermal management strategies can significantly affect the probability of thermal throttling. For instance, advancements in materials such as phase change materials and heat sinks have been promising to improve heat dissipation. Additionally, dynamic power management techniques allow devices to adjust their power consumption based on thermal load, so they do not risk overheating. As discussed in recent studies, methods like Constant Deviation Plans (CDPs) and Proportional Deviation Plans (PDPs) can be effectively employed in energy management systems to control the temperature outputs, and this has a direct correlation with the operational stability of devices. This points to a critical aspect of thermal management: not only must devices be designed to control heat output well but must play an active role in load management in order to avoid conditions likely to lead to throttling (Figure 4.10).

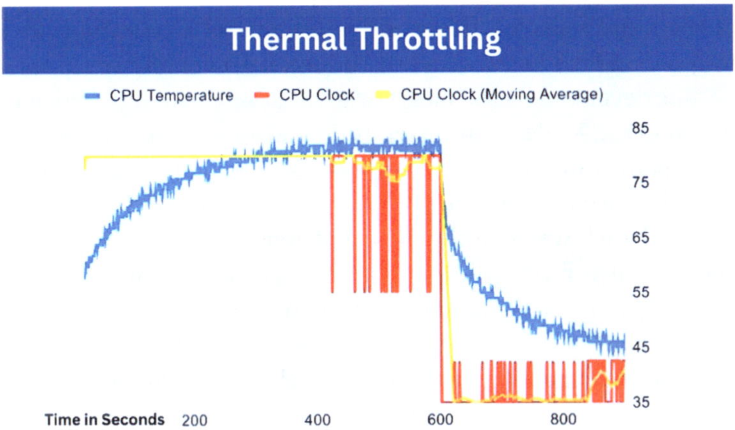

Figure 4.10. Thermal throttling.

Source: https://computermesh.com/what-is-thermal-throttling-cpu/

Furthermore, the impacts of thermal throttling are not just short-term effects on performance but are also long-term in terms of device reliability and lifespan. Devices subjected to repeated thermal throttling will have reduced lifetimes since high temperatures can accelerate material degradation and impact semiconductor reliability. In practical applications, incorporation of more intelligent thermal management is paramount for equipment subjected to extreme thermal loads, e.g., equipment used in aerospace, automotive, or high-energy physics applications. As outlined by ongoing research, implementation of such advanced thermal management techniques not only will prove effective in mitigating thermal throttling risks but also improve overall energy efficiency in electronic devices, driving innovation in the design of next-generation applications that are more tolerant to thermal problems (Cowan et al., 2025; Hassan et al., 2015).

4.4.3. Long-Term Effects on Component Lifespan

The long-term effect of heat production within electronic devices presents severe challenges that influence component life and device reliability. High heat over extended periods can lead to materials used in electronic components, including semiconductors and circuit boards, degrading over time. High operating temperatures have a tendency to accelerate processes of aging, leading to physical and chemical changes that weaken the integrity of materials. For instance, thermal cycling can create microfractures, delamination, or other forms of structural failure in components exposed to temperature changes. This type of degradation is not an abstract concern; it influences the rate of component failure and necessitates more frequent repair or replacement, raising operating costs and reducing the effective lifetime of electronic devices. It is crucial for producers to know about these long-term consequences, as this can determine the development of more heat-enduring materials and innovative cooling techniques. Aside from material degradation, heat can also affect the electrical properties of electronic components, hence affecting their lifespan. The electrical resistance in semiconductors rises with higher temperatures, which can lead to lower current flow and reduced device efficiency. High temperature can also further enhance the quality of noise and signal integrity problems, triggering miscommunication between components in complex electronic systems. For example, voltage continuity problems happen whenever temperature phenomena change voltage levels and harmony within power supplies, which eventually impact the overall system's reliability (Desmet et al., 2015). By solving such problems, engineers can prolong the working life of electronic devices to make them operational in extreme conditions. The fine balance of heat generation, electrical operation, and component lifespan makes sure that the need for ongoing R&D of thermal management methods remains. Furthermore, the long-term effect of heat on the lifespan of components transcends individual devices to affect bigger systems like networks and grids. For instance, electromagnetic compatibility and power quality stability become increasingly significant regarding heat generation in electronic systems (Dittmer et al., 2025). The consequence of thermal failure of certain elements of a larger system can lead to cascading failures affecting nearby devices or even networks. Such failure indicates the need for incorporating thermal management solutions during the electronic system design phase, particularly in planning phased incorporation of new technology. Proper design techniques greatly reduce the threat posed by heat generation, resulting in overall system longevity. Ultimately, a combined understanding of the long-term effects of heating on device life is important to the advancement of individual device reliability as well as survivability of larger electrical systems.

4.5. EMERGING TECHNOLOGIES IN HEAT MANAGEMENT

The increasing prevalence of electronic devices requires innovative approaches to the issue of generating heat, a challenge in the arena of modern technology. Emerging technologies, particularly thermal management, have the potential to enhance efficiency and longevity of the devices. Graphene and carbon nanotubes are materials with exceptional thermal conductivity that can significantly augment the capability of electronic components to remove heat. Further, advances in phase change materials (PCMs) make it possible to achieve effective thermal management with heat absorption and release during the solid-to-liquid phase transformation.

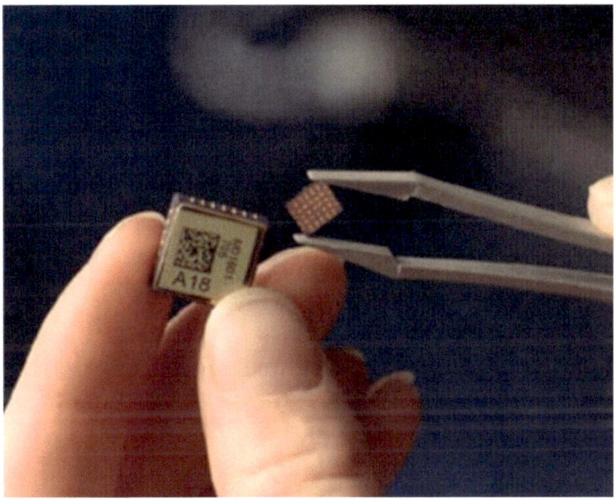

Figure 4.11. MEMS systems.

Source: https://control.com/technical-articles/understanding-microelectromechanical-system-mems-sensors/

Not only do these phases avoid overheating but also help increase overall electronic device efficiency, thereby overcoming one of the main constraints in contemporary electronics. As research goes on, the use of such advanced materials will certainly bring tangible improvements in device reliability and energy efficiency. Another area of heat management that may be pursued is the development of microelectromechanical systems (MEMS), which can be utilized for high-accuracy thermal monitoring and control within electronic devices. MEMS technology can be used to develop miniature sensors that can provide real-time feedback on temperature fluctuations, allowing for real-time dynamic adjustments in cooling approaches. This capability becomes particularly vital in applications where electronic components undergo rapid thermal load fluctuations. Further, the advent of smart grid systems emphasizes the need for efficient energy distribution, which intersects with heat management in electronic components (Figure 4.11).

Through the integration of MEMS in intelligent systems, device manufacturers can optimize energy use and enhance the thermal management capacity of their devices. Such advancements contribute not only to device efficiency but also to worldwide efforts toward sustainability and energy efficiency in the technology sector, as supported by the goals embedded in the ongoing development of smart grids (Axon et al., 2014). Utilization of renewable energy technologies like photovoltaics is very promising in enhancing the heat management of electronic devices. As the technology of solar energy advances, it will be able to compete with other end uses in residential and industrial applications. The efficiency of photovoltaic systems can be boosted through the integration of thermal management methods that utilize the waste heat generated in the process of converting solar energy. It is such integration that depends mainly on advanced power electronics converter systems, which help make energy generation and consumption interact in a harmonious way. Blending solar energy systems with combinations of new heat management technologies can lead to the production of a more robust and efficient energy system. By utilizing these new methods, the electronics industry can greatly reduce the production of heat in energy conversion processes. This not only minimizes heat stress on devices but also contributes to attempts at sustainability, ultimately resulting in a cleaner energy future (Abu-Rub et al., 2017).

4.5.1. Advanced Materials for Heat Dissipation

The ongoing miniaturization of electronic devices has created significant challenges in thermal management of heat generation, necessitating innovative solutions to thermal management. As electronic devices are operated at higher frequencies and power levels, the demand for advanced material with good heat-dissipation properties grows. Traditional cooling methods falter, prompting researchers to look for materials with improved thermal conductivity and heat-distribution capabilities. For instance, application of materials such as graphene, carbon nanotubes, and metal matrix composites is being widely used as these materials have excellent heat transfer properties. These materials not only address the thermal management issues but also come with the purpose of enhancing the reliability and performance of electronic systems, particularly as the demand for increased performance continues to rise. Central to the development of such next-generation materials is the ability to realistically model their behavior. Advanced simulation techniques, as described in recent studies, indicate how these materials behave under various conditions of operation, helping improve more efficient cooling technology. As an example, using a newly developed three-dimensional simulator, scientists can simulate thermal transport phenomena efficiently and predict heat dissipation behaviors of electronic components. Such a capability is central to comprehending the material's behavior in resistive random-access memory (RRAM) devices like Si-rich silica (SiOx), very pertinent in the present day's world of electronics. The significance of such modeling cannot be overstated; it provides a foundation for the creation of more effective thermal management systems critical to guaranteeing device performance and longevity (Asenov et al., 2015). As the telecommunication industry continues to shrink devices, the role of advanced materials in heat dissipation increases in importance. The higher levels of sophistication of electronics demand materials that can sustain higher operating temperatures without loss and with efficiency. Emerging material technologies such as ferroelectrics and high-temperature superconductors are on the horizon for their respective heat characteristics. Through the use of sophisticated measurement techniques and accurate modeling, engineers are capable of designing innovative circuit structures that minimize thermal management issues. This is particularly important in the case of nonlinearities introduced by excessive heat, which impacts the performance of the overall device. Ongoing research on circuit modeling of microwave devices focuses on the requirement for advanced heat dissipation materials, eventually to enhance the reliability and efficiency of the system as a whole, thereby meeting the demanding requirements of modern electronics (Cantenys et al., 2011).

4.5.2. Innovative Cooling Solutions (e.g., Thermoelectric Coolers)

Heat reduction in electronic devices is essential for superior performance and device life. Thermoelectric coolers (TECs) are one of the newer cooling technologies that are efficient and multi-purpose. Through the principle of the Peltier effect, TECs provide effective heat removal from the electronic elements to the outside world. This unique mechanism allows precise temperature control, a fundamental need for high-performance computing and numerous applications, for example, portable electronics and aerospace technology air conditioning. Research has established that the performance of these systems can be maximized through effective manipulation of operating parameters, such as input power and working fluid velocity, to maximize the coefficient of performance (COP) and total heat transfer efficiency (Kabeel et al., 2023). By the integration of advanced materials and structures, TECs are to be in the vanguard of future thermal management technologies. In recent advances, specific focus has been on the use of nanofluids as a coolant in conjunction with TECs. These fluids, developed to contain suspended nanoparticles, exhibit higher thermal conductivity compared to conventional coolants, providing greater heat removal in electronic equipment. Experiments show that the performance of thermoelectric devices, particularly the rates of heat transfer and COP, can be enhanced by the application of nanofluids, resulting in a better cooling system (Kabeel et al., 2023) (Figure 4.12).

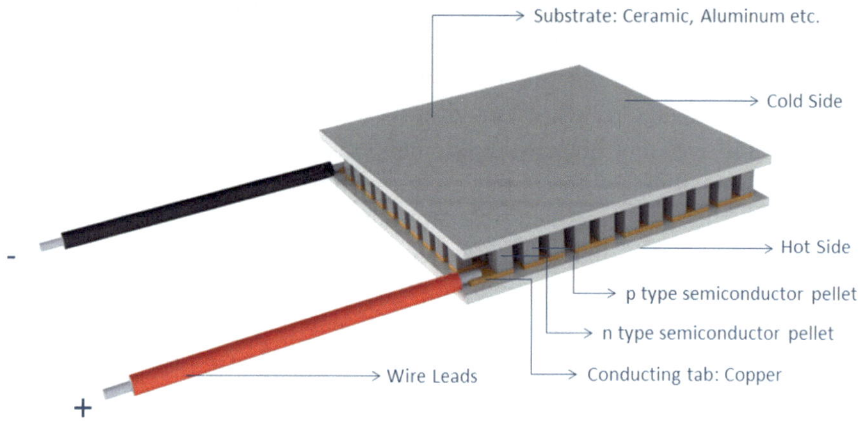

Figure 4.12. Thermoelectric cooler (TEC).

Source: https://blog.ozeninc.com/hubfs/image-png-Jan-27-2025-05-09-03-7521-PM.png.

The pairing of TECs with these high-tech fluids is a classic example of the creative solutions currently in the works to counter the mounting challenges of heat creation in ever smaller and more potent electronic devices. In this way, this synergy is not only maximizing performance but is also advancing the general aims of energy efficiency in manufacturing and operating electronics. Second, the contribution made by federal support to small businesses engaged in such new cooling technologies cannot be overlooked. Programs like NASAs Small Business Innovation Research (SBIR) have spurred many projects with the aim of exploring new cooling technologies, e.g., new-generation thermoelectric technologies. By spurring the contact between small businesses and research institutions, these projects help to accelerate the development of new technologies with the potential to effectively check heat generation in electronic devices. The SBIR program has brought into the spotlight many successful projects that demonstrate the potential of TE systems to address thermal management issues in various areas, emphasizing the necessity of innovation in enhancing the performance and reliability of electronic devices. The ongoing pursuit of novel materials, arrangements, and coolant strategies holds promise, and innovative cooling schemes keep pushing forward in tandem with advances in electronics.

4.5.3. Integration of Heat Management in Device Design

The integration of effective heat management strategies into device design has become part of the solution to making electronic devices reliable and durable. As devices shrink in size but get more powerful, heat dissipation becomes increasingly complicated. This necessitates the use of new thermal management strategies that can be integrated at design time. For instance, thermally conductive elements such as copper and graphene are increasingly being utilized to permit heat to be transferred away from sensitive components in an efficient manner. Additionally, innovative design techniques, such as heat sinks and phase-change materials, can help retain operating temperatures at optimal levels within electronic devices. Moreover, strategic component placement on printed circuit boards can further improve thermal spreading so that device-failure-causing hot spots are prevented. Consequently, integrated heat management principles in the design process not only enhance performance but also decrease the likelihood of thermal failures. Aside from structure and material design, the role of simulation technologies in heat management cannot be overlooked. Software for computational fluid dynamics (CFD) and thermal modeling allows engineers to model heat generation and dissipation patterns within a device, a critical process for the identification of potential thermal hotspots prior to them being a problem. With these simulation tools, designers are able to iterate their prototypes more effectively, fine-tuning component placement and thermal interfaces from criteria that go beyond trial and error. This is especially beneficial in high-density electronic designs, where conventional cooling forces might fall short. As components become more integrated, the art of thermal management needs to evolve to accommodate complex heat analysis between several components. This shift emphasizes the need for a comprehensive solution to heat management that addresses whole systems, rather than discreet devices, and ensures the efficiency and reliability of sophisticated electronic assemblies. The implications of integrating heat management methods extend beyond transient device operation; they impact directly upon the overall operation of smart systems. Identification of the intricate interdependence between thermal efficiency and energy efficiency is essential, particularly in the design of smart grids based on complex communication networks. In energy management, tuning of the thermal solution can reduce the consumption of energy without compromising the operational performance to a considerable degree. For example, applications of smart metering technologies can facilitate real-time temperature sensing of equipment, creating a feedback system that controls energy usage as well as the performance of the entire system. As determined by recent research, coordinated effort in standardization will be imperative to standardize such technologies, which will lay a good foundation for future innovations in thermal design and energy management (Chin et al., 2011; N/A, 1976). As the capabilities of electronic devices keep evolving, the integration of heat management methodologies will also remain a cornerstone of efficient, sustainable device design.

SUMMARY

In correlating the findings with regard to heat generation in electronic devices, there is an undeniable fact that successful thermal management is instrumental in realizing optimal performance as well as reliability. The study of next-generation cooling systems, such as the use of nanofluids in heat sinks, represents a significant advance in techniques of heat dissipation. Experiments have shown that the application of nanofluids, particularly in higher volume fractions, can enhance the rate of heat rejection, which is important as far as operation efficiency in devices subjected to severe thermal loading is concerned. In particular, the range of ~2% to ~17% of performance enhancement in terms of varying nanoparticle concentrations indicates that the right choice of cooling regimes can lead to substantial thermal regulation enhancements and consequently extend device life and performance stability (Ahmadi et al., 2020). Furthermore, electronic component development, particularly high-frequency devices, points to the necessity for ongoing betterment in heat management strategies. The evolution of low-noise InP Gunn devices achieving impressive levels of RF power is a perfect illustration of how upgraded cooling systems are able to exploit better performance without losing thermal stability. With the devices increasingly used at higher and higher frequencies, the interplay between heat generation and heat management becomes even more critical. The need for new doping profiles and best material designs is apparent, thereby implying that future advancements will have to balance performance and thermal control (Eisele et al., 2004). Bilateral thinking will ensure devices not only perform better but also operate safely within optimal thermal conditions. In summary, heat creation in electronic devices remains an ongoing priority that needs holistic solutions across all fields of technology. The addition of innovative cooling technologies, such as engineered nanofluids and advanced thermal designs, represents one possible solution to overcoming thermal-related issues. As the performance capabilities of electronic devices continue to rise, the demands for advanced thermal management systems will only become more insistent. The power to perfect the art of efficient heat dissipation will pave the way for the next generation of electronics that enhance performance and device life while preventing failure through over-thermal stress. In an era where electronic innovation is the key, thermal management is not only an engineering necessity, but also a necessity for developing sustainable technology.

The discussion of heat generation in electronic devices is centered around the crucial relationship between design and heat management. Revolutionary advancements made in material science have shed light on the significance of nanostructured materials, particularly in semiconductor technology. Silicon nanopillar research is an ideal case study where dynamics of heat generation and propagation are so critical. According to the discovery, visible light can be employed to selectively crystallize silicon pillars' innermost regions, which will improve materials' thermal and electric properties. This suggests that by localized heating with light, electronic devices' performance has a chance to be significantly improved. These observations point to the importance of integrating thermal management strategies at the early design stage to enable devices to operate optimally without succumbing to overheating, which is prevalent in modern electronics (Alessandri et al., 2018). Furthermore, power quality and electromagnetic compatibility (EMC) issues suggest an evident correlation with heat generation in electronic systems. The interference caused by voltage changes, harmonics, and transient electromagnetic fields generates inefficiencies that have a tendency to manifest as excess heat. It is required to monitor and understand how these power quality disturbances can contribute to increased thermal buildup in equipment. Careful examination of voltage fluctuation and harmonics not only highlight the restrictive operating conditions but also direct approaches towards improving cooling systems. The aggregation of results from various studies brings to the forefront the importance of proper monitoring systems that can detect and avoid thermal buildup and emphasize avenues of how electronic equipment can be designed more sustainably and reliably based on current electrical safety guidelines (Desmet et al., 2015). Lastly, the inclusion of computational tools in heat generation modeling emphasizes a new development in the field of study. The application of advanced simulations allows the detailed observation of thermal dynamics in electronic devices to make predictions on the patterns of heat generation for various time intervals. The new computational techniques proposed in recent studies have the potential to alter paradigms of light-matter understanding, paving the way for increased innovation in electronic design and manufacturing processes. This capability is not only beneficial in enhancing performance but also assists engineers in designing systems to withstand thermal stresses. As electronic devices increase in their complexity and strength, the incorporation of such computational tools will be essential in ensuring reliability and lifespan through effective thermal management (Alessandri et al., 2018).

REFERENCES

1. Abu-Rub, A., León Galván, J. I., & Malinowski, M. (2017). Solar photovoltaic and thermal energy systems: Current technology and future trends. https://core.ac.uk/download/286563376.pdf (accessed on 23rd October 2025).
2. Adachi, H., Ishida, M., Kikkawa, T., et al. (2016). Thermoelectric generation based on spin Seebeck effects. http://arxiv.org/abs/1604.00477 (accessed on 23rd October 2025).
3. Adam, J. (2018). A human powered micro-generator for charging electronic devices. https://core.ac.uk/download/159310421.pdf (accessed on 23rd October 2025).
4. Aeberhard, U. (2013). Simulation of nanostructure-based high-efficiency solar cells: Challenges, existing approaches and future directions. http://arxiv.org/abs/1304.1957 (accessed on 23rd October 2025).
5. Ahmadi, M. H., Jilte, R., Kalamkar, V., et al. (2020). Cooling performance of a novel circulatory flow concentric multi-channel heat sink with nanofluids. https://core.ac.uk/download/304220209.pdf (accessed on 23rd October 2025).
6. Alessandri, I., & Danesi, S. (2018). Using optical resonances to control heat generation and propagation *in silicon* nanostructures. http://arxiv.org/abs/1811.10908 (accessed on 23rd October 2025).
7. Allahverdyan, A. E., Callen, H. B., Chua, L. O., Curzon, F. L., De Vos, A., et al. (2004). Quantum, cyclic and particle-exchange heat engines. http://arxiv.org/abs/cond-mat/0407508 (accessed on 23rd October 2025).
8. Asenov, A., Gerrer, L., Sadi, T., et al. (2015). Physical simulation of Si-based resistive random-access memory devices. https://core.ac.uk/download/46559103.pdf (accessed on 23rd October 2025).
9. Avalos, D., Huynh, V., Lyons, D., et al. (2016). Design strategy of a thermoelectric evaporative refrigerator for safe food storage. https://core.ac.uk/download/72853297.pdf (accessed on 23rd October 2025).
10. Axon, C., Balta-Ozkan, N., Cipcigan, L., et al. (2014). Scenarios for the development of smart grids in the UK: Literature review. https://core.ac.uk/download/155776124.pdf (accessed on 23rd October 2025).
11. Azcorra, A., & Ucar, I. (2015). Deseeding energy consumption of network stacks. http://arxiv.org/abs/1511.08167 (accessed on 23rd October 2025).
12. Capata, R., & Saracchini, M. (2018). Experimental campaign tests on ultra micro gas turbines, fuel supply comparison and optimization. https://core.ac.uk/download/154949724.pdf (accessed on 23rd October 2025).

13. Chen, E., Feng, Y., Grimes, C. A., Gurrum, S. P., Huang, Y., Jaworski, C. M., et al. (2019). Thermal analysis of lithium-ion battery-equipped smartphone explosions. https://core.ac.uk/download/289095013.pdf (accessed on 23rd October 2025).
14. Chin, W. H., Efthymiou, C., Fan, Z., et al. (2011). Smart grid communications: Overview of research challenges, solutions, and standardization activities. http://arxiv.org/abs/1112.3516 (accessed on 23rd October 2025).
15. Copeland, R., Crespi, N., Rehmani, M. H., et al. (2019). Internet of Things-aided smart grid: Technologies, architectures, applications, prototypes, and future research directions. http://arxiv.org/abs/1704.08977 (accessed on 23rd October 2025).
16. Cowan, K., Lewis, R., Mislinski, P., et al. (2025). Self-unloading, unmanned, reusable lunar lander project. https://core.ac.uk/download/pdf/42813756.pdf (accessed on 23rd October 2025).
17. Desmet, J., Heimbach, B., & Renner, H. (2015). Power quality and electromagnetic compatibility: Special report, session 2. https://core.ac.uk/download/55822716.pdf (accessed on 23rd October 2025).
18. Dittmer, L. N., Drews, M. E., Hoehn, A., et al. (2025). A lunar base reference mission for the phased implementation of bioregenerative life support system components. https://core.ac.uk/download/pdf/42813737.pdf (accessed on 23rd October 2025).
19. Eisele, H., & Kamoua, R. (2004). Submillimeter-wave InP Gunn devices. https://core.ac.uk/download/58722.pdf (accessed on 23rd October 2025).
20. Faiz, A., Sinha, K., Varma, A., et al. (2025). Automotive air pollution: Issues and options for developing countries. https://core.ac.uk/download/pdf/6644979.pdf (accessed on 23rd October 2025).
21. Fisher, D., Hand, J., Liesen, R., et al. (2011). Screening of energy efficient technologies for industrial buildings' retrofit. https://core.ac.uk/download/9033399.pdf (accessed on 23rd October 2025).
22. Fraser, A. G., McCulloch, M. D., Márquez-Fernández, F. J., et al. (2018). A novel topology of high-speed SRM for high-performance traction applications. https://core.ac.uk/download/157811823.pdf (accessed on 23rd October 2025).
23. Gill, W. L., Keough, M. B., & Rippey, J. O. (2025). Technology evaluation of heating, ventilation, and air conditioning for MIUS application. https://core.ac.uk/download/pdf/42891762.pdf (accessed on 23rd October 2025).
24. Gilman, J. A., Paige, J. B., & Schwenk, F. C. (2025). NASA SBIR abstracts

of 1990 phase 1 projects. https://core.ac.uk/download/pdf/42808250.pdf (accessed on 23rd October 2025).

25. Goupil, C., Müller, E., Seifert, W., et al. (2011). Thermodynamics of thermoelectric phenomena and applications. https://core.ac.uk/download/4890584.pdf (accessed on 23rd October 2025).
26. Hassan, N. U., Khalid, Y. I., Tushar, W., et al. (2015). Customer engagement plans for peak load reduction in residential smart grids. http://arxiv.org/abs/1502.03908 (accessed on 23rd October 2025).
27. Kabeel, E., Mousa, M. G., & Elsayed, M. M. A. (2023). Theoretical study of thermoelectric cooling system performance. https://core.ac.uk/download/588392347.pdf (accessed on 23rd October 2025).
28. Kelly, B. T., Li, B., Lifshitz, E. M., Liu, G. R., Hone, J., Wang, J. S., Kubo, R., et al. (2004). Thermal conduction of carbon nanotubes using molecular dynamics. http://arxiv.org/abs/cond-mat/0402616 (accessed on 23rd October 2025).
29. Lysenko, D. (2018). Sustainable technology for processing waste electrical and electronic equipment. https://core.ac.uk/download/168412847.pdf (accessed on 23rd October 2025).
30. Markides, C. N., & Smith, T. C. B. (2011). A dynamic model for the optimization of oscillatory low grade heat engines. https://core.ac.uk/download/77011366.pdf (accessed on 23rd October 2025).
31. Muralidharan, B., & Singha, A. (2017). Incoherent scattering can favorably influence energy filtering in nanostructured thermoelectric. http://arxiv.org/abs/1708.03904 (accessed on 23rd October 2025).
32. N/A. (2013). Korea's energy standards & labeling. https://core.ac.uk/download/51181909.pdf (accessed on 23rd October 2025).
33. Rocas Cantenys, E. (2011). Nonlinear mechanisms in passive microwave devices. https://core.ac.uk/download/81575732.pdf (accessed on 23rd October 2025).
34. Saraogi, M., & Srivastava, S. (2007). Electronic chip cooling in horizontal configuration using fluent-gambit. https://core.ac.uk/download/53188790.pdf (accessed on 23rd October 2025).
35. Talbot, P., & Williams, M. (2019). Design features in multiple generations of electronic cigarette atomizers. https://core.ac.uk/download/286729283.pdf (accessed on 23rd October 2025).
36. Temnov, V. V. (2012). The marriage of plasmonic, magnetism, acoustics and ultrafast optics. http://arxiv.org/abs/1207.2624 (accessed on 23rd October 2025).

CHAPTER 5

THERMAL MODELING AND SIMULATION

Contents

INTRODUCTION

Thermal modeling is a fundamental part of the investigation of heat transfer phenomena in various physical systems. Thermal modeling essentially involves the application of mathematical and computational procedures to simulate the manner in which temperature change influences system behavior. This is particularly critical in energy engineering, in which modeling can lead to very considerable enhancements of efficiency and sustainability. For instance, the integration of material properties and heat transfer can be quantified in order to predict thermal performance of next-generation systems. As elaborated in research, providing a platform addressing physical components and their interactions as first-class citizens allows for a deeper extent of knowledge of heat transfer phenomena in thermal systems, a belief highlighted by Lepri et al. (2010). Lastly, this approach fosters conditions for precise analyses that enhance practical applications in environmental engineering and management. The contribution of computer methods to thermal modeling cannot be overstressed in the era of the prevalence of advanced simulations. With further improvements in computation capabilities, uses for approaches like Computational Fluid Dynamics (CFD) in thermal modeling become enormously significant. CFD has emerged as a critical tool in energy engineering, enabling detailed simulations of heat transfer and fluid dynamics in systems ranging from power generation to combustion. The contributions of CFD advancements over the past two decades illustrate its importance in transforming how engineers approach thermal challenges. By continuously improving modeling capabilities, researchers can better understand the intricate behaviors within thermal systems. This has been observed in the repeated surveillance of CFD programs, where emphasis is placed on the inclusion of numeric methods that allow for improved understanding and application of thermal phenomena in complex environments, as documented by Paricio et al. (2019). In addition, the inclusion of thermal modeling within broader simulation paradigms is a paradigm shift in the way engineers create and test systems. Using approaches such as Real-Time Maude, researchers are now capable of addressing challenges in modeling continuous behaviors within objects and thus improve system designs and efficiencies across different areas. Such approaches not only demonstrate the capability to simulate complex thermal behaviors but also allow for incremental testing and validation against realistic parameters. The value of such an approach is realized in its ability to facilitate model checking analysis, which provides essential insight into system response within varying contexts. The use of these advanced modeling techniques by the present research exhibits a strong shift towards increasingly adaptive and dynamic simulations within thermal modeling, reflecting a landmark shift in engineering approaches in favor of accuracy and real-time control.

Definition of Thermal Modeling

Thermal modeling is an essential engineering process of the simulation of temperature distributions and heat transfer in a system. It helps the engineers to forecast the thermal performance under various conditions so that the engineers can conduct optimal design and assess performance. Thermal modeling, using mathematical equations and computer methods, gauges how heat travels in materials and how systems respond to thermal loads. This modeling is a basis for the analysis of performance in systems, particularly in high-technology sectors such as aerospace and electronics, where temperature is very important. Proper thermal models prevent potential overheating issues and ensure that components perform within their provided thermal limits, thus extending their lifespan and enhancing overall system reliability. The process of defining the thermal model is to create comprehensive descriptions that include both steady-state and transient heat transfer conditions. This involves conducting system functional analyses in order to get top-level requirements through to detailed specifications such that all aspects of thermal dynamics are addressed within the design process. The integration of quality function deployment (QFD) as a component of the thermal modeling process contributes another level of richness to this definition by relating customer needs to important thermal performance characteristics. By alternative system concepts, engineers can study the performance of various designs against set standards (Figure 5.1).

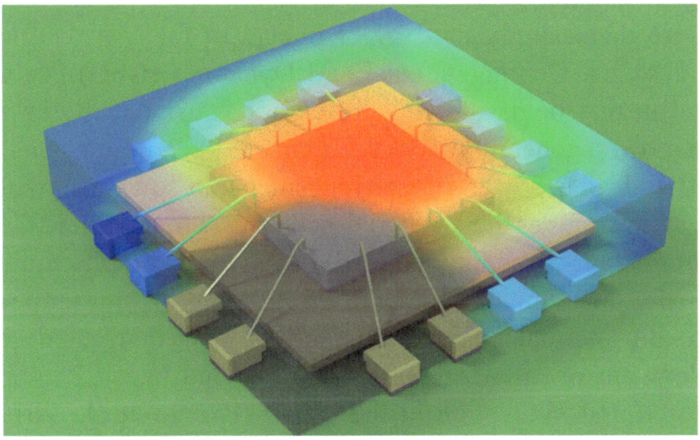

Figure 5.1. Thermal modeling in electronics.

Source: https://images.ctfassets.net/17si5cpawjzf/4f4xpJwKokko24jtyvk7Fd/e8159d1651c0798a62c39593ff7c4aff/ComponentThermalWebinar2020_LandingPage_1920x1080_tcm27-77012.jpg.

These rigorous tests become essential while making choices regarding strengths and limitations of different thermal management techniques, particularly in complex systems where traditional models cannot precisely simulate behavior. Therefore, understanding and establishing thermal modeling becomes a must while advancing thermal management solutions. Thermal modeling advances today also consider dynamic operating environments upon which emerging technology is observed, such as multicore processors. Notably, temperature control has also been researched through dynamic voltage-frequency scaling with advanced feedback controls to achieve fast convergence under model uncertainties. Such new solutions show the integration of thermal modeling into broader system control theories and their applicability beyond the traditional engineering design arena. For instance, nonlinear, time-varying models of plants, in addition to enabling prediction of temperature variation, also enable robust real-time calculations. This combination of thermal modeling and control engineering paradigms is evidence of a growing trend in the field where precise temperature control is ever more essential for guaranteeing system integrity and performance, exemplified in the simulation results demonstrating accurate tracking of temperature reference values (Rao et al., 2015).

Importance in Engineering and Science

The importance of thermal simulation and modeling in science and engineering cannot be overemphasized, especially for energy engineering. With the increasing need for energy efficiency and sustainability through technologies, simulations are increasingly used by engineers to forecast thermal behavior in different systems. Computational Fluid Dynamics (CFD) has emerged as a critical tool in this field, allowing for comprehensive analyses that inform decisions in power generation, combustion, and renewable energy sources such as wind and solar power. The advancements in software capabilities and computational power over the last two decades have transformed the landscape of thermal simulations, enabling detailed modeling of complex fluid behaviors and heat transfer phenomena. As identified in recent literature, the integration of CFD into the thermal engineering toolbox has contributed significantly to the development of efficient energy systems and hence its importance in driving research and innovation in the sector (Paricio et al., 2019). Beyond energy engineering, thermal modeling finds application in other areas such as electronics and materials science. In electronics, effective thermal management is the key to the performance and reliability of integrated circuits. Research on new materials such as carbon nanotubes as interconnects has gained momentum with the limitations of traditional copper interconnects. Carbon nanotube interconnects are projected to be more efficient in performance and energy efficiency with the support of aggressive modeling and simulation techniques that predict circuit-level behaviors. The application of such advanced materials closes the gap between theoretical modeling and practical applications and exhibits a multidisciplinary function where thermal simulations are critical in guiding engineering practice. The combination of materials science and circuit design also illustrates the overall applicability of thermal modeling in driving scientific inquiry and practical engineering solutions (Amoroso et al., 2017). Moreover, the function of effective thermal modeling has an impact on environmental sustainability and public safety. Enhanced thermal simulations can be used to design systems that minimize waste and carbon footprints, a consideration in a time where climate change is a pressing concern. For instance, accurate prediction of thermal distribution in buildings can be used to guide energy-efficient designs where heating and cooling are regulated to conserve energy and improve occupants' comfort. Similarly, for automotive and aeronautical engineering, the thermal stress on materials is an important region to be studied to be able to provide performance and safety under varying conditions of operation. With sophisticated thermal modeling techniques, designs may be optimized not only for functionality but for environmental sensitivity as well. While the world continues to fight acute environmental issues, the contributions of thermal modeling and simulation in engineering and science become more vital in influencing innovations towards a cleaner future.

Overview of Applications in Various Fields

Thermal simulation and modeling are presently at the core of numerous varied disciplines, bridging past confines to allow for creative applications. In materials science, for instance, they are applied to predict thermal properties and behavior of materials, particularly in metal-organic framework (MOF) science. Researchers employ computational chemistry methods to predict crystal structures and characterize thermal and mechanical material properties in an effective manner. With screening of possible MOFs facilitated on a large scale through extensive thermal modeling, industries are able to enhance material design suited for processes such as gas storage and catalysis, thus encouraging advancements towards sustainability and energy efficiency (Coudert et al., 2015). Thermal modeling is also crucial in the energy sector, particularly in designing and optimizing thermal systems. The applications range from increasing the solar panel efficiency to increasing the power plant heat exchangers. Using simulation tools, engineers can model heat transfer processes under various environments, thereby conserving energy and improving efficiency. In addition, with renewable energy use, thermal management becomes increasingly vital, with simulations driving thermal energy storage system efficiency. All of these advancements not only increase operating efficiency but also contribute to the overall reduction in greenhouse gas emissions, hence bringing about a cleaner future. Furthermore, the role of thermal modeling even extends beyond the realm of electrical engineering, where the management of thermal effects on electrical equipment is critical. As the industry trends toward smaller, more powerful units, the challenge of heat dissipation becomes even more significant. Simulation tools allow engineers to predict thermal performance in circuits, their design optimizing against overheating and ensuring reliability.

Convergence of thermal modeling and simulation facilitates innovation in the design of future electronics to operate reliably under various thermal loads. The intricate interplay of electrical performance and heat transfer still drives innovation in this field, both in theoretical studies and actual engineering practice (Ebert et al., 2008).

5.1. TYPES OF THERMAL MODELS

Among the various types of thermal models, continuum models are one of the most common techniques of thermal analysis. In continuum models, materials are approximated as continuous media rather than discrete particles. This allows differential equations to be used to describe thermal conduction, convection, and radiation in a system. By invoking classical heat transfer equations, continuum models allow for cogent analyses of thermal behavior on a variety of scales and geometries. A particular advantage of such modeling is its ability to offer analytical solutions or control computationally demanding schemes with relatively little effort. Such versatility is critical across industry sectors, including aerospace and automobile industries, where accurate thermal prediction can optimize performance and improve safety. Thermal models based on continuum have been improved by advancements in software, which allow for simulations with complex geometries and transient behavior. The second crucial category of thermal models is discrete models, which deal with the idiosyncratic characteristics and responses of particles or elements in a thermal system. This approach is especially suited to materials where the continuum hypothesis cannot be applied, e.g., granular media or nanostructured material. Discrete element modeling (DEM) is a significant approach that is typically applied, taking into account particle-particle interactions and their related thermal implications. Such models are able to provide minute details of heat transfer mechanisms which cannot be accounted for by continuum models, e.g., particle-to-particle conduction and spot heating. Secondly, the combination of DEM and CFD gives broad knowledge of thermal processes whose actions are dictated by particle motion and fluid flow. The combination broadens thermal modeling applications in material sciences and energy production so that researchers can achieve greater precision in simulations (Paricio et al., 2019). Hybrid thermal models are in-between continuum and discrete modelling approaches, being sensitive to the richness of thermal phenomena in nature. Hybrid models have both methods' advantages in treating systems with both continuous and discrete characteristics. This is particularly relevant to phase-changing applications such as melting and solidification processes where the material and thermal properties undergo dramatic changes. The versatility introduced by hybrid modeling implies that bulk areas can be modeled by continuum equations and discrete models at the same time, in surface interactions or particulate flows. Moreover, the advancement in computational techniques implies that simulation of such complex interactions can be accomplished in real time, which enhances the predictive ability of thermal models. The use of hybrid models in research and industrial applications makes major contributions to thermal engineering disciplines since they enable a comprehensive understanding of complicated thermal responses in intricate systems (Lepri et al., 2010).

5.1.1. Analytical Models

The use of analytical models in thermal modeling and simulation is critical to successfully analyzing the thermal profiles of complex systems. Analytical models possess a significant advantage over numerically based approaches in that they are capable of providing quick estimations without sacrificing significant accuracy. One of the most important domains where analytical models excel is in the simulation of very-large-scale integration (VLSI) structures. Traditional commercial thermal analysis programs struggle with intricate designs and lengthy computation times; therefore, applying analytical methods based on electrical resistance solvers is a viable solution. Through the development of sophisticated closed-form thermal models that account for isotropic and anisotropic characteristics, scientists have been able to reduce simulation complexity while maintaining thermal profile accuracy, enhancing the prospects of successful product development and silicon realization (Banerjee et al., 2010) (Figure 5.2).

Figure 5.2. Very-large-scale integration (VLSI).

Source: https://www.rglobits.com/assets/img/technologies/7a.jpg.

Besides application in VLSI systems, analytical models also play a critical role in explaining thermal dynamics in permafrost regions. As frozen ground conditions have a large impact on local hydrology and energy balances, effective model simulation of ground thawing and freezing depths (GTFDs) is required. Various algorithms have been employed to simulate GTFDs with varying degrees of complexity and type of parameterization, emphasizing the need for rigorous testing of these analytical procedures. Through comparative studies, it has been found that analytical algorithms can be superior to semiempirical schemes under given circumstances, particularly when surface and bottom thermal forcing are taken into consideration. Besides, analytical model precision can also be greatly improved through best input selection and specific soil parameter configurations, demonstrating the flexibility and actual application potential of analytical models in clarifying complex thermal processes (Abbey et al., 2008). Analytical model development and improvement under thermal simulation highlight their ability in bridging the theory application gap. Not only do these models allow for quick assessments needed in iterative design cycles, but they also create avenues for more accurate predictions in complex environments. With successful application across various applications—ranging from VLSI thermal management to permafrost science—analytical models are assuming increasingly prominent roles in thermal modeling. With the growing need for effective thermal solutions, the fusion of analytical methods and advancement in computational methods may revolutionize the course of thermal management techniques in engineering and environmental science. Through the marriage of theoretical modeling with real-world application, researchers are able to enhance the efficiency and applicability of their studies, ultimately paving the way to more sustainable engineering practices.

5.1.2. Numerical Models

Numerical models are a central component in the practice of thermal modeling and simulation, with important tools to investigate and predict heat transfer behavior in complex systems. By discretizing physical models into concrete components, numerical methods enable researchers to identify complex behaviors that might be unrecognizable through analytical solutions alone. The application of numerical models in thermal systems enables the simulation of a variety of conditions such as transient heat conduction, convection, and radiation that are typical of most engineering applications. Moreover, the models are helpful in evaluating system performance under various operating conditions and thus forming the basis for the design of more efficient thermal systems. The increasing sophistication of numerical techniques reflects the urgent need for advanced techniques capable of accommodating real-life complexities, making new inroads in the areas of thermal analysis and simulation. The use of advanced numerical techniques is significantly benefited by developments in computation technologies, such as high-performance computing and advanced algorithmic techniques. The rapidly increasing computation capability at hand has enabled scientists and engineers to undertake simulations with higher accuracy and at unimagined scales. For instance, Computational Fluid Dynamics (CFD) is now an indispensable tool that integrates numerical methods into fluid flow and heat transfer analysis, revealing critical information regarding thermal management systems in various energy sectors. Recent studies indicate that developments in the application of CFD have enhanced thermal modeling strengths, providing dense data for optimization in applications like combustion, renewable energy systems, and optimum turbine designs. As outlined in recent critiques, such advances underscore the central place of numerical modeling in solving sophisticated thermal engineering issues in many industries, representing a change of paradigm in research approaches. Furthermore, research on hybrid numerical models represents an innovative frontier in thermal simulation that best links continuum and discrete modeling techniques. Such models represent a fulcrum for depicting the continuous and discontinuous behaviors characteristic of most thermal systems. For example, the use of systems like Real-Time Maude allows for the simulation and modeling of physical phenomena, whereby interactions as well as components are treated systematically. Current research identifies that numerical methods, namely for the simulation of continuous behaviors in object-oriented hybrid systems, offer contexts to consider thermal interactions more dynamically. Such methods enhance the stability of simulation and accommodate more advanced model checking analyses in thermal engineering applications. Thanks to hybrid numerical models, researchers are able to build a more accurate picture of thermal dynamics that can, in turn, contribute to enhancing energy systems and applications, which play a significant role in sustainable engineering practices (Lepri et al., 2010; Paricio et al., 2019).

5.1.3. Hybrid Models

Thermal modeling and simulation have evolved to increasingly apply hybrid models, which pair different techniques and methods for enhanced computational efficiency and accuracy. Hybrid models tend to pair continuum-based approaches, such as finite element analysis, with discrete approaches such as molecular dynamics. The pairing allows scientists to model the thermal properties of materials at various scales, thus bridging phenomena that are difficult to describe using either technique alone. For instance, as one explores micro- and nano-scale systems, faith in purely conventional modeling can fail to capture key interactions at the atomic level. Recent studies focus on hybrid approaches, whereby simulations can be used to give complete insight required for designing smart materials and optimization of thermal systems, putting a case forward for adaptable modeling approaches in contemporary thermal analysis (Asproulis et al., 2014). In engine management systems, hybrid models have been particularly helpful since they are able to combine various modeling strategies with meta-heuristic algorithms. Through the use of evolutionary algorithms and swarm intelligence strategies, researchers have been able to successfully optimize engine performance as well as thermal efficiency. The versatility of these algorithms allows them to work through a collection of optimization challenges in parallel, such as engine calibration and fault detection, leading to improved engine robustness. The combination not only eases modeling but also increases predictability, which is critical to meeting increasing regulatory requirements and performance expectations. In turn, the application of hybrid models promotes innovations in thermal management in motor vehicle engineering that led to advancements that are central in modern engine production (Tayarani et al., 2015). Additionally, the application of hybrid models in thermal simulation crosses over beyond automotive systems to cover a wide range of applications, including microelectronics and heat exchangers. Given the complexity that is associated with micro- and nano-fluidic systems, as presented in recent review papers, a hybrid model approach enables scientists to comprehensively study fluid dynamics and thermal effects at various scales. Through fusing atomistic and mesoscale simulations, researchers can successfully estimate temperature distribution and fluid flow characteristics that are of paramount significance in developing efficient thermal systems. This paradigm not only enhances understanding of heat transfer processes but also trains one on the design of new materials and technologies with improved thermal performance. Lastly, reliance on hybrid modeling contributes directly to progress in thermal simulation applications in diverse scientific and engineering disciplines (Asproulis et al., 2014).

5.2. SIMULATION TECHNIQUES

In thermal modeling and simulation, a broad spectrum of simulation techniques is employed to forecast the thermal performance and system response. At the heart of such techniques is Computational Fluid Dynamics (CFD), which has developed into a crucial tool in energy engineering. In the past 20 years, advances in software capability and hardware performance have significantly contributed towards the capacity of CFDs to simulate heat transfer, fluid flow, and energy conversion processes. The ability to interpret complex thermal interactions quantitatively allows researchers and engineers to design more efficient systems in many applications ranging from power generation to electronic device thermal management. The broad scope of applications serves to emphasize CFDs invaluable contribution to thermal system optimization and guides future research direction within engineering practices and methods leading to the discipline as a whole as presented by Paricio et al. (2019). Another critical section of simulation techniques in thermal modeling is the use of future materials, such as carbon nanotubes for interconnects. Research on carbon nanotube interconnects examines the need for improved thermal and electrical conductivity compared to traditional copper interconnects, in order to address intrinsic scaling issues in microelectronics. With modeling and simulation, scientists can contrast circuit-level performance and predict behavior under various conditions. Theoretical modeling and experimental characterization of such materials form a foundation of understanding their application in integrated circuits. In the wake of continuing energy-efficient demands, such research highlight the importance of harmonious cooperation between simulation methods and material science development, providing a paradigm to innovate thermal management and circuitry design (Amoroso et al., 2017). Effective thermal modeling also necessitates a systems-level approach to integrating simulation techniques and real-world applications. This entails capitalizing on the results of CFD and carbon nanotube research to inform design choices and operating procedures in thermal systems. By integrating cooperation among engineers, materials researchers, and computer specialists, thermal modeling's inter-disciplinary emphasis can deliver breakthroughs in efficiency and effectiveness. For example, the information obtained through simulation can help facilitate more environmentally friendly energy solutions, optimizing the thermal efficiency of power plants and renewable sources. Thus, such simulation techniques not only contribute to immediate technological advancements but also ensure long-term energy engineering sustainability, proving their central role in deciding the future of thermal modeling and simulation.

5.2.1. Finite Element Method (FEM)

The finite element method (FEM) is a flexible computational method used extensively in thermal modeling and simulation to assist engineers and researchers in investigating complex thermal behavior within materials and structures. FEM allows for temperature distributions and heat transfer effects to be simulated with high accuracy by discretizing continuous systems into smaller, manageable pieces. This method is of significant value in cases where analytical solutions are difficult or even impossible to obtain, such as in components with complicated geometries or non-uniform heat properties. Therefore, FEM is nowadays the standard in engineering disciplines, helping in making decisions that depend on the understanding of thermal dynamics within a design. One of the major advances in FEM is the development of the vademecum-based Generalized Finite Element Method (V-GFEM), which enhances computational efficiency in thermal simulations. The technique uses precomputed parametric solutions as enrichment functions stored in memory to be utilized rapidly for analyses. Offline, precomputed information reduces the computational expense typically associated with FEM but preserves high accuracy in transient analysis. Applications of the procedure are also of most relevance to manufacturing operations, such as welding, where local thermal behavior heavily influences materials quality and performance. The V-GFEM is indicative of how the incorporation of new methods into FEM can help guide more effective thermal modeling practices (Ammar et al., 2016) (Figure 5.3).

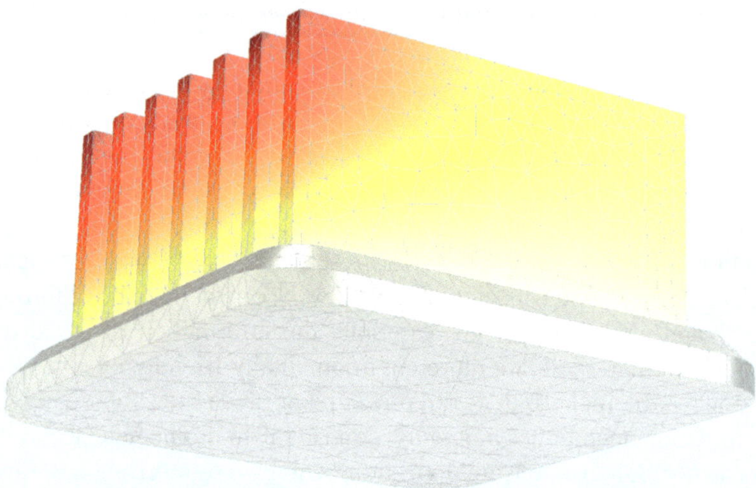

Figure 5.3. Finite element method (FEM).

Source: https://cdn.comsol.com/cyclopedia/finite-element-method/temperature-field-heat-sink-model.png.

Aside from advancements in V-GFEM, simpler alternatives such as one-dimensional reduction methods also demonstrate the versatility of FEM in thermal simulation. For instance, one of the recent methods utilizes finite difference methods (FDM) in modeling temperature rise in electrical equipment and demonstrates how one-dimensional analysis can produce highly accurate results with less computational effort. By breaking down a three-dimensional geometry into one-dimensional substructure, crucial thermal parameters are evaluated efficiently without sacrificing fidelity. This methodology holds substantial potential for utilization in substation connector design; wherein thermal performance needs to be minimized for highest operating efficiency. The flexibility of the application range from the intricate generalized formulations to the more straightforward ones underlines the significance in thermal modeling and simulation across diverse applications (Rubio et al., 2017).

5.2.2. Computational Fluid Dynamics (CFD)

The field of computational fluid dynamics (CFD) is at the forefront of thermal simulation and modeling, offering invaluable information on complex fluid behaviors and interactions. CFD is a branch of fluid mechanics that uses numerical algorithms and methods to solve fluid flows, and it presents a safe environment for simulating how thermal energy is transferred in various systems. This capability is crucial, particularly in energy engineering, where optimal thermal performance can significantly enhance efficiency and sustainability. According to recent reviews, considerable advancements in software modeling and computing power over the last two decades have generated a surge of interest in CFD applications across diverse domains, including power generation and renewable energy technologies. They demonstrate the pivotal function of CFD to further advance thermal modeling as scientists strive to further explain heat transfer processes and develop more effective energy solutions, as noted in a special issue on CFD application to energy engineering research (Paricio et al., 2019) (Figure 5.4).

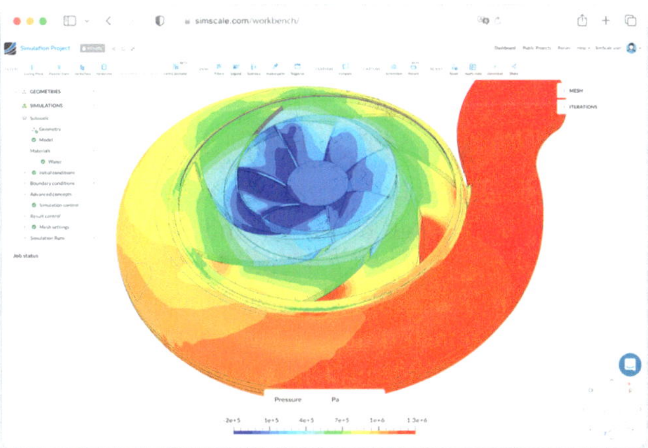

Figure 5.4. Computational fluid dynamics (CFD) simulation software.

Source: https://www.simscale.com/wp-content/uploads/2023/11/product-page-header-fluid-dynamics.png.

Further, application of CFD in thermal process simulation entails the assessment of many engineering challenges, such as those in heat exchangers and combustion systems. CFD provides the capability to make sophisticated simulations that account for the intricacy of thermal interactions, such as conduction, convection, and radiation effects. With this technology, parameters of design can be optimized and thus the risk of thermal inefficiencies that can lead to failure or energy losses can be minimized. From the literature, CFD has not only helped establish a better understanding of core thermal phenomena but has also made innovative design with higher performance characteristics in thermal systems a possibility. In this context, CFD stands as a central tool for modeling existing systems as well as forecasting the not-so-distant future energy systems that are not only efficient but also eco-friendly (Paricio et al., 2019). The combination of CFD with heat modeling is further enhanced through the integration of smart materials and techniques such as Molecularly Imprinted Polymers (MIPs). These materials, which are noted for their selective molecule recognition characteristics, may be potentially optimized through CFD simulations to allow for a better understanding of their thermal behavior across different operating conditions. Computational modeling of such smart materials can open up possibilities for innovations in sensor technology and targeted applications. With the application of an integrated approach that combines simulation data with experimental results, scientists can refine by iteration the properties of MIPs for specific thermal applications and hence enhance efficiency. Thus, the marriage of CFD with advanced materials is a multi-dimensional thermal modeling approach with the potential to revolutionize energy systems and make them more dependable in operation.

5.2.3. Boundary Element Method (BEM)

Boundary element method (BEM) is now widely recognized for its competence in solving problems of thermal modeling and simulation, particularly in cases involving complex geometry and boundary conditions. Compared to conventional methods such as Finite Element Method (FEM), BEM reduces the problem dimension, transforming a three-dimensional problem into a two-dimensional boundary problem. This simplification not only decreases calculations but also decreases the memory requirement significantly, making BEM particularly effective for simulating intricate heat transfer processes on surfaces. With the application of BEM in thermal analysis, engineers are able to calculate precise estimates of temperature distributions and boundary heat fluxes, making more efficient design options and optimization processes available across fields ranging from material science to mechanical engineering. Apart from this, BEM can handle nonhomogeneous materials, which is essential in a wide array of thermal simulations where material properties can vary significantly with location. For instance, Antonio et al. (2017) illustrates the application of BEM in the simulation of propagation of elastic waves in fluid-infilled boreholes, with the surrounding material exhibiting a high level of nonhomogeneity due to multiphase construction. By combining BEM with other numerical solutions, such as the meshless local Petrov–Galerkin scheme or finite element schemes, even greater accuracy can be achieved in fluid and solid domain modeling. The beauty of this flexibility lies in the fact that it enhances the predictive ability of thermal simulations further to simulate heat transfer in materials with intricate geometrical and material characteristics. The marriage of BEM with other computational software also opens up opportunities for new optimization techniques. Having the ability to effectively model thermal performance allows scientists to apply optimized algorithms for performance enhancement across many fields. In wind power use, for instance, wind turbine performance optimization through thermal analysis is crucial to overall efficiency improvement (Figure 5.5).

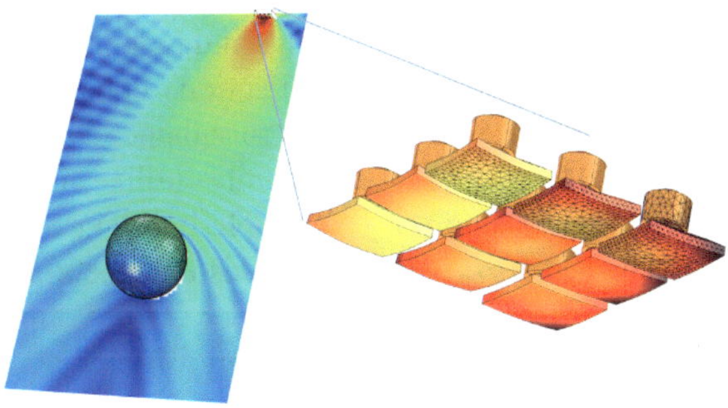

Figure 5.5. An example of a transducer array located far from a scattering object. This type of problem is very difficult—or even impossible—to solve with a purely FEM-based approach due to the large memory requirements.

Source: https://www.comsol.com/blogs/how-to-use-the-boundary-element-method-in-acoustics-modeling.

As Haghnegahdar et al. (2019) finds in the use of optimizing wind turbines through Computational Fluid Dynamics (CFD), achieving a general understanding of thermal interactions is of essential significance to turbine design. Thus, the use of BEM contributes not just to increased thermal simulation but also to the broader objective of performance optimization within engineering disciplines. This blend of strategies reflects the strength of BEM to drive innovation and effectiveness in thermal modeling and simulation strategies.

5.3. SOFTWARE TOOLS FOR THERMAL SIMULATION

Software programs in the thermal simulation and modeling industry are the key to enhancing designers' capabilities to predict and optimize the thermal performance of buildings. Software enables specialists to create detailed simulations that incorporate different building materials and weather conditions, leading to more efficient energy designs. Leading software platforms generally employ finite element analysis (FEA) or computational fluid dynamics (CFD) to model thermal behavior with precision. The complexity of building models means there is a growing necessity for high-level software to verify the performance of various components, such as insulation and façade systems. Thus, software that supports whole-building modeling can be used to calculate energy savings and improve comfort levels for occupants, forming a vital connection between thermal simulation software and sustainable design methodology, critical in meeting today's energy efficiency standards. Further, coupling new technologies with thermal simulation software has significantly enhanced the analysis of energy use in buildings and indoor climatic conditions. A good example of such integration is the application of phase change materials (PCM), which hold and release heat energy to enhance comfort and reduce cooling energy needs. However, to effectively utilize PCM in building envelopes, numerical computation through whole-building energy simulation software is crucial. This type of software enables designers and engineers to model the behavior of PCM-treated elements under various climatic conditions and have a firm grasp of their impacts on energy efficiency. It is especially fascinating, for instance, that PCM system performances are very sensitive to design parameters and environmental conditions, suggesting the necessity of precise simulation in order to use them effectively in green buildings (Cabeza et al., 2017). Finally, the future of thermal simulation software is potentially in its capacity to adapt with new technology advancement and keep pace with increasing complexity of energy-efficient design requirements. The designing of smart, responsive façades represents an excellent example of this new trend because they apply real-time information and end-user input to optimize energy efficiency. These systems can be capable of operation on dynamic simulations that take into consideration variable weather patterns and occupied statuses. For instance, advanced control systems employing physics-based equations and math solvers can significantly reduce electricity bills by dynamically controlling shading and daylighting elements. In California, their integration has also been linked with considerable energy and peak demand savings, especially pointing out the central role of software tools in fueling innovation in thermal modeling and simulation (Breshears et al., 2020).

5.3.1. Overview of Popular Software (e.g., ANSYS, COMSOL)

In thermal modeling and simulation, program packages such as ANSYS and COMSOL have become the focal tools that make complex analyses easy. ANSYS, with its high-computational power, enables engineers to perform sophisticated thermal simulations predicting heat distribution and thermal stresses in various materials. Its ability to conduct fluid dynamics analysis extends its field, and it finds special application in fields like electronics cooling and power generation systems. ANSYS also uses advanced algorithms to analyze transient thermal analysis and has the capability of real-time response to thermal loads that change in an instant. Due to this, this platform is highly used in industry requiring high-fidelity simulations for enhancing product performance and reliability, more firmly establishing itself as an elite software in thermal modeling. Similarly, COMSOL Multiphysics features a versatile platform for thermal modeling through its unique multiphysics capabilities. The program enables simultaneous simulation of heat processes together with other physical processes, such as structural mechanics and electromagnetics. A simple-to-use interface supports seamless model configuration and adjustment to accommodate a broad range of engineering applications (Figure 5.6).

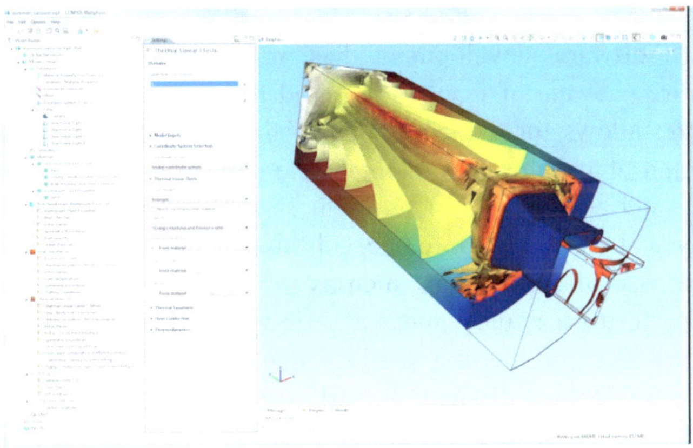

Figure 5.6. COMSOL software.

Source: https://img.laserfocusworld.com/files/base/ebm/lfw/image/2021/07/10_LFW_2_comsol.60f70cfd4ff34.png?auto=format,compress&fit=fill&fill=blur&w=1200&h=630.

Additionally, COMSOL offers support for a number of modes of heat transfer, i.e., conduction, convection, and radiation, for stable analysis of intricate thermal interactions. Integrating thermal simulations with other physical effects demonstrates COMSOL's ability to provide solutions to challenging engineering challenges where multidimensional behavior is the focal point. These features enable designers to innovate without allowing development cycles to get in the way, solidifying COMSOLs position as a sought-after product for high-end thermal analysis. Besides, the applications of these computer tools infiltrate specialized fields, justifying their efficiency in addressing contemporary engineering challenges. For example, employment of Computational Fluid Dynamics (CFD) within ANSYS has been cited as essential in energy engineering because it simulates thermal phenomena in procedures like combustion and heat exchangers (Paricio et al., 2019). Alternatively, COMSOLs electrothermal coupling modeling is a crucial function in semiconductor device design, towards sluggish thermal response dynamics influencing device performance (Bonani et al., 2008). The above examples illustrate the critical role of high-end software solutions towards thermal modeling procedures, where precision is assured for prediction functionality and insight into thermal processes. As such, both ANSYS and COMSOL continue to play a pivotal role in the development of thermal simulation, following the industry's continued quest for innovative and effective designs.

5.3.2. Features and Capabilities of Simulation Tools

Simulation software development has revolutionized thermal modeling into an entire package of features required for accurate analysis and design. The software allows for the simulation of thermal processes using advanced computational algorithms that facilitate the simulation of heat transfer, fluid flow, and material properties. One of these high points is the integration of various physical phenomena into one, cohesive model, and this becomes totally vital in the case of applications such as microgrid design. Because projects involve a performance study under a variety of operation modes, simulation software provides for thermal performance to be examined by simulating temperature fields, heat dissipation, and efficiency. This predictive capability allows engineers to make smart design decisions, making the systems robust against environmental stresses while realizing maximum performance and reliability (Bao et al., 2018). Secondly, the generic nature of modern simulation tools provides the facility to tailor and add to their analytical capabilities. Most software platforms allow new modules and functions to be incorporated, which provides the means to satisfy specific requirements in their thermal models. For instance, the Integrated Analysis Center (IAC) illustrates such versatility through the incorporation of a number of diverse analytical modules, for instance, thermal and structural analysis, which co-exist harmoniously within the larger frame of the tool (Vos et al., 1982). This is particularly handy within interdisciplinary projects where different technical disciplines need to co-exist harmoniously with one another. Such unification guarantees balanced vision of system behavior under all circumstances and therefore enhances precision of predictions by simulation tools. Engineers then can design optimally based on comprehensive insights from simulations that are realistic. Simulation tools also include robust visualization capabilities beyond modeling, which enhance interpretation and communication of data among stakeholders. Visualizing the thermal simulations assists engineers in identifying problematic areas, checking results in real-time, and presenting findings to clients or stakeholders. Since the software is able to produce detailed plots, 3D geometries, and interactive visualizations, the discrepancy between the numerical complexity of the result and stakeholder knowledge is bridged. In addition, most of these platforms come pre-configured with templates and easy-to-use interfaces, which allow the less technical user to instantly appreciate the software's potential. This access to simulation tools contributes to wider use across a variety of domains, fostering a culture of innovation and continuous improvement in thermal modeling practice (Bao et al., 2018). With the increasing complexity of the environments in which engineers operate, these capabilities allow them to enhance system designs and ensure success in operations.

5.3.3. Comparison of Software Performance

In thermal modeling and simulation, cross-comparison of software performance is essential in ensuring accuracy and efficiency in forecasting thermal behavior. Software programs have varying methods of simulating thermal dynamics, and platform selection can decisively influence both the outcome of the process and useability. While some software excels significantly at computational speed and user interfaces, others are coded to have finer or specialized features. Careful analysis enables practitioners to determine which packages effectively process complex calculations without taking much computational time, ultimately enabling sound decision-making in thermal management schemes. Further, as emerging technologies become available, it becomes more imperative to determine how software integration with cutting-edge system components can be used to support greater overall performance. For instance, simulation packages with real-time data feedback mechanisms can reduce the operation of thermal systems by adjusting parameters to changing conditions. Insights gained through such combined simulations guide innovative design solutions to maximize efficiency in energy usage. Projects like those outlined by Breshears et al. (2020) require adaptive software packages that respond suitably to real-time needs in energy-saving conditions with increased fortification of the interrelational aspect of software performance and actual thermal consumption. In addition, broadening the scope of comparison requires comparing cost-effectiveness with performance measures. As in the case of construction and ground-stabilizing technologies, such as the new use of rice husks described by Mokhtar et al. (2011), economic viability usually plays a determining role in software selection for modeling activities. Thermal simulation software with full capabilities at a modest fee can facilitate easier access to practitioners in various industries, opening the doors to green technology development and energy efficiency. This balance between cost considerations and solid performance is a critical aspect in the decision-making process while selecting thermal modeling software to guide the direction of sustainable engineering practice.

5.4. VALIDATION AND VERIFICATION OF MODELS

Verification and validation of models are major contributors to the success of thermal modeling and simulation. These processes ensure that the developed models correctly replicate the physical systems they are meant to model. Verification refers to the process of ensuring that the model has been developed correctly according to provided requirements, while validation assesses whether the model sufficiently represents the phenomenon in the real world it is meant to model. Sound validation and verification processes are essential, particularly in high-risk fields such as manufacturing and aerospace where system failure would have catastrophic consequences. For instance, as explained by Beck et al. (2025), validation is essential for those technologies ranked with the highest priority under national space missions, which sets a high value on trustworthy model assessments in this context. Lack of participation in extensive validation and verification can lead to wrong predictions, ultimately compromising system performance and safety. In thermal modeling, a number of methodologies are used to validate and verify simulations to make them believable. One of such common practices involves the comparison between the results of the simulations and empirically observed data obtained through experiments. This process not only enhances the reliability of the model but also promotes stakeholders' faith in its forecasting capability. Furthermore, validation techniques, such as sensitivity analysis, help determine the most significant parameters accountable for the dramatic effect on model behavior, which is crucial for sim updating and accuracy enhancement. Advanced tools and frameworks, such as those listed by Blech et al. (2014), provide infrastructure for systematic testing. These methods help in visualization and validation of industrial automation cyber-physical systems, demonstrating how numerous applications can benefit from strict validation processes. Using these methods, engineers and researchers are able to limit uncertainty and effectively develop thermal modeling innovation. A mix of validation and verification processes also encourages collaboration among different sites that participate in thermal modeling and simulation.

By establishing a comprehensive model evaluation framework, individuals are able to share knowledge and best practices and thereby create enhanced methodology and outcomes in a lot of areas. With more involvement of thermal modeling in cyber-physical systems, the involvement is even more important. Improvements in simulation and visualization techniques, for example, allow remotely operated teams to validate models from various geographical locations, as presented by Blech et al. (2014). This capability is transforming the terrain of thermal modeling, enabling more thorough and multi-faceted verification and validation processes. With the processes further developing, they will likely provide more precise thermal simulations in the long run, which will ultimately lead to technological progress across many industries.

5.4.1. Importance of Model Validation

In the field of thermal modeling and simulation, there can be no overestimation of the significance of model validation. Model validation is an invaluable tool to ensure that simulations accurately represent physical phenomena, something of crucial concern in research touching on energy and thermal systems. When models are verified using experimental data, practitioners can ascertain their prediction accuracy, thus allowing informed decision-making in such domains as combustion and fluid dynamics. Not only does this verification process certify the numerical models' function, but it also helps in identifying potential differences between predicted and real behavior. Due to the inherent intricacies involved in thermal processes, solid validation frameworks are essential to prevent the pitfalls of extending defective models into practical applications, resulting in enhanced safety and efficiency in various engineering applications. Further, the advancements in computational capabilities and software sophistication have also hastened the importance of proper model validation in thermal simulations. With the addition of sophisticated aids such as Computational Fluid Dynamics (CFD), scientists can simulate complex physical phenomena and interactions. But with this complexity comes the necessity for strict validation protocols to ensure that they simulate reality as accurately as possible. A case in point is validation against data obtained from experiments, which allows scientists to refine their models by adding corrections that enhance precision in simulating phenomena such as turbulence and heat transfer. Knowledge obtained through validation not only provides a vision of tested procedures in the past but also paves the way for improvements in energy engineering in the future. As emphasized in recent reports, the continuous evolution of CFD applications in thermal and energy engineering emphasizes the necessity for ongoing validation so that there are scientific cohesion and technological improvement (Paricio et al., 2019). Lastly, the application of model validation transcends academic circles and is important in terms of actual real-world application in energy-saving technology and conservation of the environment. Thermal model validation allows engineers to better estimate performance outcomes, which in turn aids in designing better performing and more efficient systems with reduced energy consumption and emissions. This use is particularly vital in fields such as combustion engineering, where correctly advised, changes can significantly contribute to sustainability. The ability to validate models not only enhances the effectiveness of operations but also ensures public trust in new technologies that rely on complex thermal simulations. As is evident in research on effervescent sprays, the continued advancement of modeling techniques requires emphasis on validation so that models remain effective and relevant to current and future energy requirements (Broukal et al., 2014). Therefore, the emphasis on model validation in thermal modeling is critical in order to enable technological innovations and promote sound engineering practices.

5.4.2. Techniques for Verifying Simulation Results

Simulation output verification is important in thermal modeling because it ensures predictive outcomes are accurate and reliable. An effective method involves comparing simulated results and experimental data. It is a standard employed in assessing the performance of thermal models, and it is employed particularly for verifying assumptions during modeling. Differences between simulated and real values can indicate where fine tuning is needed or highlight the model's limitations. Precise experimental settings are necessary, as these produce high-quality data against which the simulation is validated. As is demonstrated by recent studies, utilizing two-parameter models for analyzing complicated geometries, the validation process can be improved with a higher degree of accuracy, particularly where traditional materials are not able to deliver ideal thermal and mechanical properties. By consistently comparing simulation results to experimental data, scientists can greatly improve the validity of their thermal models. Another essential method associated with verification is sensitivity analysis, which evaluates the impact of variations in input parameters on simulation results. By systematically changing these parameters, scientists can learn which inputs most significantly impact the thermal results. This capability also assists model refinement in pointing to the most significant factors that must be controlled at high accuracy or defined more precisely. Sensitivity analysis also contributes to a deeper understanding of the system under investigation, allowing one to determine the probable defects of a model during its construction. In the case of advanced materials such as nanotrusses, for which there are complex interdependencies between structural properties and thermal response, employing sensitivity analysis can be particularly beneficial. It helps to establish key parameters governing thermal conductivity and mechanical performance, thus facilitating more informed design choices and material selection (Dou et al., 2015). The findings from such a study help to foster overall faith in simulation predictions. With the inclusion of advanced modeling techniques, such as Monte Carlo simulations, the verification process can be enhanced further in thermal modeling. These simulations allow for the study of a broad range of conditions, often producing statistically significant observations about thermal behavior across various conditions. For instance, Monte Carlo methods can easily tackle the problem of intricate geometries like carbon nanotube interconnects, which are typically employed in silicon technology. Through the combination of results from multiple runs, scientists are able to have a strong conclusion of the effects of particular configurations and conditions on system performance and reliability (Amoroso et al., 2017). The probabilistic nature of Monte Carlo simulations thus provides an auxiliary verification tool, allowing scientists to compare the output with voluminous theoretical and experimental data. This unified paradigm not only converges the findings but also offers means for further innovation in thermal modeling and material science.

5.4.3. Case Studies Demonstrating Validation Processes

In thermal modeling and simulation, validation processes come to the forefront in an effort to validate that the models represent the physical processes being studied. An Electron Beam Welding (EBW) case study illustrates rigorous validation processes both involving numerical simulation and experimental validation. In this case, a multi-physics finite element (FE) program was utilized for simulating thermal, mechanical, and metallurgical processes characteristic of EBW interactions. The model attempted to mimic the heat transfer with a specially designed ad hoc moving heat source and was able to represent the thermal behavior experienced in welding. Validation of the model proceeded through direct comparisons of numerical solutions to empirical observations, realistically measuring temperature development and quantifying distortions and residual stresses in the welded material. This comprehensive approach highlights the importance of validation for achieving feasible and accurate thermal model outcomes, which are essential to aerospace and other sensitive sectors (Agelet et al., 2016). Similarly, application of Computational Fluid Dynamics (CFD) to energy engineering presents a wealth of case study scenarios that highlight validation processes in thermal simulation. During the last two decades, advances in computer capabilities and software abilities have greatly enhanced the credibility of CFD modeling across a wide range of applications in fields ranging from combustion and renewable energy technologies to thermal management systems. The history of applications of CFD contains many instances in which serious validation procedures have been performed to ensure that simulated conditions closely approximate real-world operational conditions. Through comparison of simulation results with experiments, researchers have been able to advance their models so that predictions can be made more accurately. It is imperative that simulation and experimental activity be in such a symbiotic relationship because it encourages the advancement of complex thermal engineering solutions while confirming that computer results are accurate (Paricio et al., 2019). Finally, these case studies of validation procedures in thermal modeling and simulation likewise work to highlight the critical function of accuracy in the engineering process. By either EBW or CFD approaches, incorporating validation into the modeling process works not simply to enhance predictive result confidence but also to assist in improvement to the models themselves. Iterative process of this sort is necessary to bridge the gap between theory prediction and practice in application, particularly in high-impact areas like aerospace, energy production, and manufacturing. As these case studies show, proper validation procedures are a cornerstone for continued advances in thermal modeling, ensuring engineering practice remains in harmony with empirical findings and operational requirements. This ongoing emphasis on verification later leads to more durable and innovative engineering solutions (Agelet et al., 2016; Paricio et al., 2019).

5.5. CHALLENGES IN THERMAL MODELING

The application of thermal modeling is faced with an extensive variety of challenges that strongly impact its accuracy and efficiency. One main challenge lies in the complexity of heat transfer behavior, which occurs in various forms such as conduction, convection, and radiation. The larger and more complex the systems become, the harder it is to represent these diverse thermal interactions. These are further exacerbated by variations in material properties and external environments that can influence thermal performance. Scientists and engineers thus have to develop sophisticated models that can treat these variables, often relying on computational fluid dynamics (CFD) as a means of enhancing simulation precision. This approach has seen extensive growth, but its effective use in thermal modeling is contingent upon overcoming the system interaction and environmental complexities, which are of extreme significance in existing research activities (Paricio et al., 2019). Another key challenge inherent in thermal modeling is the need for precise boundary conditions and input values. The accuracy of thermal simulations relies to a great extent on the quality and precision of input parameters, including material properties, temperature distributions, and heat generation rates. In practice, obtaining these parameters can be difficult, especially in sophisticated systems or in new materials where empirical information is unavailable or limited. Variations in experimental measurements can also result in significant differences in model predictions, which can translate to potential inaccuracies in thermal management approaches. Thus, scientists must provide the utmost importance to developing standardized test procedures and data collection techniques that help ensure input data reliability in thermal modeling. This pursuit not only makes the model more faithful but also assists in enhancing the transferability of simulations to real situations in which variability occurs (Amoroso et al., 2017). Finally, the computational loads associated with high-level thermal models provide an additional degree of complexity. As simulations become more sophisticated and higher spatial and temporal resolutions must be computed in order to properly reproduce transient behavior, the utilization of computational resources is significantly increased. This provides real-world limitations, particularly in terms of simulation time and available hardware. High-performance computing capabilities are often required to efficiently perform these simulations, which may not be accessible to all. The challenge is also given a boost by the cross-disciplinary requirement since thermal modeling increasingly crosses materials science and electrical engineering domains. Future research will have to focus on more efficient algorithms and optimization techniques that reduce computational costs without sacrificing model accuracy to address these challenges. This challenge is key to enabling broader application of thermal modeling across industries (Paricio et al., 2019).

5.5.1. Complexity of real-World systems

The subtleties of real-world systems are achieved by the complexity of interaction between elements in thermal simulation and model environments. They are characterized by tens of thousands of dependencies and interactions which lead to non-linear result, making predictions difficult. For instance, in thermal applications, modeling of heat transfer in a building involves considering a number of factors, including properties of the material, environmental factors, and habits of occupants. When consumed individually, one tends to underestimate their overall impact on heat dynamics. Interdisciplinary modeling techniques need to be employed to capture this complexity. Through advanced statistical techniques and algorithms, e.g., in multiple-input multiple-output (MIMO) systems, part of the performance degradation due to residual limitations in heat systems can be mitigated, hence enhancing predictive accuracy in simulations (Burg et al., 2010). In the evolving world of thermal modeling, integrating dynamic systems adds additional complexity. Buildings with sophisticated attributes such as dynamic facades or adaptive environmental controls are the epitome of this complexity because they introduce variables that can respond to internal and external inputs. A buildings operation, for instance, requires an energy demand model that incorporates real-time adjustments due to weather, occupancy, and user choice. This kind of detail is vital towards guaranteeing correctness in energy consumption and load demand simulation. The necessity for robust dynamic models is evident, as seen in Californias quest for efficiency in utilizing technology that allows load management strategy—ultimately, the utilization of these dynamic properties aids in enhancing both sustainability and electrical grid management (Blum et al., 2020). Furthermore, the implications of understanding the complexity of practical systems extend beyond individual constructs to influence urban settings. They are compounded when considering approaches such as demand response (DR), which is founded on the collective behavior of numerous buildings to achieve resource abundance in electrical grids. The marriage of multiple modeling approaches, such as Modelica for thermal simulation and Radiance for daylighting analysis, demonstrates how holistic approaches can lay bare the systemic effects of energy management devices. From integration of such models, scientists can develop optimization plans that effectively balance energy expenditures and resource allocations during high-demand periods. To this end, the variability in real systems can be utilized to make urban infrastructure systems more resilient and efficient, with a focus on advanced thermal modeling as a means of solving contemporary environmental problems (Blum et al., 2020).

5.5.2. Limitations of Current Modeling Techniques

The limitations placed on the current modeling methodologies to thermal modeling and simulation are mostly a function of the physical intricacies present within the systems being simulated. The traditional methodologies are typically founded upon simplifying assumptions regarding material properties, boundary conditions, and heat transfer mechanisms, which eventually lead to predictive accuracy limitations. For instance, in electronic device simulation modeling, approximations for heat conductivity and specific heat can skew results, affecting simulation result accuracy. These simplifications become more problematic as device size reduces and heat densities increase, with the need for more refined understanding of thermal interactions. Therefore, these modeling techniques must evolve to encompass more sophisticated representations of material and its behavior under harsh conditions that remain a significant bottleneck to creating precise thermal simulations for practical purposes. Beyond the simplifications that current approaches fall prey to, there is also the issue of computational efficiency. Sophisticated modeling techniques, such as those involving complicated finite element analysis, are likely to require great amounts of computing effort and time, thereby limiting their application in real-time simulation. Recent literature has indicated that new algorithms like hybrid optimization techniques have shown promise in addressing these efficiency problems but their incorporation in standard workflows is not uniform (Restituto et al., 2005). Furthermore, such sophisticated models can demand expertise that can exclude them from widespread use. Hence, the balance between accuracy and efficiency of thermal simulations remains a pressing concern that requires innovative solutions to achieve maximum utilization of these modeling techniques. In addition, the ability to authenticate and validate numerical models is crucial for ensuring their dependability in thermal simulations. Most of the available methods do not have a strong validation against experimental results, so the predicted model results may significantly differ from actual experiences. The issue is brought up in existing literature since materials processing and growth need to be correctly represented in order to enable trustworthy circuit-level performance prediction (Amoroso et al., 2017). This shortcoming lowers the confidence of researchers and engineers in employing the models for design and optimization tasks. Thus, developing comprehensive strategies integrating experimental outcomes and theoretical frameworks is vital not only for the evolution of thermal modeling but also for the enhancement of interdisciplinary interactions within materials science and circuit design. Without such advancements, application of thermal modeling is limited, and innovation capacity is maintained low in energy-efficient integrated circuits and associated technologies.

5.5.3. Future Trends and Research Directions

As thermal modeling and simulation evolve, there are some key future developments and trends emerging that have the potential to further enhance computational accuracy and efficiency. One of the more advanced developments is coupling Computational Fluid Dynamics (CFD) with thermal analysis, offering more comprehensive simulation environments. This interaction will address different topics of energy systems, including heat transfer, fluid flow, and thermodynamic cycles, thus broadening the application of thermal models to more complex engineering issues. Furthermore, increasing computational power one can mobilize facilitates more intricate and high-resolution simulations that can mirror the details of heat behavior in real time and finally attain the desired convergence of theoretical research and engineering applications for aerospace and energy production sectors, as noted in evolving curricular requirements in aeropropulsion research (Glassman et al., 2025). Another leading trend in the field is the emphasis on sustainability and efficiency of thermal systems. As environmental concerns increasingly become more prominent, researchers are also placing a high level of priority on models that not only predict thermal behavior but also towards developing efficient energy systems as well as carbon footprint reduction. The trend reflects the growing influence of external pressures, such as economic competitiveness and environmental legislation, on research emphasis in thermal modeling. New techniques, such as innovative heat exchangers and phase change materials, are being integrated into simulations to test their impact on system performances. Through the integration of sustainable design in simulated environments, engineers can create new solutions that not only complement global efforts to combat climate change but also ensure that ongoing research streams are not just technically skilled but socially responsible as well (Glassman et al., 2025). Finally, interdisciplinary research collaboration is increasingly becoming more vital in thermal modeling and simulation research. Future trends will witness various scientific fields—materials science, computational mechanics, environmental engineering, among others—converging to design stronger model schemes. Enhanced collaboration can result in the development of wholesome models that consolidate broad aspects of thermal dynamics, from micro-scale impacts to macroscale implementations. With the advancement in CFD technologies, there will be more requirements for industries and researchers to collaborate so that the tools can tackle real challenges effectively. This collaborative effort will pave the way for coupled simulations that provide better understanding of thermal behavior and energy systems and assist in the designing of innovative solutions for future energy issues.

SUMMARY

The comprehension gained in exploring thermal modeling and simulation justifies the significance of sophisticated modeling techniques in various applications, with emphasis on energy efficiency and resource minimization. Development and validation of advanced models have played an instrumental role in anticipating thermal conduct accordingly in complex systems, as shown in the research results in various studies of effervescent sprays and high-efficiency energy systems. For instance, Broukal et al. (2014) illustrates the ability of high-fidelity spray models to model accurately the characteristics of effervescent sprays, detailing how increased sophistication in modeling can produce significant increases in combustion efficiency and overall system performance. It is these findings that highlight the necessity for increased research and development of modeling technologies to address emerging thermal management needs. Furthermore, thermal modeling is used outside combustion equipment and encompasses new façade arrangements minimizing energy use throughout shifting climatic conditions. The outcome of what is accomplished on those projects that are seeking greenhouse gas reductions, such as the design of an R-5 insulating glass unit, illustrates the instantaneous impact of simulation-based design on modern architecture. As recognized by Breshears et al. (2020), such systems achieve not just substantial gains in energy efficiency but also signify the potential of smart building designs to respond dynamically to energy requirements. These innovations demonstrate the catalytic role thermal simulation has in enabling sustainable practice in the built environment. Lastly, the future of thermal simulation and modeling lies in the continued integration of multi-disciplinary approaches, reconciling progress in computational techniques with experimental verification for enhanced predictive power. Synching theoretical models with empirical data is paramount to the credibility and usefulness of thermal simulations to real-world applications. Next-generation research needs to attempt to craft new methodologies that take into account the complexity of thermal interactions and make the most of the generalizability of such models across various industries. By addressing such challenges, we can turn thermal modeling into an advanced technology that can bring efficiency, sustainability, and flexibility to our ever-complex energy systems.

In the overall framework of thermal modeling and simulation, one of the overall concerns has been atmospheric states' interactions with the resulting thermal consequences, particularly in coastlines. Recent work delineates the intricate character of convection, indicating its dependence on large-scale atmospheric processes decreases near coastlines compared to open ocean environments. This is the result of the extensive influence of meso-scale land-sea interaction and topography that characterize coastal areas. Inability of standard cumulus parametrization schemes to resolve these complex dynamics means that coastal convection is often not properly simulated. Closing this gap, researchers propose an idealized modeling approach that uses a decision algorithm to decide on conditions for coastal convection and thereby enhance the simulation of such events in global models. Besides, combining stochastic models of clouds with a trigger function makes it possible to simulate in greater depth deep convection events, particularly for atmospheric conditions typical of the coastal zone. Empirical results demonstrated that a hybrid model based on such a combination is capable of reconstructing spatial and temporal structure of clouds, patterns confirmed by observations. Because coastal convection is notoriously hard to model, these advances represent a paradigm shift towards the improvement of the predictive capacity of global numerical weather prediction. Not only does this advance promise to improve representations of clouds and convection within models, but it has implications for understanding wider-ranging climatic impacts within coastal regions, thus improving the field of thermal modeling. Apart from atmospheric convection, thermal dynamics inherent in modern-day multiprocessor systems must also be dealt with seriously. As technology continues to evolve, the use of heterogeneous cores in system-on-chip implementations has become widespread with the goal of providing better energy efficiency. While this innovation brings significant challenges in coping with thermal management in that power consumed by individual cores is linear with temperatures on the chip floorplan, thermal dynamics inherent in modern-day multiprocessor systems must be dealt with. These extreme temperatures also enhance leakage power, creating a nonlinear feedback loop with the potential to destabilize thermal properties. Some latest research has suggested a paradigm that accurately foretells power-temperature paths, thus keeping operational parameters in thermally safe limits. With the help of a predictive model, developers can ascertain stable fixed points for the use of power, avoiding thermal hazards that threaten system integrity and performance.

REFERENCES

1. Abbey, D., Alexeev, A., Alexiades, V., Andersland, O., Anderson, J., Anisimov, M., Becker, T., et al. (2008). Evaluation of the algorithms and parameterizations for ground thawing and freezing simulation in permafrost regions. https://core.ac.uk/download/143689532.pdf (accessed on 23rd October 2025).
2. Abomailek Rubio, B. C., Capelli, F., Casals Torrens, P., Riba Ruiz, J., et al. (2017). Transient thermal modelling of substation connectors by means of dimensionality reduction. https://core.ac.uk/download/81571924.pdf (accessed on 23rd October 2025).
3. Agelet de Saracibar Bosch, C., Cervera Ruiz, M., Chiumenti, M., et al. (2016). Numerical modeling of the electron beam welding and its experimental validation. https://core.ac.uk/download/81578299.pdf (accessed on 23rd October 2025).
4. Ajayamohan, R. S., Arakawa, A., Arritt, R. W., Bergemann, M., Birch, C. E., et al. (2017). Coastal tropical convection in a stochastic modeling framework. http://arxiv.org/abs/1708.06868 (accessed on 23rd October 2025).
5. Ammar, A., Aquino, W., Babuška, I., Bellman, R., et al. (2016). Vademecum-based GFEM (V-GFEM): Optimal enrichment for transient problems. https://core.ac.uk/download/81571350.pdf (accessed on 23rd October 2025).
6. Amoroso, S., Asenov, A., Brown, A. R., et al. (2017). A survey of carbon nanotube interconnects for energy efficient integrated circuits. https://core.ac.uk/download/82958608.pdf (accessed on 23rd October 2025).
7. Antonio, J., Galvín, P., Romero Ordóñez, A., et al. (2017). Modeling elastic wave propagation in fluid-filled boreholes drilled in nonhomogeneous media: BEM–MLPG versus BEM–FEM coupling. https://core.ac.uk/download/161255112.pdf (accessed on 23rd October 2025).
8. Asman, S. (2015). Adsorption behavior of molecularly imprinted-beta-cyclodextrin polymers prepared by reversible addition-fragmentation chain transfer (RAFT) polymerization for selective recognition of benzylparaben. https://core.ac.uk/download/12006860.pdf (accessed on 23rd October 2025).
9. Asproulis, N., Drikakis, D., & Frank, M. (2014). Advances and challenges in computational research of micro and nano flows. 4th Micro and Nano Flows Conference (MNF2014). https://core.ac.uk/download/29139821.pdf (accessed on 23rd October 2025).

10. Banerjee, K., Caron, A., Deutsch, A., Jiang, L., et al. (2010). A thermal simulation process based on electrical modeling for complex interconnect, packaging, and 3DI structures. https://core.ac.uk/download/37962018.pdf (accessed on 23rd October 2025).
11. Bao, Y., Feng, W., Jin, M., Liu, X., et al. (2018). A review of microgrid development in the United States–A decade of progress on policies, demonstrations, controls, and software tools. https://core.ac.uk/download/323310623.pdf (accessed on 23rd October 2025).
12. Beck, S. M., Breckenridge, R. A., Clark, L. G., Collier, J. W., et al. (2025). Technology for the future: In-space technology experiments program, part 2. https://core.ac.uk/download/pdf/42817270.pdf (accessed on 23rd October 2025).
13. Bhat, G., Gumussoy, S., & Ogras, U. Y. (2017). Power-temperature stability and safety analysis for multiprocessor systems. http://arxiv.org/abs/1806.06180 (accessed on 23rd October 2025).
14. Blech, J. O., Peake, I., Schmidt, H., Spichkova, M., et al. (2014). Cyber-virtual systems: Simulation, validation & visualization. http://arxiv.org/abs/1410.1258 (accessed on 23rd October 2025).
15. Blum, D. H., Gehbauer, C., Lee, E. S., Wang, Y., et al. (2020). An assessment of the load modifying potential of model predictive controlled dynamic facades within the California context. https://core.ac.uk/download/287622187.pdf (accessed on 23rd October 2025).
16. Bonani, F., Camarchia, V., Cappelluti, F., Donati Guerrieri, A., et al. (2008). When self-consistency makes a difference. https://core.ac.uk/download/pdf/11399348.pdf (accessed on 23rd October 2025).
17. Breshears, J., Curcija, C., Deisler-Moroder, D., Fernandes, L., et al. (2020). High-performance integrated window and façade solutions for California. https://core.ac.uk/download/286728862.pdf (accessed on 23rd October 2025).
18. Broukal, J. (2014). Effervescent breakup and combustion of liquid fuels: Experiment and modelling. https://core.ac.uk/download/30311594.pdf (accessed on 23rd October 2025).
19. Burg, A., Studer, C., & Wenk, M. (2010). MIMO transmission with residual transmit-RF impairments. http://arxiv.org/abs/1002.0406 (accessed on 23rd October 2025).
20. Cabeza, L. F., Gracia Cuesta, A. de, Saffari Tabalvandani, M., Ushak, S. N., et al. (2017). Passive cooling of buildings with phase change materials using whole-building energy simulation tools: A review. https://core.ac.uk/download/304371814.pdf (accessed on 23rd October 2025).

21. Corban, R. (2025). NTP comparison process. https://core.ac.uk/download/pdf/42806827.pdf (accessed on 23rd October 2025).
22. Coudert, F. X., & Fuchs, A. H. (2015). Computational characterization and prediction of metal-organic framework properties. http://arxiv.org/abs/1506.08219 (accessed on 23rd October 2025).
23. Delgado Restituto, M., Rodríguez Vázquez, Á. B., Rosa Utrera, J. M., Ruiz Amaya, J., et al. (2005). Simulation-based high-level synthesis of Nyquist-rate data converters using MATLAB/SIMULINK. https://core.ac.uk/download/286565261.pdf (accessed on 23rd October 2025).
24. Dou, N. G., & Minnich, A. J. (2015). Heat conduction in multifunctional nanotrusses studied using Boltzmann transport equation. https://core.ac.uk/download/216229066.pdf (accessed on 23rd October 2025).
25. Ebert, U., & Sentman, D. D. (2008). Streamers, sprites, leaders, lightning: From micro- to macroscales. https://core.ac.uk/download/301646016.pdf (accessed on 23rd October 2025).
26. Glassman, A. J., Povinelli, L. A., & Reichert, B. A. (2025). Trends in aero propulsion research and their impact on engineering education. https://core.ac.uk/download/pdf/42811690.pdf (accessed on 23rd October 2025).
27. Haghnegahdar, M. A., Kolios, A., Lander, M., Mussetta, M., et al. (2019). State of the art in the optimization of wind turbine performance using CFD. https://core.ac.uk/download/228157082.pdf (accessed on 23rd October 2025).
28. Iranzo Paricio, J. A. (2019). CFD applications in energy engineering research and simulation: An introduction to published reviews. https://core.ac.uk/download/288003442.pdf (accessed on 23rd October 2025).
29. Lepri, D., Lien, E., Villani, E. A., Lee, E. A., Ábrahám, E., Meseguer, J., Hoffman, J. D., et al. (2010). A rewriting-logic-based technique for modeling thermal systems. http://arxiv.org/abs/1009.4263 (accessed on 23rd October 2025).
30. Mokhtar, M. (2011). Mechanical properties of soft clay stabilized with cement-rice husks (RH). https://core.ac.uk/download/12007624.pdf (accessed on 23rd October 2025).
31. Rao, K., Song, W., Wardi, Y., Yalamanchili, S., et al. (2015). Temperature regulation in multicore processors using adjustable-gain integral controllers. http://arxiv.org/abs/1507.06357 (accessed on 23rd October 2025).
32. Tayarani-N, M. H., Xu, H., & Yao, X. (2015). Meta-heuristic algorithms in car engine design: A literature survey. https://core.ac.uk/download/132585049.pdf (accessed on 23rd October 2025).

CHAPTER 6

ACTIVE COOLING TECHNIQUES

Contents

INTRODUCTION

The research into active cooling methods is more and more applicable as the temperatures on the planet increase and energy efficiency takes center stage in all walks of industries. Generally, in environments like data centers, the issue of keeping temperatures at an optimal level is essential for running efficiency and reliability. As demand continues to grow for data storage and processing, existing cooling methods are being questioned for their efficiency and sustainability.

Traditional cooling systems may not be sufficient anymore, necessitating a transition to more sophisticated alternatives. Active cooling techniques, such as a range of engineered structures used to actively remove heat, represent an appealing path toward enhancing system performance while restricting energy usage. The purpose of this chapter is to untangle the various approaches of active cooling and their implications on energy efficiency and economic viability. Among active cooling technologies, heat exchangers, particularly those incorporating heat-pipe technologies, have emerged as most prominent in terms of enhancing efficiency of cooling. Recent studies suggest that heat-pipe based heat exchangers can reduce energy expenditures in data centers considerably by offering efficient solutions for what has been referred to as "free cooling." This approach leverages natural conditions in the environment, which keeps dependence on mechanical cooling systems at a minimum. By quantifying current practices in the sector and recognizing the technical and legislative limitations faced when data storage is involved, we are able to make educated assumptions about areas of possible improvement. The impact of utilizing such technology can lead to up to 75% savings in energy, thereby proving to be not only economically worthwhile, but ecologically too towards greater environmental causes (Jouhara et al., 2014). With businesses shifting increasingly toward retrofitting existing buildings to achieve improved energy performance, incorporating energy-efficient active cooling practices becomes the area of focus. The study of a variety of technologies brings forth a picture with abundant promising technologies that can revolutionize indoor climate control design. Active cooling practices have the double benefits of comfort and sustainability; therefore, their dissemination becomes a requirement for industrial application in the modern world. By analyzing performance data in the past and employing theoretical models, stakeholders can better implement effective strategies that can optimize energy requirements as well as care for the environment. As explained in this chapter, pursuing efficiency in energy management is not a technical issue but also a vital aspect of future-proofing industrial architecture (Fisher et al., 2011).

Definition of Active Cooling Techniques

Active cooling techniques refer to mechanical devices employed to extract heat from indoor environments, hence creating a favorable temperature condition. Active cooling techniques usually entail energy-based devices such as air conditioners, chillers, and fans employed to actively regulate indoor temperatures. Unlike passive cooling, which relies on natural mechanisms such as ventilation and shading by building features, active cooling necessitates constant energy supply. According to recent studies, the quantification of energy performance in active cooling systems became more advanced, with an increasing focus on the significance of key performance indicators (KPIs) that facilitate performance measurement at the system level rather than at the equipment level. This integrated approach is crucial to achieve maximum utilization of energy and reduce deviations in cooling performance for increased sustainability in the operational strategy of buildings (Hong et al., 2020). Active cooling techniques must be incorporated not just for comfort but also to assist in maintaining indoor air quality and enable occupant productivity. While traditional active cooling products are based on temperature reduction, technological advancements have made way for more sophisticated ones, like variable refrigerant flow and heat pumps, that maximize efficiency and control. As the architectural industry grapples with tackling the pressing issues of climate change and energy consumption, there is a growing acknowledgment of the need to balance active cooling strategies with environmental imperatives. As mentioned in recent architectural writings, a more introspective use of cooling techniques can avoid excessive reliance on energy-intensive systems. Emphasis on an integrated approach to thermal comfort can lead to new design processes bringing together both active and passive methods for higher performance and sustainability (Napier et al., 2015) (Figure 6.1).

Figure 6.1. Active cooling.

Source: https://www.talius.com/wp-content/uploads/2024/08/SAVE-scaled.jpg.

In the case of the definition and impact of active cooling techniques, consideration should be given to their role in today's building practice and urban planning. Active cooling systems can be rendered highly efficient through the assistance of innovative design techniques, such as the use of appropriate building materials and thermal insulation that can reduce overall cooling loads. Furthermore, the implementation of system-level KPIs, as argued in the recent literature, offers valuable insights on performance for various cooling systems in various building types and climates. Through the energy consumption, peak load, and user satisfaction considerations, architects and engineers can promote an active cooling strategy that is not only technically proficient but also responsive to the environmental concerns of contemporary society. Ultimately, a trained comprehension of active cooling approaches affirms the potential potency of innovation and collaboration in shaping sustainable futures (Hong et al., 2020; Napier et al., 2015).

Importance of Cooling in Various Applications

Cooling systems are essential in many applications, as control of temperature is crucial for peak performance and device life. In industries, for example, inadequate cooling leads to overheating, reduced efficiency, and catastrophic failure. Assessment of energy-efficient cooling technologies is essential in retrofitting current systems in industrial buildings, so their working capacity is maximized while energy intake is minimized. The selection of appropriate active cooling techniques not only negates hazards posed by excessive operating temperatures but is also a component of green operation. Learning the importance of cooling is thus a priority in keeping up with efficiency and embracing green philosophy in industry, as seen from the findings of Fisher et al. (2011). Beyond the industrial world, so important is cooling that even specialized industries such as aerospace are accorded benefit from it. Sophisticated cooling systems are critical for spacecraft, where they are exposed to extreme thermal conditions; proper heat control keeps the electronic parts of the spacecraft at safe temperatures.

Programs like MISAT show developments underway, where micro satellites can be designed with sophisticated cooling solutions for inboard machinery. Through the application of microsystems technology, these solutions provide lower energy consumption and higher functionality. Consequently, the evolving landscape of active cooling techniques in aeronautics not only propels engineering frontiers but also emphasizes the broader implications of effective cooling strategies in diverse domains, as reported by Gill et al. (2007). Furthermore, active cooling is critical to industries such as data centers and electronics, where heat control is critical to performance and reliability. Servers and data center rooms overheating result in throttling, failures, and long-term damage, which makes it imperative to incorporate effective cooling systems. They regulate temperatures through techniques like liquid cooling, which provides higher heat extraction compared to traditional air-cooling technology. The transition from passive to active solutions is a great leap forward towards providing optimum conditions in high-performance computing environments. As more companies rely on data centers for their operations, the adoption of new coolants will be vital in making operations efficient, sustainable, and durable, reaffirming the underlying need for active cooling in modern applications.

6.1. PRINCIPLES OF ACTIVE COOLING

Active cooling techniques utilize diverse principles to ensure reliable control of indoor temperatures, thus promoting comfort and energy efficiency. A fundamental principle of active cooling is thermal dynamics control with mechanical support. This incorporates uses like air conditioning, where indoor spaces are cooled by refrigerants, which transfer the heat outdoors. By creating a temperature gradient, such systems enhance comfort in addition to energy efficiency by their ability to focus on specific building thermal zones. Further, the use of energy-efficient technologies upon the retrofitting of industrial buildings is crucial towards the attainment of optimum with such systems. Existing designs are focused on reducing energy loss with effective temperature management, highlighting the importance of adopting a holistic approach to the implementation of active cooling techniques (Fisher et al., 2011). Moreover, development of dynamic façades also highlights yet another basic concept of active cooling. These advanced systems incorporate external climatic conditions, which significantly reduce the usage of traditional air conditioning. By adjusting their attributes as per real-time information, dynamic façades can regulate solar heat gain, daylight admission, and even occupant views, thus facilitating an interiorized balance. The effectiveness of such advanced systems to a great extent depends on the incorporation of model predictive control. With the use of algorithms, this approach forecasts the optimal state of dynamic façades so that energy usage is coordinated with real-time demands. Thus, the integration of technology and environmental factors presents scope for the maximum reduction of peak loads and total consumption of energy in office buildings to develop an active multi-dimensional strategy for cooling that is effective and sustainable (Blum et al., 2020). The philosophy of active cooling technologies, nonetheless, also emphasizes the importance of adaptability and control in modern systems. With increasing sophistication in energy-efficient technologies, integrated automated control plays an increasingly important role. Modular control system's light-weight agents, for instance, can adjust cooling outputs in response to the customized needs of buildings and enhance life-cycle performance at less expense. Dynamic strategy supports modulating to user desirable levels as well as changing ambient conditions so that active cooling devices can stay responsive and efficient. The development of low-cost controllers further democratizes access to such advanced cooling technologies for broader application across different building types. Thus, the principles behind active cooling lie in part not merely in the progress of technology but also in a consideration of sustainable approaches which are responsive to both occupant needs and to environmental constraints (Figure 6.2).

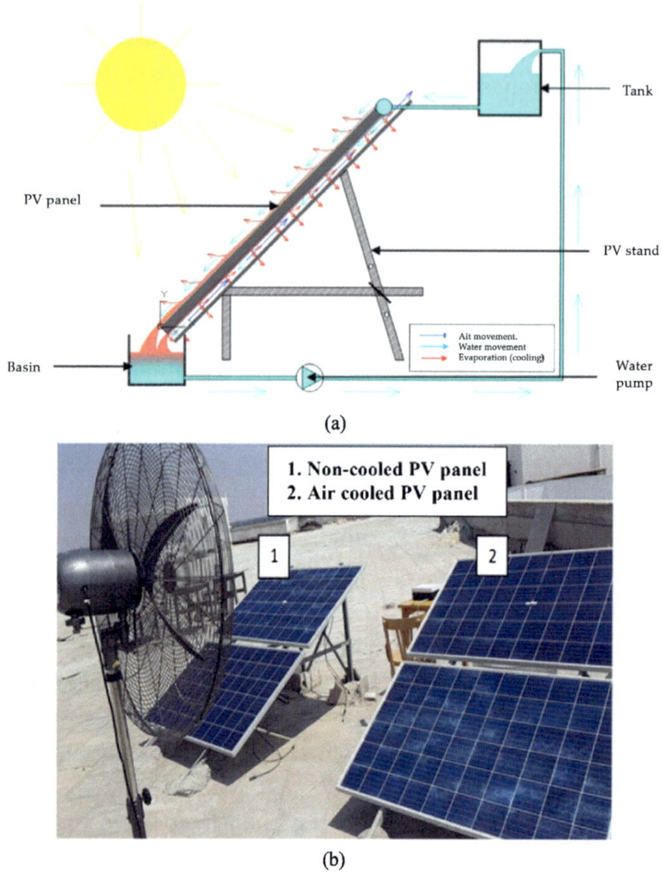

Figure 6.2. Active cooling techniques: (a) Schematic view of an application using water flow on the front and back sides of the PV panel; and (b) photograph of an application using forced convection for active cooling.

Source: https://www.sciencedirect.com/topics/engineering/active-cooling.

6.1.1. Thermodynamics and Heat Transfer

The fundamental laws of thermodynamics and heat transfer are critical to the design and utilization of active cooling techniques. Active cooling relies on the deliberate dissipation of heat from systems, and a grasp of conductive and convective heat transfer mechanisms is necessary. Thermodynamics gives the energy balance inside such systems, allowing for the optimization of cooling systems tailored to specific applications. For instance, industrial settings, where there is a high heat load, understanding the thermodynamic properties of different cooling fluids can optimize the performance of cooling towers and heat exchangers. Further, control of air flow and the design of the surface for the highest heat rejection are just some of the direct applications of such principles in achieving desired thermal outcomes and thus improving system performance and reliability.

The state to which working cooling systems respond has a drastic impact on their efficiency and effectiveness, particularly under reduced gravity. Pioneering experiments aboard space shuttles have confirmed the overarching significance of thermodynamics in various gravitational settings. Fluid flow, heat transfer mechanisms, and capillary phenomena, to name a few, can vastly differ in appearance when confronted with reduced gravity. Abramson et al. (2025) emphasizes the importance of investigating such phenomena with earth-based experiments under space-mimicking conditions to accurately predict thermal management issues in space missions. Investigating such a peculiar environment enhances our knowledge of heat transfer mechanisms, which can potentially lead to the development of advanced cooling techniques efficient not only in space but also in terrestrial applications where efficiency is also a key factor. Apart from the fundamental laws of thermodynamics, mathematical modeling is also a successful method in predicting heat transfer phenomena in active cooling systems. Mathematically highly sophisticated models allow for high-resolution simulation of heat transfer from underground pipes and other devices critical in many industrial processes. For instance, Babcock et al. (1975) offers models for analyzing heat transfer behavior of buried piping systems transporting cooling water. The models involve variables such as soil temperature profiles and energy balances to determine the effectiveness of heat exchange processes. The prognostic character of the models enables engineers to optimize active cooling design, making feasible solutions that enhance energy efficiency at lower environmental impacts. This union of thermodynamics with predictive modeling highlights the foundational significance of rigorous scientific approaches in advancements in cooling technologies across numerous areas.

6.1.2. Types of Active Cooling Systems

Active cooling systems are needed for maintaining the optimum temperatures in various environments, particularly data centers and commercial buildings where heat can be responsible for decreased performance and safety. The most prevalent active cooling is vapor-compression refrigeration systems. This system is dependent on the principles of thermodynamics and phase change, utilizing compression of refrigerants that absorb heat from an environment and direct it outside. These systems are greatly valued for their reliability and efficiency but need strict management to minimize the energy expense of refrigerant cycles. As the energy demands of systems such as these continue to rise, there is a desperate need for innovations that make them more sustainable while still maintaining performance, as is seen in the critical data center challenges of power management and energy efficiency (Mittal et al., 2014) (Figure 6.3).

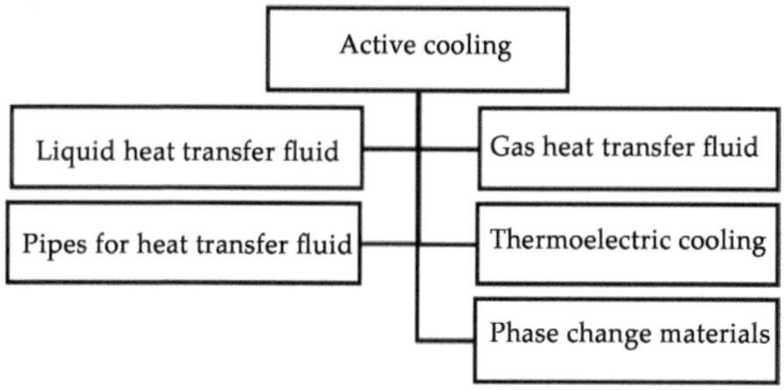

Figure 6.3. Types of active cooling.

Source: https://www.researchgate.net/publication/339468275_Review_of_cooling_solutions_for_compact_electronic_devices.

The second major method of active cooling is the liquid cooling system, which is becoming relevant with increasing computation intensity and data center densification. Liquid cooling systems utilize water or other suitable liquids to transfer heat from critical components with high efficiency. The use of liquid eliminates some of the disadvantages of traditional air cooling, such as lower efficiency in dense conditions. Systems are as simple as direct-to-chip to large-scale installations with chillers and cooling towers. Application of liquid cooling solutions not only promotes thermal control but is also crucial when it comes to energy efficiency, thus eliminating environmental concerns. Researchers are continually testing liquid cooling with green energy sources to formulate a more sustainable method of cooling in high-demanding situations (Hong et al., 2020). Thermal storage systems are another active cooling innovative solution that can be critical when it comes to increasing energy efficiency during peak demand. Thermal storage systems work by generating ice or chilled water during off-peak hours, which can be utilized for cooling during periods of greater energy demand. By shifting cooling loads to off-peak hours, these systems help to reduce the electrical grid load and total energy cost. They also help improve cooling system reliability during heatwaves since the stored thermal energy can act as a lifesaver when conventional systems can be subjected to strain. Because energy saving increasingly becomes priceless due to climate variability and scarce resources, the combination of thermal storage with conventional cooling techniques is a progressive strategy towards efficient and sustainable cooling (Hong et al., 2020; Mittal et al., 2014).

6.1.3. Efficiency Metrics for Cooling Systems

The efficiency of the cooling system is a significant factor in the case of active cooling systems, particularly in the context of energy consumption and sustainability. Efficiency is quantified in terms of several metrics, including the Coefficient of Performance (COP) and the Energy Efficiency Ratio (EER), which are among the most used. The COP is the useful cooling output delivered per unit of energy consumed, and the EER measures the cooling output delivered per unit of electricity consumed under a specific test condition. Building operators and engineers can make rational evaluations of the operating efficiencies of their cooling plants with the knowledge that they would not only select equipment sufficient to provide the cooling load, but also one that saves on energy expenses and environmental impressions. With the requirement for sustainable operations increasing, these performance measures are great tools to assist in guiding investment and operational decisions in the cooling sector. In addition to traditional levels of efficiency, there is also heightened recognition of the need for holistic approaches that consider the overall environmental impact of cooling systems. Data center energy consumption, for instance, is one of the largest sources of carbon output, and this requires innovative action to reduce energy usage as well as the environmental footprint of cooling systems in general. Sustainable data centers can only be created by means of combined solutions that enable synergistic management of energy, as indicated by recent research. By taking on metrics that include both energy efficiency and environmental sustainability, for example, lifecycle assessment and carbon footprint, organizations can make more effective assessments of the long-term consequences of their cooling solutions and make educated decisions that are in harmony with the world's sustainability objectives (Fern et al., 2016). The similarity of energy efficiency and thermal resilience is another parameter that is important to cool system performance, particularly in more stringent climate conditions. New studies indicate that a building's HVAC system must be evaluated not only for their capacity to save energy but also for their capacity to deliver adequate internal temperatures during adverse weather conditions. For example, active measures such as on-site power generators are required to maintain cooling systems operating even when grid power is lost due to hurricanes or extreme heat waves. Integration of passive and active measures is critical in optimizing energy efficiency while guaranteeing enhanced thermal resilience (Hong et al., 2020). Assessing the efficiency of cooling systems based on these holistic approaches allows stakeholders to favor solutions that save not only on energy costs but also safeguard against potential risks due to climate change.

6.2. REFRIGERATION SYSTEMS

Refrigeration systems represent a central element of the active cooling technologies in the broader environment of industrial and commercial applications. They operate according to thermodynamics using refrigerants to absorb heat from areas of interest and subsequently discharge it into the surrounding environment. By effectively governing the phase changes within a closed system, refrigeration technologies regulate temperatures according to requirements, thus enabling the best conditions for various processes. As the growing demand is to be energy efficient, modern refrigeration systems employ cutting-edge technologies to enhance efficiency while minimizing the use of energy. Use of energy-efficient technologies in industrial buildings, particularly during retrofitting operations, is important to counter environmental challenges as well as economic viability, as emphasized by Fisher et al. (2011) (Figure 6.4).

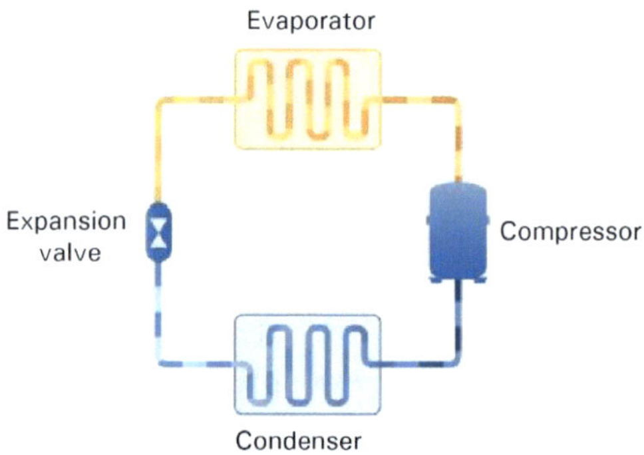

Figure 6.4. Basic configurations of a refrigeration system.

Source: https://www.intarcon.com/en/refrigeration-system/

Refrigeration system design and engineering demand a strong understanding of mechanical engineering principles along with the special requirements of the given application. Various design projects within the area demonstrate the necessity to use a systematic process encompassing all stages of the design process, ranging from problem definition to alternative comparison. This involves the meticulous planning of sustainable methods since effective refrigeration systems are capable of conserving energy and minimizing waste. Students of engineering can benefit from learning these design principles when used in refrigeration systems to thereby acquire the necessary skills in the designing and implementation of effective cooling solutions. The conclusions of design-focused literature help fill the gap between theoretical knowledge and practical implementation, thereby making them all the more applicable for future engineers (N/A, 2017). Additionally, refrigeration technology continues to advance with responsiveness to regulatory pressures as well as market needs toward sustainability. Innovations like magnetic refrigeration or thermoelectric cooling represent potential solutions over traditional vapor-compression cycles, toward more environmentally friendly approaches. These new technologies can also significantly reduce greenhouse gas emissions from the utilization of conventional refrigerants, making them appealing to environmentally conscious industries. Hence, stakeholders keen on active cooling technologies should be knowledgeable of these advancements. Modern research and development of refrigeration technologies are indicative of the need to cross-apply disciplines, as such developments trigger both performance enhancement and sustainability, which are in compliance with the strategic priorities of modern engineering practice (Fisher et al., 2011; N/A, 2017).

6.2.1. Vapor-Compression Refrigeration (VCR)

Vapor-compression refrigeration (VCR) has been a popular technology for contemporary active cooling systems due to its efficiency and reliability. VCR works on a cycle where a refrigerant is compressed, taking in heat from the ambient environment and releasing it elsewhere. The main components of a VCR system include a compressor, condenser, expansion valve, and evaporator. In the compression phase, the refrigerant is compressed, increasing its temperature, which is then cooled in the condenser and discharged into the atmosphere in the form of window heat. The refrigerant then expands and cools in the evaporator where it warms to the space in question. The process actually maintains the ambient temperatures lower, and hence VCR systems prove to be extremely valuable in several industries, including residential cooling, commercial refrigeration, and industrial processes, thereby showcasing their application in the current energy-consumer-oriented world. Increased demand for cooling systems has brought interest in innovative developments within vapor-compression refrigeration systems. Recent advancements are aimed at improving energy efficiency and reducing the environmental impact associated with traditional refrigerants. Changing to low-global warming potential refrigerants is one aspect of this activity, which is complemented by system design innovation such as two-phase cooling technologies. Such technologies, in this instance, micro-evaporation, can improve the energy efficiency of cooling systems by higher operating temperatures, which can result in increased waste heat recovery. As proposed by Blutt et al. (2025), the use of these novel cooling principles not only optimizes the refrigeration cycle but also makes possible the recovery of waste heat in applications and enhances system-wide energy efficiency (Figure 6.5).

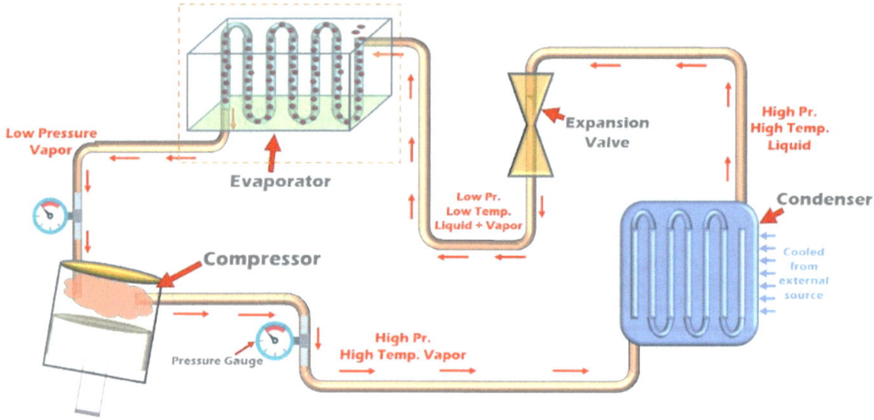

Figure 6.5. Vapor-compression refrigeration (VCR) system.

Source: https://i.ytimg.com/vi/PjcdqAkP0UA/maxresdefault.jpg.

Against the background of rising data centers, the role of vapor-compression refrigeration technologies has become much more prominent. Data centers will have over 1,00,000 servers with high power demands, which translates to high heat dissipation for which conventional air conditioning units are not sufficient. Overcoming this challenge requires high-end cooling solutions that can adapt in line with the fluctuating thermal dynamics of data centers. Research by Ding et al. (2011) emphasizes the necessity of exploring alternative cooling cycles, such as mini-compressor systems, which are able to recover additional exergy. Such innovations, besides reducing energy consumption, also enhance the cooling system's controllability, which improves a more sustainable way of dealing with heat in high-density computing setups. Evolution of VCR technologies evidence sustained interest in improving cooling performance across a broad range of applications both for operating performance and environmental reasons.

6.2.2. Absorption Refrigeration

Absorption refrigeration is an essential technology within the field of active cooling techniques, defined by its specific operation principles and environmental consequences. Different from conventional vapor-compression technology based on mechanical compressors, absorption refrigeration uses heat energy to facilitate a process of cooling. Usually comprised of a refrigerant and an absorbent, the process absorbs refrigerant vapor into a liquid form, releases heat and facilitates cooling. This is particularly useful in applications where waste heat is readily available such that a renewable means of cooling may be achieved. In the rapidly growing concern over climate change, with the effectiveness of energy consumption becoming increasingly critical, absorption refrigeration is a green technology that offers an environmentally friendly alternative aligned with global efforts at energy conservation and environmental technologies. Most importantly, system efficiency during operation in absorption refrigeration heavily relies on accurate design and material selection. Modern advances in the technology have put more emphasis on improved heat exchangers and absorbent materials that facilitate increased absorption with a decrease in energy loss. For instance, research directed at optimizing system performance has realized potential benefits through new materials and system configurations. These developments can lead to systems operating dependably over a broader set of conditions, making their application possible in a broad range of environments—from home cooling systems to large-scale industrial processes. These developments highlight the needed convergence of absorption refrigeration with modern engineering practices toward optimizing performance and energy consumption (Figure 6.6).

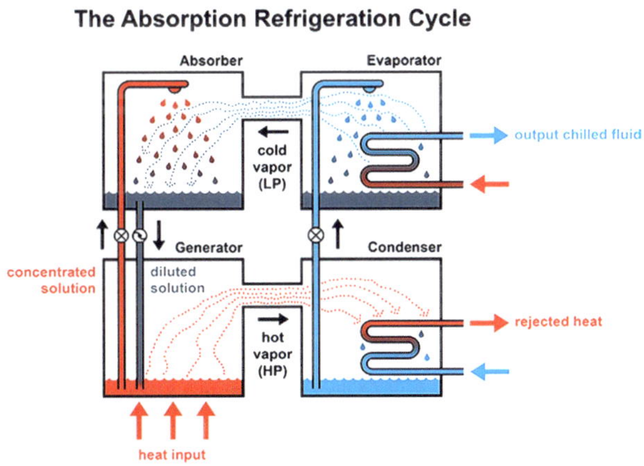

Figure 6.6. The absorption refrigeration cycle.

Source: https://blogger.googleusercontent.com/img/a/AVvXsEh34Oq8oBI6EJh-6Hifv6HuxmVVjjYlQcevm-U8RHG2mtJ-eqgp_ty44FnT0OmexBff8tYZy-KH_GokrWmCAB3W8mqZE0ljH3JuYDijTZLJOEgDIEc6CKXWYVXM5u6-83T_w7MMdVZxJYvJiZtAtHYG6J1CZwYkYUqs0ma6dEmcuPGQEY_RSJBb5WUO-p=w640-h360.

Furthermore, research of absorption cooling technology has also important consequences for its use in the future in space and remote conditions. The pioneering work analyzing vapor-absorbing refrigerator systems irrespective of gravity demonstrates the overall potential of this technology for non-traditional environments. Cooling in microgravity using thermal sources has the potential to benefit space missions wherein traditional refrigeration is not feasible. The development of such systems would then enhance the efficiency of life support technology, enabling extended space flight. With the advancement of research, it will become imperative to continue fine-tuning the systems, creating robust, dynamic solutions that cater to earth-based and space-based applications of cooling (Agnew et al., 2025; Blutt et al., 2025).

6.2.3. Thermoelectric Cooling

Thermoelectric cooling, which is a component of active cooling techniques, applies the Peltier effect to create a temperature difference between two surfaces of a thermoelectric module. The effect occurs when an electric current flows through a loop of two dissimilar conductors, enabling the heat to be absorbed on one surface and discharged from the other side.

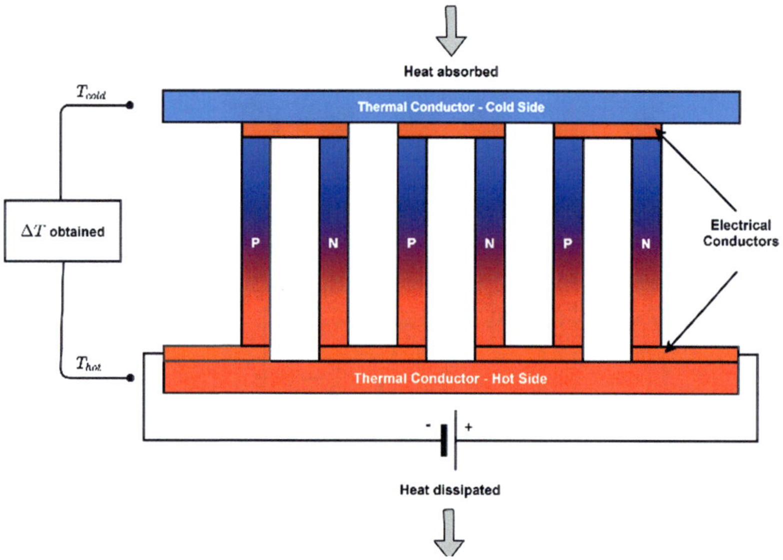

Figure 6.7. Schematic of a thermoelectric cooling system.

Source: https://www.researchgate.net/profile/Abdul-Aabid/publication/360644590/figure/fig2/AS:1158952767242241@1653327280525/Schematic-of-a-thermoelectric-cooler.png.

One of the most compelling arguments for thermoelectric cooling is its small size and lack of moving parts, making it quiet and needing less maintenance. With more applications in electronics, automotive, and HVAC arising, thermoelectric devices have been featured with using them in energy-efficient cooling due to being appropriate in environmentally friendly solutions, representing a green option for environmental issues. The versatility of thermoelectric cooling technology is observed in a number of applications from portable refrigerators to industrial uses. Specifically, their use in consumer devices serves to mitigate the increasing heat output in miniaturized devices where traditional cooling may be cumbersome. The use of thermoelectric modules in solar panels also enhances their efficiency by minimizing heat, thereby optimizing efficiency. Such thermoelectric cooling–PV energy synergy provides dual benefits in energy management, as demonstrated in recent studies emphasizing the function of thermoelectric elements to increase the overall efficiency of solar energy applications, which is consistent with today's demands for energy conservation and sustainability (Bagarella et al., 2016) (Figure 6.7).

The promise of thermoelectric refrigeration technology is extended further with ongoing research in new materials and systems. Advances like the construction of new materials with high figures of merit for thermoelectric properties, and the application of nano-structuring, are under consideration in the research pipeline to increase the performance level. Furthermore, as environmental impact considerations shape the next wave of technologies, thermoelectric systems could be the key to reducing dependence on conventional refrigeration techniques often tied to greenhouse gas emissions. Research activities, including those investigating system properties and application prospect, cite the focal importance of thermoelectric cooling in not just promoting energy efficiency but also addressing regulatory thrust for sustainability across industries (Copeland et al., 2025).

6.3. AIR CONDITIONING SYSTEMS

Air conditioning systems are a big advance over passive cooling systems, providing vital climate control for houses and companies. Air conditioning systems employ a refrigeration process to force heat from interior spaces to the exterior, thereby reducing interior temperatures. Frequently consisting of multiple significant components—compressors, condensers, evaporators, and expansion valves—air conditioning systems can be built to accommodate specific cooling needs. The efficiency of these systems is typically measured by their Seasonal Energy Efficiency Ratio (SEER), indicating the extent to which they can maximize energy usage to cool a given area. Increasingly, technological innovations aim to optimize these efficiencies while keeping environmental impact to a minimum in order to meet sustainability goals. Furthermore, smart integration of technology allows for more control and monitoring with energy-saving potential and greater comfort for the user as temperatures can be regulated from distant locations. During recent years, there has been growing interest in optimizing air conditioning systems for efficiency and performance.

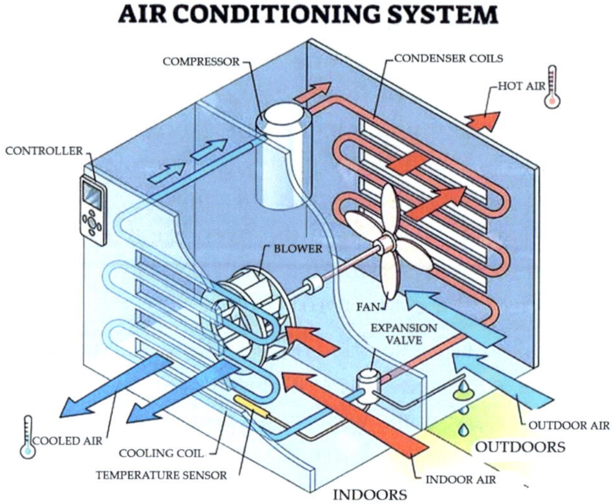

Figure 6.8. Air conditioning systems.

Source: https://external-content.duckduckgo.com/iu/?u=https%3A%2F%2 Fvectormine.b-cdn.net%2Fwp-content%2Fuploads%2Fair_conditioning_system_outline_diagram-1.jpg&f=1&nofb=1&ipt=7617da6366f3f8713e5b76b3abd04487458cc7bd7d3092f250347d32eb069360.

This transition is especially important in cooling-dependent industries like data centers, where the conventional cooling methods tend to result in an overconsumption of energy and high operational costs. Utilizing advanced cooling systems, including heat-pipe based systems, can play an important role in improving the efficiency of air conditioning systems in such applications. Research has demonstrated that through heat-pipe technologies, data centers are able to conserve up to 75% of their energy compared to conventional cooling methods (Jouhara et al., 2014). The systems exploit natural thermal forces to transfer heat around and thereby espouse the greater agenda of limiting energy consumption while lessening the carbon footprint of extensive cooling operations within an increasingly data-driven world. In addition, retrofitting current industrial buildings with energy-efficient air-conditioning technologies is a step of high significance towards making urban locations sustainable. Advances in technology are rendering energy-efficient design feasible, enabling building owners to reduce their energy consumption without sacrificing performance. Screening of various energy-efficient technologies, including variable refrigerant flow systems and improved insulation, has intensified as industries aim to satisfy the strict environmental regulations while still being economically acceptable (Fisher et al., 2011). With their implementation by organizations, long-term benefits emerge not only in operational cost savings but through the added employee comfort and productivity as well. Hence, the evolution of air conditioning systems continues to remain a critical aspect of discussion for active cooling technologies, showing technology, efficiency, and regard for the interactive roles of the environment (Figure 6.8).

6.3.1. Central Air Conditioning

Central air conditioning systems are a core component of contemporary climate control, providing consistent cooling across multiple rooms in residential and commercial spaces. This active cooling process consists of the movement of cooled air through a network of ducts, which allows for efficient temperature control across different zones. Contrary to window or stand-alone air conditioning, where everyone requires special attention and service, central air conditioning systems offer a more integrated solution that has the capability of cooling an entire building simultaneously, thereby ensuring user comfort and enhanced productivity. Accordingly, the upfront investment in central air conditioning can be offset by its eventual benefits of long-term energy efficiency and consumer benefit, demonstrating its status as a cornerstone of modern environmental control systems. In addition, ongoing energy efficiency debates have highlighted the importance of integrating central air conditioning systems with renewable technologies. The marriage of energy-saving technologies and active cooling technologies presents opportunities for enhanced thermal resilience, especially with increasing extreme weather conditions. For instance, buildings installed with state-of-the-art air conditioning technologies are more resilient to elevated temperature demands, especially where the climate shifts suddenly, as research evaluated building infrastructures' resilience. Through the integration of this technology with passive cooling techniques such as improved insulation and natural ventilation, individuals can achieve peak performance and sustainability, validating central air conditioning as something greater than a utility but a vital component of a holistic climate response strategy (Fisher et al., 2011; Hong et al., 2020) (Figure 6.9).

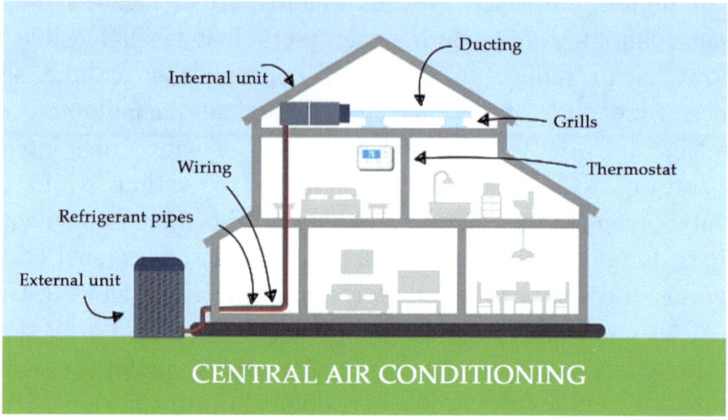

Figure 6.9. Central air conditioning system.

Source: https://www.d-air-conditioning.co.uk/wp-content/uploads/2019/09/central-air-con-diagram-600.jpg.

There is also the innovation of central air conditioning systems in the form of converting them to using renewable energy sources, thereby supporting the visions of reducing environmental imprints. The integration of solar power into air conditioning systems is now an attractive alternative, with buildings capable of drawing on renewable energy while reducing the use of traditional energy grids. This transition not only conserves operating costs but also meets broader environmental goals aimed against climate change. As the intersection of energy efficiency and thermal resilience gains increasing importance, understanding where central air conditioning fits within this dynamic is crucial. The effectiveness of next-generation cooling systems to operate in harsh conditions will ultimately determine their viability and attractiveness within the realm of expanding climate concerns, shedding light on the role of active cooling techniques within the future of building design (Fisher et al., 2011; Hong et al., 2020).

6.3.2. Ductless Mini-Split Systems

Mini-split ductless systems represent a significant innovation in active air-cooling technology, especially suitable for home and office use. Ductless systems are a departure from conventional central air conditioning in that they provide a more accurate and energy-saving way of temperature control. A system consists of an indoor air-handling unit and an outside compressor, which work together to provide low-temperature air where it is most needed. This feature not only enhances the efficiency of energy use but is also convenient, as people have the option of cooling one room rather than the entire space. Advancement in this technology indicates a higher level of concern towards environmental needs, particularly with the rising global temperatures and the necessity of green measures in energy consumption, in accordance with the guidelines advocated in contemporary energy reviews (Engelmann et al., 2013). In evaluating ductless mini-splits, energy consumption emerges as an overarching consideration. The technology enjoys a high level of energy efficiency compared to conventional systems, and this makes it extremely popular among environmentally conscious consumers. Interestingly, studies have determined that ductless systems use less energy without sacrificing indoor comfort levels. They operate by means of variable speed compressors, which adjust their output in relation to the existing cooling load, thus maximizing the consumption of energy and reducing costs over the long term. This is most essential in an era of increased demand for active cooling due to global warming and increased summer temperatures (Engelmann et al., 2013) (Figure 6.10).

Figure 6.10. Ductless mini-split systems.

Source: https://images.squarespace-cdn.com/content/v1/5fdbfc63eeb0d96bf9fb0833/0a6c7a46-5fb7-439a-835b-3ae1323dff7b/Ductless+Mini+Split+Ac+System.jpg.

These performance indicators suggest that ductless mini-splits can be an effective solution to the application of net zero-energy homes and, simultaneously, better indoor air quality. Furthermore, installation and maintenance of ductless mini splits are much less intrusive compared to traditional ducted systems. Lack of having much ductwork not only simplifies installation but also reduces the disadvantages pertaining to duct losses, which account for a significant amount of heat loss in conventional HVAC systems. Furthermore, advances in technology ensure that the mini-split systems come equipped with programmable thermostats and smart technology compatibility, enhancing convenience and user control even further. While active cooling requirements are on the rise, examining the systems as a total solution to cooling, energy saving, and comfort for occupants grows increasingly relevant. With these factors in mind, ductless mini splits have a compelling case for incorporation into ongoing discussions of efficient cooling in a hotter world (Alamoodi et al., 2023).

6.3.3. Portable Air Conditioning Units

Portable air conditioners have become a significant component of active cooling systems, particularly where traditional HVAC systems are not practical or possible. Portable and convenient, these units can be transported easily to offer cooling solutions in any given space, whether home, office, or temporary workspace. Portable air conditioners' portability ensures the consumer can cool specific areas or rooms without having to undergo long installation processes, making them ideal for renters as well as those who want affordable and versatile systems.

Figure 6.11. Portable air conditioning units.

Source: https://m.media-amazon.com/images/I/71DSU5Q5bmL._UF894,1000_QL80_.jpg.

Ultimately, hybrid strategies compromising between passive and portable active cooling methods are required to address the growing challenges of climate adjustment. Also, energy consumption of portable air conditioners should be noticed, especially as occupancy patterns fluctuate and demand shifts. Energy-saving portable units currently feature advanced technologies, such as variable speed compressors and programmable timers, intended to maintain overall power level low. Research indicates that application of optimization techniques, such as those used in traditional HVACs, can also be optimized to maximize more efficient utilization of such portable devices towards maintaining optimal comfort levels while lowering energy consumption. Model predictive control techniques can be employed for the monitoring of occupancy patterns, projecting usage to modulate the cooling output in a more efficient manner (Jain et al., 2018). This versatility not only conserves energy costs but also reduces the environmental footprint of air conditioning technology. Therefore, as the climate crisis intensifies, optimizing portable air conditioners by both innovative engineering and smart energy management can be a decisive leap towards comfort and sustainability indoors.

6.4. COOLING IN ELECTRONICS

Within the context of electronics, heat dissipation management is increasingly significant with the constantly increasing device performance, shrinking, and power density. Specifically, active cooling has emerged as a key solution to ensure peak performance and longevity of electronic devices. Active cooling can provide accurate temperature control even in harsh applications through the use of mechanisms such as fans, pumps, and heat exchangers. This technology is compared to passive cooling methods that rely on natural convection or conduction alone. As electronic systems progress towards miniaturization, the need for active cooling technology has expanded exponentially, significantly enhancing thermal management. These advancements facilitate the integration of high-power electronics into various applications, ranging from consumer devices to complex computing systems, transforming the way cooling technologies are perceived and utilized in the field of electronics. Moreover, recent active cooling technologies have provided opportunities for cost-effective and efficient electronic package designs. A novel solution for example, integrating directly injected cooling with traditional package types such as ball grid arrays, has demonstrated outstanding performance attributes at lower production expense. Not only does this technique involve fewer components—enabling ease of manufacturing processes—it facilitates better design scalability across multiple pieces in an electronic device as well. Therefore, designers are capable of delivering uniform and simultaneous cooling, thereby enabling greater design freedom with the best-performing results. These advancements demonstrate the strength of interdisciplinary integration of engineering in developing next-generation cooling solutions, thereby proving the central position occupied by active cooling techniques in developing next-generation electronic systems (Brok et al., 2008).

Further, the growing application of big data centers due to the explosion of big data and urgent demand for computing power has induced the advancement of energy management methods, particularly in cooling systems. The high-power consumption of such facilities generates substantial heat, which is difficult to manage to function the equipment accordingly. Successful power management techniques are vital to improve energy efficiency and reduce operational expenses (Mittal et al., 2014). A number of architectural solutions have been proposed to address these problems, allowing for more effective thermal solutions that are data center targeted. This sustained need for novel coolants and energy-saving designs represents a wider industry movement towards sustainability, emphasizing the importance of sustained research and development of electronic application cooling technologies.

6.4.1. Liquid Cooling Solutions

Liquid cooling technologies are a sophisticated form of heat management, particularly where high-heat dissipation applications are required, such as in electric vehicles (EVs). The advantage of liquid cooling over air-based techniques is multi-faceted. For instance, liquids possess a higher thermal capacity compared to gases, implying that they are able to absorb and carry heat more efficiently. This efficiency is especially significant for the most vital parts of EVs, including electric traction motors and power electronic components (PECs), where thermal output management directly affects performance and life. Recent advancements in direct slot cooling (DSC) and phase change materials (PCMs) further enhance operating efficiencies, bearing witness to the pivotal position held by liquid cooling solutions in new thermal management strategies (Gharehghani et al., 2025). Also, the use of liquid cooling systems minimizes the thermal challenges surrounding the current electronic devices. The increasing electronic component density generates a large amount of heat, and as such, there is a necessity for advanced methods of cooling that will avoid overheating and proper functionality. Liquid-cooled systems employ microchannels with high heat transfer coefficients, which effectively dissipate heat from critical zones. It has been proven through research that if appropriately engineered, they are capable of greatly outshining the traditional methods by increasing heat exchange rates and ensuring even temperature distribution between parts (Ding et al., 2011) (Figure 6.12).

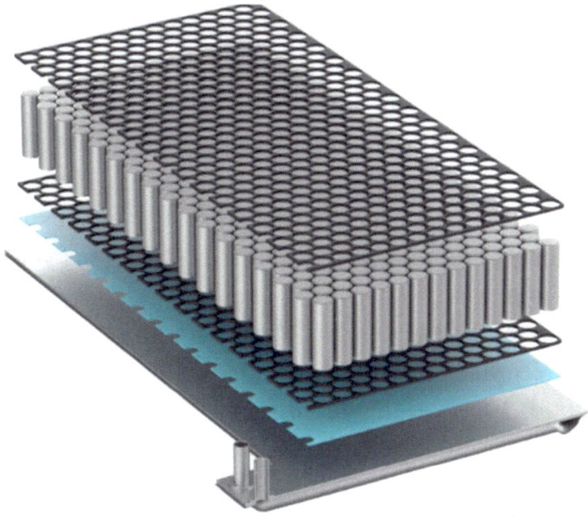

Figure 6.12. Battery liquid cooling system.

Source: https://www.trumonytechs.com/wp-content/uploads/2022/08/datwyler.com_battery_thermal-management-768x639.jpg.

The ability of these cooling solutions to be customized further enhances their cross-industry applicability for different industrial applications, while emphasizing efficiency and dependability in energy-intensive applications. Not only are liquid cooling solutions efficient, but their design complexity and associated challenges also define them. Designing efficient liquid cooling solutions poses complexities that have to balance cost, material science, and integration into existing systems. Optimizing flow paths and heat sinks for thermal transfer necessitates meticulous planning and advanced computational methods. Thermal-fluid distributions in these systems are commonly exposed by simulations, enabling engineers to apply design adjustments prior to implementation. The material technology breakthroughs, for instance, the application of copper in heat sinks, show how advanced engineering continues to improve the performance of liquid cooling technology as well as ensure that such technology enhances not only performance but also general sustainability in modern electronic technology (Ding et al., 2011; Gharehghani et al., 2025).

6.4.2. Heat Sinks and Fans

The history of electronics has necessitated innovations in cooling technology, particularly because small form factors have a tendency to lead to overheating. Out of all active cooling technologies, heat sinks and fans are the most efficient. Heat sinks, made primarily of high thermal conductivity materials such as copper or aluminum, are designed to dissipate heat from critical components, thereby preventing thermal buildup. They are more effective when combined with fans, which directly blow air across the surface of the heat sink. The end result is far greater exposure of the surface to cold air, which permits more effective cooling. These strategies are critical in modern electronics, as devices are requesting improved performance with less thermal issues, impacting both device life and operation, as newer thermal solutions such as piezoelectric fans are being notable for their compact form and effectiveness (Shaker et al., 2014). In relation to heat sinks and fans in synergy, the shape and functionality in consideration are what lead the design. The layout must offer the most effective airflow, as even the most effective heat sink will not work at its best if the air flow is not good. Fans offer dynamic cooling by moving air through the heat sink, enhancing natural convective currents that would be too weak under closely spaced conditions of portable devices like smartphones and laptops. This intersection of technology and design allows for flexibility in diverse applications, including high-performance computing and gaming systems, where thermal output is large. Growing performance requirements have mandated progress in fan and heat sink technology, justifying the justification of good thermal management practice in consumer goods and reinforcing the pertinence of ongoing research into new materials and architectures (Han et al., 2012) (Figure 6.13).

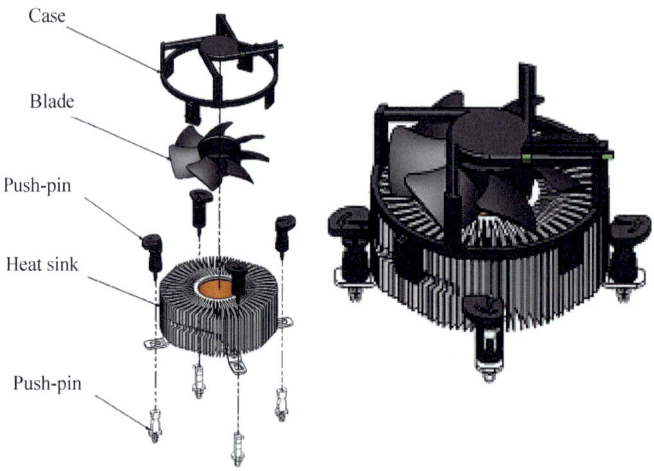

Figure 6.13. Schematic diagram of heat pipe embedded with heat sink and cooling fan units.

Source: https://www.researchgate.net/profile/Anumut-Siricharoenpanich/publication/336396773/figure/fig2/AS:814171121807365@1571124930908/Schematic-diagram-of-heat-pipe-embedded-with-heat-sink-and-cooling-fan-units.png.

New technological advancements in materials science are bringing forth improved heat sink and fan technologies, respectively, with the aid of metal foams and metal matrix composites (MMCs). These new materials possess designed thermal properties that enhance heat transfer efficiency in heat sinks and heat exchangers. Metal foams, due to their permeable structures, provide increased surface area, which allows for improved cooling while simultaneously reduced weight, a factor of paramount concern in mobile devices. As the world keeps trying to merge performance with power efficiency, such materials applied to active cooling methods represent an advancement. Manufacturing and scalability issues must be overcome before such material can become universally applicable. Therefore, investigations of practical applications remain an important function in realizing the full potential of nascent thermal management technologies (Han et al., 2012).

6.4.3. Phase Change Materials (PCMs)

In active cool structures, phase change materials (PCMs) are a great technology in offering enhanced thermal comfort for buildings. Using the latent heat storage characteristic of PCMs, PCMs are able to absorb and release thermal energy efficiently in the process of phase transition from solid to liquid state. This allows PCMs to maintain indoor temperature within a desired range and reduce reliance on mechanical air-conditioning systems. As buildings account for a large proportion of global energy consumption—particularly to maintain thermal comfort—the application of PCMs is an innovation measure to minimize cooling energy requirements. Theoretical evaluation of the performance of PCMs in operation requires strong numerical modeling by whole-building energy simulation software, which can help architects and engineers simplify their integration into building envelopes to achieve energy efficiency and user satisfaction (Cabeza et al., 2017). Use of PCMs has its drawbacks, primarily controlled by climatic factors and the specific application of the material. Operation of PCMs relies heavily on such parameters as melting temperature and thermophysical characteristics of the space to be filled. Although passive PCM systems are beneficial in numerous ways, their energy-saving properties can be non-uniform, meaning it is extremely essential to carry out comprehensive simulations to evaluate performance in different scenarios. Such simulations yield the essential resource used to maximize PCM selection and placement, ultimately ensuring these materials' capacity to deliver on their potential for greater energy efficiency and comfort (Cabeza et al., 2017). Therefore, ongoing study of natural night ventilations impact on PCM performance could potentially contribute more to the maximum overall performance of such cooling systems. In addition, the use of heat-pipe technology is a suitable complement to PCMs' role, particularly in single-purpose applications like data centers where cooling demands are stringent. The systems provide effective heat transfer by taking heat away from high-value equipment, hence reducing temperature fluctuations and maintaining optimal operating conditions. Combining heat-pipe exchangers with PCM systems is in some cases capable of achieving significant energy savings, up to 75%, particularly in environments with favorable weather for free cooling. On this combination of PCMs and heat-pipe technology rests a new trend toward hybrid cooling systems tailored to specific environmental and operational needs. As the research advances, understanding and maximizing how such technologies combine will be key to addressing the increasing energy demands involved in contemporary building operations and climate management.

6.5. INDUSTRIAL COOLING TECHNIQUES

In terms of industrial cooling technologies, optimization of heat exchangers is at the apex of achieving maximum energy efficiency in industrial processes. Research emphasizes the utmost relevance of mixed-integer nonlinear optimization such that systems are optimized to optimize both equipment design and utility cost. By considering heat recovery systems and stream interplay, industrial plants can effectively decrease their energy demand, which, as indicated by current statistics, constitutes about 73% of industry energy use. This approach utilizes deterministic global optimization algorithms, which are meant to evaluate large-scale networks and thereby ensure exhaustive analysis of different configurations that may enhance overall plant efficiency (Baltean-Lugojan et al., 2017). As companies move towards sustainable operations, such advanced computer techniques become the focal point in maximizing cooling processes and achieving energy conservation targets on buildings. Another crucial component of industrial cooling techniques entails using energy-saving technologies through the retrofitting of industrial buildings. The transition not only enhances the capacity of existing systems for cooling but also aids in larger-scale efforts in minimizing carbon footprints. Through filtering and combining innovative cooling technologies, businesses can achieve tremendous energy savings without compromising efficiency. Methods could include application of high-efficiency chillers, variable speed drives, and intelligent controls that enable adaptive cooling processes to adapt according to varying operational loads (Fisher et al., 2011). The comprehensive examination of aforementioned technologies makes it possible for industries to systematically advance their cooling systems, necessitating ongoing evaluations in order to keep up with shifting standards and performance levels in energy efficiency. Lastly, technology's convergence with industrial cooling techniques is also embodied through advancements in algorithmic design that maximize process systems engineering. These advances solve difficult optimization issues, namely nonconvex quadratically-constrained optimization issues that arise in cooling network designs. With the understanding and utilization of the topological structures inherent in such networks, industries are not just capable of improving the efficiency of liquid cooling processes but can also overcome combinatorial issues. The learning so obtained provides improved understanding of the systematic integration of active cooling techniques into industrial processes, moving significantly towards improving better energy utilization in industrial processes of any type. Realization of the potential for these algorithms to transform industrial processes highlights the role of technology, research, and utilitarian application in determining industrial future cooling solutions (Baltean-Lugojan et al., 2017).

6.5.1. Chillers and Cooling Towers

The effectiveness of active cooling technologies, particularly in commercial and industrial use on a large scale, is greatly reliant on chillers and cooling towers. Chillers operate by removing heat from a liquid using a vapor-compression or absorption refrigeration cycle to facilitate indoor climatic control.

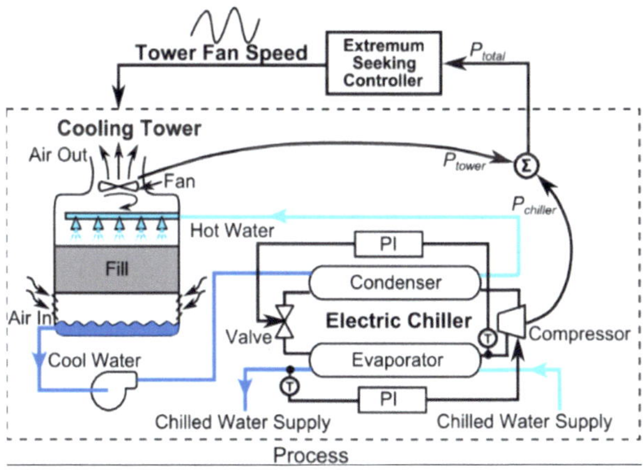

Figure 6.14. Tradeoff between energy consumption of chiller and cooling tower.

Source: https://www.researchgate.net/profile/John-Seem-2/publication/261281692/figure/fig5/AS:668695051374596@1536440732079/Tradeoff-between-energy-consumption-of-chiller-and-cooling-tower.png.

It is a very energy-consuming process, typically making chillers account for more than 50% of a building's total energy consumption when they are included in heating, ventilation, and air conditioning (HVAC) systems. In order to be more energy-efficient, technologies such as supervisory optimization control have gained popularity to maximize the system according to real cooling loads and weather conditions. However, the in-built assumption of a steady cooling load has the potential to limit energy savings, with specific focus on predictive optimization methods that adapt for anticipated changes in cooling demand (Feng et al., 2019). Cooling towers complement chillers in their ability to reject heat efficiently to the atmosphere, considerably optimizing the system's efficiency as a whole. By utilizing water as a heat exchange medium, cooling towers cool such water before it is circulated back to the chiller. Such a process is not only accountable for optimizing energy in chillers but also for regulating excessive humidity levels in a building. This inter-relation between chillers and cooling towers emphasizes the importance of overall system performance determination (Figure 6.14).

Key performance indicators (KPIs) play an important role in this context, providing a methodical means to benchmark and compare system component operating efficiency from HVAC systems to service water heating systems. Specific KPIs allow tracking of energy performance and fuel efforts towards conserving energy in building operations (Hong et al., 2020). Besides, the integration of newest monitoring technologies enhances the efficiency of chillers and cooling towers by measurement of performance indicators in real time. As more building construction work adopts sensor technology, the value of improved information gathering is greater than before. These advances enable close monitoring of interactions among different energy systems in a building, with the result enhancing integrated management of energy. Enhanced performance monitoring can warn when maintenance is needed or recognize inefficiencies which can be rectified. Finally, the implementation of a formal process to quantify the efficiencies of cooling towers and chiller, as well as precise KPIs, promotes not only energy cost reductions but a more sustainable running of HVAC systems in all building types. It is imperative to integrate these steps for maximizing the performance of active cooling techniques in today's energy-conscious age (Feng et al., 2019; Hong et al., 2020).

6.5.2. Process cooling in manufacturing

In production, effective cooling of processes is a part of ensuring operation efficiency and product quality. Equipment and processes create much heat, so proper temperature is required to be sustained to prevent overheating and equipment breakdown. This demand fuels the implementation of various active cooling systems that employ fluid, air, or phase change material to effectively transfer heat. The implications of these techniques extend beyond straightforward protection of equipment; they also make a massive contribution towards saving energy. For instance, the application of new cooling technology can realize extremely high savings in energy, something increasingly relevant with industries endeavoring to establish sustainability targets and maintain levels of productivity. Emerging process cooling technologies are revolutionizing manufacturing environments, particularly with innovations such as heat-pipe type heat exchangers.

Figure 6.15. Design and construction of process cooling systems.

Source: https://external-content.duckduckgo.com/iu/?u=https%3A%2 F%2 Faquachill.co.in%2Fassets%2Fimg%2FIMG_3038.JPG&f=1&nofb=1&ipt=fda918af5ec16a1690beb125069c1d2e1c9d8d67711ee39971d1dbe5c3297fe3.

These systems benefit from the efficient heat transfer through the change of phase, with improved cooling performance and less energy use. One study indicates that the integration of such technologies into cooling systems has the potential to record up to 75% energy savings in some applications, particularly in environments like data centers where the cooling loads are significant (Jouhara et al., 2014). As manufacturing industries continue to evolve, energy-efficient systems gain further significance not just for the sheer effectiveness of operations, but also to satisfy stringent environmental policies aimed at reducing carbon emissions. Incorporation of cutting-edge cooling technologies in the production sector, however, does not happen without its share of challenges. Mass investments in research and development, along with experiential checks for effectiveness in various manufacturing environments, are necessary in order to develop reliable and cost-effective solutions. Attention must also be paid to understanding the life cycle costs associated with these cooling systems, especially when adapting concepts for high-bypass turbofans that promise significant energy efficiencies (Baughman et al., 2025). Continued monitoring of performance indicators and the possible operations benefits of these systems will ultimately decide whether they are feasible and used in long-term procedures in manufacturing in the future. As manufacturers are becoming engaged in such intricacies, a combined focus on innovative cooling methods will prove vital in maximizing overall efficiency and competitiveness in a tighter and tighter marketplace (Figure 6.15).

6.5.3. Waste Heat Recovery Systems

The integration of waste heat recovery systems into active cooling technologies is a milestone in achieving energy efficiency in various industrial and commercial processes. Waste heat recovery systems are used to harness excess thermal energy generated through processes such as manufacturing or electricity generation to increase the performance of the system. By converting waste heat into useful energy, these systems reduce the consumption of conventional cooling processes that consume high electrical energy. Invention in the process of creating new technologies, such as the Organic Rankine Cycle (ORC), is an ideal representation of this trend. ORC systems utilize low-grade heat sources to produce power and, in effect, transform waste heat into an asset rather than a liability. Strategic waste heat utilization does not only make the system more efficient but also facilitates cost savings and emissions reduction, highlighting its significance in contemporary energy management practices. Moreover, recent advancements in the fields of thermodynamic models, such as Artificial Neural Networks (ANN), have optimized waste heat recovery systems to their fullest potential by enhancing their efficiency and feasibility. ANN's approach allows for precise understanding and optimization of performance parameters in ORC applications, according to recent studies. Utilizing ANN for solving difficult nonlinear optimization problems, engineers can arrive at cost-minimizing and energy-maximizing operating conditions. The experiments performed with R1234yf as the working fluid illustrate how the application of high-performance computational techniques can assist in developing more efficient designs and operational modes in waste heat recovery systems. This intersection of waste heat recovery with advanced modeling is a reflection of the revolutionary nature of combining novel technologies and traditional energy systems to improve their performance. Moreover, the technical advancements in high-efficiency engines, such as those done by NASA-GE, identify the importance of recovering waste heat in order to maximize the efficiency of high-bypass turbofans. The engines produce significant amounts of waste heat that can be recuperated and transferred by using recovery systems, leading to increased operational efficiency and life cycle cost savings. Emphasizing structural design innovation of engines as well as the recovery of waste heat gives a glimpse into how a cleaner aviation and industrial world can be achieved. With industries increasingly recognizing the benefits of active cooling by means of waste heat recovery, research activities combined with technology development aim to make the world cleaner. Lastly, the integration of waste heat recovery systems is not only in line with environmental goals but also economic ones, an essential step toward a cleaner future.

SUMMARY

It follows a discussion of active cooling techniques to underscore their vitally important role in addressing the thermal challenges of today's computing systems. With data centers expanding in scale and size, cooling efficiency increasingly becomes an imperative requirement. The use of techniques like liquid cooling, phase change cooling, and high-end airflow management not only allows for optimum operating conditions but also optimizes the overall efficiency of the facilities. Through prevention of overheating, such mechanisms increase system performance and the lifetime of equipment, something that is significant in keeping operations alive in high-demand settings. Further investigations into the mechanisms of operation and efficiencies of such cooling systems will be instrumental in shaping how future innovations reduce energy consumption while providing maximum cooling efficiency, which will contribute to the sustainability of technological infrastructure. In addition to ensuring computer system longevity, active cooling techniques also play a critical role in maximizing energy efficiency for data centers. According to research into power management, the increasing demands for data capacity and storage at ever-increasing rates, driven by the phenomenal increase in big data, pose unprecedented challenges to energy consumption (Mittal et al., 2014). The addition of novel cooling technologies not only is a reaction to thermal management requirements but also reduces power dissipation, in keeping with an increasing focus in society on green computing initiatives. This combining of effective cooling and energy management highlights the necessity for designers and engineers to design innovative solutions that deal with cooling and power concerns simultaneously. Another integrated understanding of such relationships will enhance collaboration in building design, reinforcing the position of active cooling in the overall discussion of energy efficiency. Lastly, the continuous evolution of cooling technologies reflects a promising shift toward sustainability in computing environments. Improvements in active cooling, in addition to methods such as evaporative cooling, are an innovative way of remaining current with the thermal dynamics of expanding data centers. Research focusing on limitations in the attainment of quantum ground states using cooling techniques, such as those reported in work on fermionic gases

(DeMarco et al., 1999), offer opportunities to explore new applications in fields of science and engineering. As the demands for computational capability rise, ongoing innovation with cooling techniques solves near-term thermal challenges and also generates an environment in which pioneering research can thrive. Thus, the way forward necessitates an interdisciplinary solution, yielding a reflective and prescriptive text that shapes the destiny of active cooling technologies in an age defined by the pace of technological advancement.

The discourse of active cooling techniques reveals some prominent understandings that collectively are refining our comprehension about their functioning and performance. At the core of these techniques lies the need for indoor climate control maximization to achieve both energy efficiency and occupant comfort to the fullest. Foregrounded research places emphasis on the necessity of achieving proper indoor temperatures as one of the most significant factors that influence thermoregulation in built environments. Specifically, research has indicated that the operational comfort temperature of cooling modes falls between 21.9°C and 25.4°C, and therefore moderate adjustments can lead to added satisfaction without compromising energy performance. The findings confirm the premise that the integration of such optimal temperature ranges in building design is capable of facilitating anticipatory modulation of the cooling system, thereby synchronizing sustainable measures with user preference, eventually leading to improved building operation and energy management (Hong et al., 2020).

Furthermore, research on active cooling techniques extends previous heating and cooling paradigms to encompass emerging methods such as low frequency nanoflare heating within astrophysics. This technique highlights the complexity of heat processes in variable environments and stresses the importance of using multi-disciplinary models to study thermal dynamics. A close examination of observation measurements indicates a non-linear scaling of emission measures with temperature, uncovering variable plasma states influencing cooling processes. This complexity influences the effectiveness of active regions in heat management. Impregnating this with the discussion of cooling approaches reveals the potential for cross-applicability of approaches that may be applied in both Earth-based and extraterrestrial environments, paving the way for more sophisticated designs and interventions that prioritize efficient thermal management (Bradshaw et al., 2012). Finally, the convergence of findings on active cooling approaches calls for an adaptive system that is capable of responding to variable demands from the environment. One of the most important takeaways from current research is the dual focus on occupant comfort and energy efficiency, and this can be optimally met through a fine balance of thermal comfort parameters. Recognizing the overlapping ranges of variability in preferred temperature between heating and cooling modes, as in the research, building managers and architects can design systems that properly balance both of these factors. This proactive approach is essential for the sustainability of urban development over the long term, given the uncertainty of climate and energy consumption patterns. The integration of these essential issues into practical applications has the potential to drive the evolution of active cooling systems to ensure their applicability in todays and future design practice (Hong et al., 2020).

REFERENCES

1. Abramson, H. N., Angrist, S. W., Catton, I., Churchill, S. W., et al. (2025). Fluid physics, thermodynamics, and heat transfer experiments in space. https://core.ac.uk/download/pdf/42890762.pdf (accessed on 23rd October 2025).
2. Agnew, D. L., & Jones, P. A. (2025). Large Deployable Reflector (LDR) System Concept and Technology Definition Study. Volume 2: Technology Assessment and Technology Development Plan. https://core.ac.uk/download/pdf/42810764.pdf (accessed on 23rd October 2025).
3. Alamoodi, A. H., Albahri, A. S., Albahri, O. S., Coffman, T., et al. (2023). Multi-perspective evaluation of integrated active cooling systems using fuzzy decision-making model. https://core.ac.uk/download/581548966.pdf (accessed on 23rd October 2025).
4. Albers, M., Dodds, W. J., Gerstle, J. H., Hasel, R. A., et al. (1992). High-speed civil transport aircraft emissions. https://core.ac.uk/download/pdf/10437196.pdf (accessed on 23rd October 2025).
5. Andersen, T. R., Cottam, S., Elkington, D. C., Griffith, M. J., et al. (2020). Manipulating nanoscale structure to control functionality in printed organic photovoltaic, transistor and bioelectronic devices. https://core.ac.uk/download/288432664.pdf (accessed on 23rd October 2025).
6. Babcock, R. E., & Havens, J. A. (1975). Heat and moisture conduction in unsaturated soils. https://core.ac.uk/download/215462988.pdf (accessed on 23rd October 2025).
7. Bagarella, G., Lazzarin, R., & Noro, M. (2016). Advancements in hybrid photovoltaic-thermal systems: Performance evaluations and applications. https://core.ac.uk/download/84486664.pdf (accessed on 23rd October 2025).
8. Baltean-Lugojan, R., Floudas, C. A., Misener, R., Mistry, R., et al. (2017). Designing energy-efficient heat recovery networks using mixed-integer nonlinear optimization. https://core.ac.uk/download/84869350.pdf (accessed on 23rd October 2025).
9. Baughman, J. L., Coffinberry, G. A., Demel, H. F., Johnston, J. C., et al. (2025). Structures performance, benefit, cost-study. https://core.ac.uk/download/pdf/42859533.pdf (accessed on 23rd October 2025).
10. Blum, D. H., Gehbauer, C., Lee, E. S., Wang, Y., et al. (2020). Integrated dynamic facade control with an agent-based architecture for commercial buildings. https://core.ac.uk/download/287624235.pdf (accessed on 23rd October 2025).

11. Blutt, J. R., & Sadek, S. E. (2025). A gravity independent vapor absorption refrigerator. https://core.ac.uk/download/pdf/85247142.pdf (accessed on 23rd October 2025).

12. Bradshaw, S., Dahlburg, R., Guarrasi, M., Guennou, C., Klimchuk, J. A., Reep, J. W., et al. (2012). Diagnosing the time-dependence of active region core heating from the emission measure: I. Low-frequency nanoflares. https://core.ac.uk/download/pdf/10180421.pdf (accessed on 23rd October 2025).

13. Brandner, J. J., Brighenti, F., & Kamaruzaman, N. (2011). Experimental investigation on self-similar heat sinks for liquid cooled electronics. 3rd Micro and Nano Flows Conference (MNF2011). https://core.ac.uk/download/9633348.pdf (accessed on 23rd October 2025).

14. Brok, G. J. H. M., Legtenberg, R., Mannak, J. H., Wits, W., et al. (2008). Novel cooling strategy for electronic packages: Directly injected cooling. https://core.ac.uk/download/pdf/11480087.pdf (accessed on 23rd October 2025).

15. Cabeza, L. F., Gracia Cuesta, A. de, Saffari Tabalvandani, M., Ushak, S. N., et al. (2017). Passive cooling of buildings with phase change materials using whole-building energy simulation tools: A review. https://core.ac.uk/download/304371814.pdf (accessed on 23rd October 2025).

16. Ceschin, F., Vezzoli, C., & Zhang, J. (2010). Sustainability in design: Now! Challenges and opportunities for design research, education and practice in the XXI century. https://core.ac.uk/download/9633150.pdf (accessed on 23rd October 2025).

17. Copeland, R. J., & Oren, J. A. (2025). Cooling systems for satellite remote sensing instrumentation. https://core.ac.uk/download/pdf/42889921.pdf (accessed on 23rd October 2025).

18. DeMarco, B., Kaoy, C. G., Myatt, C. J., Butts, D. A., Ferrari, G., et al. (1999). Limits to sympathetic evaporative cooling of a two-component Fermi gas. http://arxiv.org/abs/cond-mat/9909014 (accessed on 23rd October 2025).

19. Engelmann, P., Roth, K., & Tiefenbeck, V. (2013). Comfort, indoor air quality, and energy consumption in low energy homes. https://core.ac.uk/download/71267048.pdf (accessed on 23rd October 2025).

20. Feng, W., Lee, E. W. M., Wang, L., Yuen, R. K. K., et al. (2019). Cooling load forecasting-based predictive optimization for chiller plants. https://core.ac.uk/download/323310630.pdf (accessed on 23rd October 2025).

21. Fiorucci, M., Marmoni, G. M., Martino, S., Mazzanti, P., et al. (2018). Thermal response of jointed rock masses inferred from infrared

thermographic surveying (Acuto test-site, Italy). https://core.ac.uk/download/159635933.pdf (accessed on 23rd October 2025).

22. Fisher, D., Hand, J., Liesen, R., & Wilson, E., et al. (2011). Screening of energy efficient technologies for industrial buildings' retrofit. https://core.ac.uk/download/9033399.pdf (accessed on 23rd October 2025).
23. Gharehghani, A., Könnö, J., Mahmoudzadeh Andwari, A., Moradi, R., et al. (2025). Advanced thermal management strategies for electric vehicles: Enhancing efficiency, reliability, and performance. https://core.ac.uk/download/653750538.pdf (accessed on 23rd October 2025).
24. Gill, E., Monna, G. L. E., Scherpen, J. M. A., et al. (2007). MISAT: Designing a series of powerful small satellites based upon micro systems technology. https://core.ac.uk/download/pdf/12941826.pdf (accessed on 23rd October 2025).
25. Guitart Fernández, J. (2016). Toward sustainable data centers: A comprehensive energy management strategy. https://core.ac.uk/download/81578280.pdf (accessed on 23rd October 2025).
26. Han, X. H., Jacobi, A. M., Park, Y. G., Sommers, A. D., et al. (2012). A review of metal foam and metal matrix composites for heat exchangers and heat sinks. https://core.ac.uk/download/55763987.pdf (accessed on 23rd October 2025).
27. Hong, T., & Wang, Z. (2020). Learning occupants' indoor comfort temperature through a Bayesian inference approach for office buildings in United States. https://core.ac.uk/download/288432592.pdf (accessed on 23rd October 2025).
28. Hong, T., Lee, S. H., Li, H., Sofos, M., et al. (2020). System-level key performance indicators for building performance evaluation. https://core.ac.uk/download/288432586.pdf (accessed on 23rd October 2025).
29. Hong, T., Specian, M., & Sun, K. (2020). Nexus of thermal resilience and energy efficiency in buildings: A case study of a nursing home. https://core.ac.uk/download/323082718.pdf (accessed on 23rd October 2025).
30. Jain, M., Kalaimani, R. K., Keshav, S., Rosenberg, C., et al. (2018). Using personal environmental comfort systems to mitigate the impact of occupancy prediction errors on HVAC performance. http://arxiv.org/abs/1810.10619 (accessed on 23rd October 2025).
31. Jouhara, H., & Meskimmon, R. (2014). Heat pipe based thermal management systems for energy-efficient data centers. https://core.ac.uk/download/30339678.pdf (accessed on 23rd October 2025).
32. Marcinichen, J. B., & Thome, J. R. (2011). On-chip micro-evaporation:

Experimental evaluation of liquid pumping and vapor compression cooling systems. 3rd Micro and Nano Flows Conference (MNF2011). https://core.ac.uk/download/9633343.pdf (accessed on 23rd October 2025).

33. Massimiani, A., Palagi, L., Sciubba, E., Tocci, M., et al. (2017). Neural networks for small scale ORC optimization. https://core.ac.uk/download/153367074.pdf (accessed on 23rd October 2025).
34. Mittal, S. (2014). Power management techniques for data centers: A survey. http://arxiv.org/abs/1404.6681 (accessed on 23rd October 2025).
35. Napier, J. D. (2015). Climate based facade design for business buildings with examples from central London. https://core.ac.uk/download/29176762.pdf (accessed on 23rd October 2025).
36. Shaker, S. F. (2014). Thermal and flow analysis of piezoelectric fans for cooling LEDS packages. https://core.ac.uk/download/32600178.pdf (accessed on 23rd October 2025).

CHAPTER 7

PASSIVE COOLING TECHNIQUES

Contents

INTRODUCTION

The continuous technological advancement in the area of electronic devices has resulted in increased reliance on efficient cooling methods in an attempt to maximize performance and reliability. As the devices reduce in size but boost power, they are generating more quantities of heat, and thus there is a higher demand for quality thermal management solutions. Traditional cooling methods utilize mechanical components such as fans and refrigeration systems, which consume power and are environmentally unfriendly. Conversely, passive cooling methods have immense potential in tapping into the natural physical laws for dissipating heat without spending active energy. This also improves device performance but, more crucially, helps contribute to sustainability objectives as passive cooling technologies tend to be less environmentally damaging and capable of conserving enormous amounts of energy. Therefore, it is important to research these techniques since technology continues to evolve and the demand for efficient cooling systems rises. One intriguing research area among passive cooling techniques is the application of heat-pipe technology, which has been well established as quite effective in heat transfer and temperature control. Heat pipes are able to transfer heat from essential components in a highly efficient way using the principles of phase change and capillary action, hence delivering thermal stability. The technology also has future uses beyond consumer electronics; for instance, it has been tried in the case of data centers, where heat management is important to maintain efficiency in running. Recent research has shown that energy usage can be saved by as much as 75% when integrating free cooling systems using heat pipes, with potential energy savings (Jouhara et al., 2014). These findings underscore the paramount value innovative passive cooling technologies add to solving both efficiency and sustainability concerns of modern electronic devices. As urgent as it is for electronic devices to have efficient thermal management strategies, learning about various passive cooling technologies is paramount to maintaining electronic device reliability and longevity. Methods such as natural convection, radiant cooling, and thermal mass can be employed for applications ranging from consumer electronics to mega data centers. As industry participants increasingly worry about green practices, the use of passive cooling solutions can decrease the environmental impact of electronic devices significantly. Moreover, the ongoing research in this field finds a complete treasure of possibilities for innovation and advancement of existing techniques, thereby establishing possibilities of more advanced cooling technologies in the future. Giving top priority to passive cooling methods, the electronics sector can contribute its part meaningfully towards energy efficiency targets with the development of innovation and technology growth (Hoell et al., 2019).

Definition of Passive Cooling

Passive cooling is a technique that employs natural means of moving heat away from electronic devices without the use of active mechanical systems like fans or refrigeration. Passive cooling relies on various physical principles, including conduction, convection, and radiation, to provide the best operating temperatures. Passive cooling in electronic devices can make them perform better and live longer by minimizing thermal stress. The effectiveness of passive cooling is dependent on design and material selection, which may include thermal interfaces, enclosures, and deliberately placed heat sinks for facilitating the dissipation of heat. Additionally, a customized geometry used can optimize air flow, enhancing the natural cooling process. These factors are worthy of consideration, especially with diminishing sizes of electronic components being squeezed in closer proximity to one another, so that enhanced thermal management becomes necessary. An additional important aspect of passive cooling is that it has environmental and cost benefits. Due to the fact that this process does not rely on power usage in its cooling mechanisms, energy savings are realized, and it therefore proves useful for battery-powered systems and devices where power supply is limited. Further, lower energy needs can mean substantial cost reductions for manufacturers and consumers. With emerging technologies and growing emphasis on sustainability, the incorporation of passive cooling solutions can complement efforts to reduce energy consumption and carbon emissions globally. Research indicates that dependable thermal management is essential in electronics to provide reliability and efficiency, emphasizing the importance of passive cooling approaches as a sustainable solution (Clark et al., 2025). These approaches not only minimize the thermal loads but also address heightened pressures toward resource conservation as well as environmental impacts on electronics manufacturing. Finally, the use of passive cooling technologies is a dominant consideration in the improvement of the reliability and life span of electronic systems. Utilization of natural heat transfer processes inherently restricts the number of moving parts employed, reducing the possibility of mechanical malfunction inherent in traditional active cooling technology. Passive cooling configurations can also be integrated very easily into every product line, adding to the widespread durability and tenacity of electronic hardware. As businesses continue to adopt these approaches, advancements in material science and design engineering are fueling innovations in thermal management technologies. The use of passive cooling is thus set to be a core component in building future electronic devices, not just in their functional operation but also in the overall objective of creating energy-efficient technology solutions in a way that meets societal needs in a responsible manner (Glasheen et al., 2025).

Importance of Thermal Management in Electronics

Proper thermal management is critical in electronic products as excessive heat can lead to performance loss, component degradation, and shorter device lifetimes. With the miniaturization and complexity of electronic circuits, heat generation has become an intrinsic challenge in their design and operation.

Particularly for high-performance products, satisfactory temperatures are paramount to ensure reliability and efficiency. The consequences of poor thermal control extend beyond outright failure; they also make thermal cycles impact the structural integrity of components and subsequently necessitate costly repairs and replacement. Due to this, passive cooling solutions become a vital method of addressing such issues because they aim to remove heat effectively without the need for additional power consumption, improving overall system performance. Among the various passive cooling strategies, the use of phase-change materials and heat pipes is a highly effective way to address thermal loads. Liquid/vapor phase-change based passive thermal spreaders such as heat pipes have been popular for use in macro-scale thermal management for decades. The developments in micro-fabrication processes in recent years have enabled one to fabricate grooved heat pipes at the micro-scale and even the nanoscale. These advancements allow for effective heat dissipation by proper capillary action despite the compact form factor of modern electronics. In this context, geometrical parameter optimization of filling ratios and aspect ratios of micro- and nano-scale heat pipes becomes mandatory to obtain maximum thermal performance achievable. Findings indicate that such optimizations have the ability to significantly enhance the efficiency of heat transport, thus enabling devices to function at safe thermal margins (Akkus et al., 2018). Moreover, the research in nano-grooved heat pipes is an area in passive cooling technology that is evolving, as can be observed with simulations verifying their operating performance across various conditions. These simulations have revealed that the operation of nanoscale heat pipes is not merely fluid dynamics but also rests on intricate geometrical issues. Through research on these parameters, researchers have opened the doors to tailored thermal management strategies that are adapted to the diverse demands of modern electronics. The novel modeling methods employed in these research articles are directed toward the capacity of computational experiments to improve thermal interface design through optimization. With continued advancement in electronics, innovation in passive cooling technology will be central to maintaining performance while limiting power consumption to its least level, consequently enhancing the performance and life of electronic systems in different applications (Hoell et al., 2019).

Overview of Passive Cooling Techniques

The concept of passive cooling is all about establishing measures that assist in maintaining electronic equipment at manageable temperatures without depending on the use of mechanical cooling measures. This is particularly significant in confined electronic environments where conventional strategies might not be effective or practical. Passive cooling measures employ natural factors such as conduction, convection, and radiation to reduce heat generated by electronic components. There are several various strategies that can significantly enhance thermal control, including ideal material choice and component placement at strategic locations. By utilizing materials with high heat conductivity, engineers enable the heat to be transferred away from heat-sensitive components effectively, preventing overheating and maintaining stable operation. Another important passive cooling method is ventilation, which utilizes natural airflow to enhance thermal management within electronic components. Adequately designed air ventilation schemes can produce a flow of air which sucks heat away from the devices surface, in effect maintaining cooler operating temperatures. This is particularly beneficial in enclosures where overheating is an issue. Employment of natural cross-ventilation, typically using the assistance of vents strategically positioned, allows cooler air to enter with hot air being expelled, a continuous airflow cycle. Besides, the design considerations like device placement in the context of ambient temperature gradients may also optimize passive cooling efficiency. They are significant factors, especially in the design of data centers, where thermal loads control is central in realizing energy efficiencies and operational feasibility (Jouhara et al., 2014) (Figure 7.1).

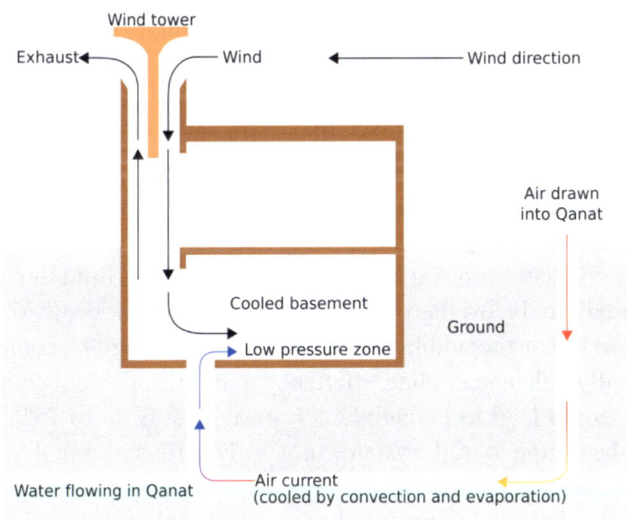

Figure 7.1. *Solar cooling design using a wind tower.*

Source: https://upload.wikimedia.org/wikipedia/commons/thumb/4/45/Wind-Tower-and-Qanat-Cooling-1.svg/1024px-Wind-Tower-and-Qanat-Cooling-1.svg.png.

New technologies such as heat pipes and phase change materials (PCMs) are classic applications of advanced passive cooling technology. Heat pipes employ a phase change within an enclosed vessel to transfer heat away from sensitive components in a highly efficient manner, typically dissolving heat at much higher levels than traditional thermal management devices. The technology is also applicable to densely packed electronic assemblies where space is limited. Similarly, PCMs can take up and discharge latent heat when transforming from solid to liquid state or vice versa, serving as a buffer for temperature change. The integration of such cutting-edge passive cooling techniques is transforming the design world of electronic products toward not only improved thermal performance but also energy efficiency because they allow devices to operate without activating additional power to cool them (Hoell et al., 2019).

7.1. PASSIVE COOLING TECHNIQUES OVERVIEW

In the electronic equipment industry, passive cooling techniques are now very prominent as a green solution to managing thermal loads independent of traditional mechanical cooling mechanisms. These techniques leverage the inherent thermodynamic properties of materials and the ambient environment to dissipate heat generated by electronic equipment. With the aid of mechanisms like conduction, convection, and radiation, passive cooling systems can maintain the optimal operating temperatures while they are energy conserving and environmentally friendly. Thus, an exploration of these processes is significant, particularly with decreasing sizes in electronic devices, which generate high amounts of heat, hence thermal management becomes a critical consideration. One of the innovative techniques of passive cooling involves heat pipe applications that have the potential to transmit thermal energy efficiently away from heat sources by utilizing phase changes in an enclosed system. Heat pipes employ condensation and evaporation of a working fluid to facilitate heat transfer with extremely low thermal resistance. It has been recently demonstrated through research that the addition of heat-pipe technology to cooling systems can dramatically enhance cooling efficiency and, in some applications such as data centers, could lead to potential energy savings of up to 75%. Therefore, research in heat-pipe based systems not only offers thermal management solutions but also helps the industry's growing concern for energy efficiency, showing the revolutionizing potential of passive cooling principles in modern electronic devices (Jouhara et al., 2014). Moreover, the concept of free cooling also describes another avenue to low-energy thermal management by leveraging ambient environmental conditions for dissipating heat. This strategy focuses on the use of naturally cold ambient air or outside bodies of water to cool equipment rather than employing energy-intensive active cooling systems. Data centers, for instance, can greatly benefit from this by using ambient air during months when temperatures are low, very aggressively reducing operating costs and carbon footprints. The study of free cooling technology combined with established technologies such as heat pipes is a formidable synergy for achieving maximum energy efficiency. The integration of these passive cooling methods enables technology stakeholders to address crucial thermal management challenges in an eco-friendly way (Hoell et al., 2019).

7.1.1. Natural Convection

Natural convection is a significant aspect in passive cooling techniques in electronic devices, essentially being the source of the heat dissipation created by their operation. As opposed to forced convection, which relies upon mechanical devices such as fans, natural convection draws its concept from buoyancy gradients.

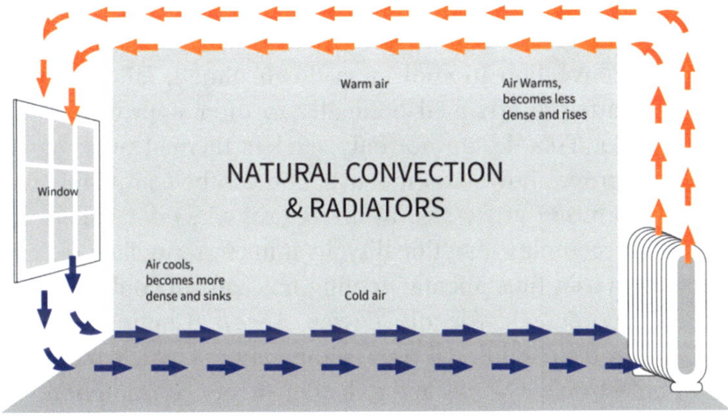

Figure 7.2. Passive cooling by natural convection.

Source: https://www.calyos-tm.com/news-article/passive-cooling-with-convection---natural-convection.

When electronic devices generate heat, the surrounding air warms and becomes lighter, and it forces the hot air to rise. This movement creates a continuous circulation of cooler air from the surroundings, in effect siphoning off excess heat. This is a passive method which is particularly beneficial in compact electronic packages where there is no room for the addition of additional cooling mechanisms. Moreover, the simplicity of relying on natural convection reduces the energy needs and noise concerns of mechanical cooling mechanisms, and as such is a desirable option for manufacturers seeking to implement energy-saving designs. Natural convection as part of electronic systems can enhance thermal control without the increased complexity of active cooling. The design of heat sinks, for instance, is created to optimize the natural convection currents; fins or porous media are typically employed to optimize heat-exchange surface area. Such systems enable increased heat dissipation, allowing the device to run safely even under maximum performance (Figure 7.2).

Additionally, the availability of new materials with greater thermal conductance can further enhance the efficiency of natural convection devices. The incorporation of these materials with natural convection mediums is also in harmony with current electronics trends, where power efficiency and miniaturization are given priority over all else. With such design practices, manufacturers are able to develop devices not only that work at an optimal capacity but are also strong in the aspect of thermal degradation. Recent developments in passive cooling technologies also include integrating natural convection with built-in electronic components, emphasizing the practical application of this technique. A remarkable example is found in hybrid stereolithography printers where temperature-responsive circuits are actually built into materials that use natural convection to cool them down. Small, efficient devices are possible where parts are arranged optimally to offer improved air flow and thermal dissipation. This design not only renders thermal management more efficient but also proves how natural convection can be compatible with state-of-the-art manufacturing processes. In the future, when consumer electronics need increasingly complex functionality in thinner form factors, employing natural convection as a fundamental cooling mechanism could be an essential component of environment-friendly device design. Continuing research in the method can lead to additional uses, ultimately extending the lifetime and functioning of electronic systems and reducing energy consumption.

7.1.2. Radiation Cooling

Radiation cooling is a cutting-edge passive cooling technology that utilizes the natural radiation of infrared waves to extract heat from electrical devices. Unlike typical active fan-based or refrigeration cycle-based cooling technologies, radiation cooling leverages thermodynamic principles of heat transfer. Its efficiency is primarily characterized by the emissivity of the material, which defines a surface's ability to emit thermal radiation. Within electronic devices, high-emissivity materials have the potential to be very effective at dissipating heat and maintaining optimal working temperatures as well as increased lifespan. Such applications of radiation cooling are particularly relevant where it is not feasible or energy-hungry to use mechanical cooling technologies, and this is commensurate with expanding demands for energy efficiency in modern electronics. Not only does the use of radiation cooling need the selection of suitable materials but also design within electronic device architecture. For example, surfaces can be structured or textured to make them have a greater effective area to improve heat radiating properties. Studies have shown that, by integrating radiation cooling with conventional methods of thermal management, better performance can be achieved, maximizing the overall thermal behavior of electronic devices. Such multimodal approaches are especially important in applications of high-performance computing, where excessive heating can cause system breakdown.

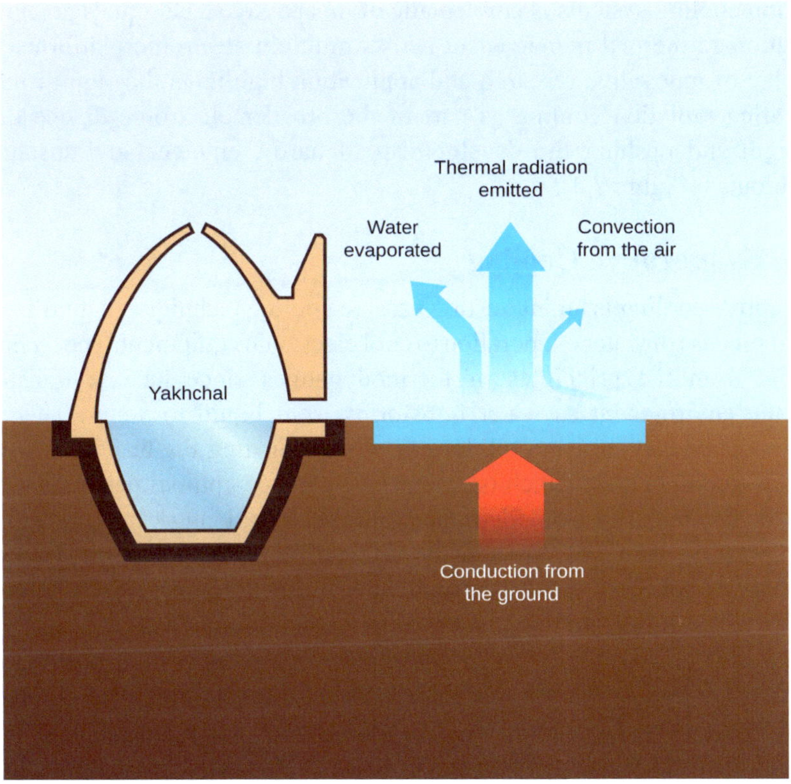

Figure 7.3. Radiative cooling energy budget in a traditional Iranian architectural element.

Source: https://en.wikipedia.org/wiki/Passive_cooling#/media/File:Yakhchal_radiative_cooling.svg.

Moreover, the overview of ongoing research in optical technology and cryogenics, as cited by Clark et al. (2025), demonstrates the interconnected nature of various passive cooling techniques, necessitating an integrated solution to the thermal management in designing the next-generation electronics. In addition, radiation cooling becomes increasingly pivotal in solving the demands posed by modern high-energy particle detectors, e.g., those employed in large-scale physics experiments at the Large Hadron Collider. These systems operate in extreme thermal environments where traditional cooling methods may not suffice, and radiation cooling thus constitutes an imperative method of thermal control, as highlighted through innovations in radiation-hard sensors and readout chips by Garcia-Sciveres et al. (2018). Ongoing development of materials and technologies to support augmented performance, life, and efficiency in radiation cooling systems is emblematic of this passive technique's potential to revolutionize thermal management across multiple electronic platforms. This interplay of innovative research and application highlights the significance of integrating radiation cooling as part of the broader electronic device design paradigm and pushing the development of hardy, efficient, and sustainable technologies (Figure 7.3).

7.1.3. Evaporative Cooling

Evaporative cooling is an interesting passive cooling technique that utilizes the natural process of water evaporation to cool electronic equipment. The technique benefits from the principles of thermodynamics since heat is transferred from the environment as water transforms from liquid to vapor. Therefore, the process of cooling can go a long way in mitigating the heat generated by electronic components, which is crucial to guarantee optimal performance and lifespan. The majority of modern electronic devices, being very tiny and with high operating demands, are mostly susceptible to overheating. The use of evaporative cooling systems can actually negate such heating problems, and the devices will still operate under stressed usage conditions. Evaporative cooling systems also mostly get designed into device implementations with relatively less redesign effort, thereby enhancing the performance of heat dissipation without increasing the energy consumption significantly. As the demand for efficient thermal management in electronics increases, the application of evaporative cooling becomes increasingly relevant. Advances in materials science provide avenues toward realizing efficient water-retention systems for enhanced evaporative cooling capacity. For instance, porous material and hydrophilic surface design may enable higher evaporation rates and thus enhance the maximum cooling capability. Furthermore, evaporative cooling technologies can be integrated with other passive cooling approaches to create hybrid technologies with increased thermal management. This cooperative approach can lead to more resilient electronic components that can operate effectively in hostile environments, as witnessed from the recent establishment of microelectronics technologies to address challenges arising from data-driven applications (Garcia-Sciveres et al., 2018). Consequently, the strength and effectiveness of evaporative cooling render it a critical strategy for electronic cooling in the future. Ecological sustainability of evaporative cooling is also noteworthy, particularly over conventional active cooling methods based on energy-demanding refrigeration cycles. By leveraging natural effects, evaporative cooling reduces the aggregate energy needed for maintaining device temperatures at their ideal values. This not only minimizes the carbon footprint of the electronics but also aligns with the growing emphasis on green technology in the current market. The added aspect that the locally accessible water resources can be employed to execute this "passive cooling" technique makes the method more appealing, with economic and environmental benefits. With shifting trends in the industry towards energy efficiency and sustainability, the implementation of evaporative cooling in device design and manufacturing processes can be the touchstone for innovation and sustainable practices (Fisher et al., 2011). For this purpose, the future of electronic devices is full of promise, with evaporative cooling as the focal point of energy-efficient passive thermal management strategies (Figure 7.4).

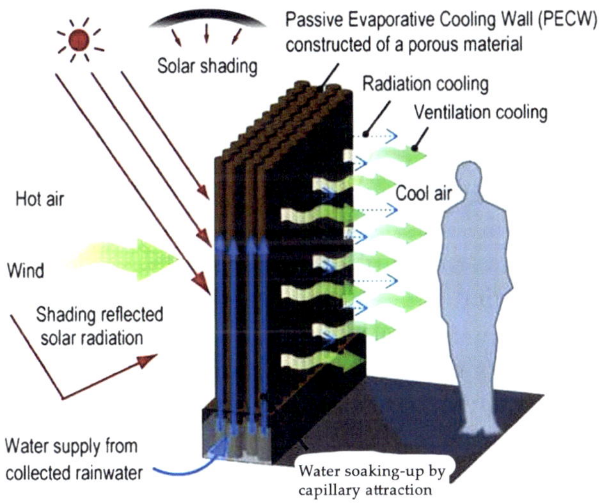

Figure 7.4. Schematic of a passive evaporative cooling wall constructed from a porous material with high water absorption capacity.

Source: https://www.researchgate.net/figure/Schematic-description-of-a-passive-evaporative-cooling-wall-constructed-of-a-porous_fig1_269728249.

7.2. MATERIALS USED IN PASSIVE COOLING

The effectiveness of passive cooling techniques of electronic components to a large extent depends on the material employed in their manufacture. Those possessing good thermal conductance, such as copper and aluminum, are especially crucial in efficient heat dissipation. These metals facilitate quick heat conduction away from components to maintain them at optimal temperatures. The incorporation of phase change materials (PCMs) in device packaging also enhances efficiency in cooling. PCMs absorb and release thermal energy during phase transitions, and this can be beneficial in buffering temperature fluctuations. Not only does the selection of materials influence thermal management but also because it addresses overall reliability and longevity of electronic hardware, which is typically exposed to alternating thermal stresses while operational. A further important feature of passive cooling is the use of insulating materials in conjunction with conductive materials. Materials such as aerogels or expanded polystyrene may be utilized to minimize unwanted heat transfer from the environment so that internal heat still gets dissipated effectively. These insulators can be tuned to reflect or absorb specific wavelengths of infrared radiation so that their cooling effect is enhanced. By examining these interactions, device manufacturers can tailor their devices for many different applications and environments. The employment of these materials supports a more integrated device design, such that electronic components remain within their temperature safety margins and have longer lifetimes. Finally, the research into composite materials also offered new prospects for passive cooling technology. For instance, the application of high-conductivity materials combined with advanced coatings can be an efficient measure to enhance heat-dissipation characteristics. Research has suggested incorporating materials like graphene into polymer matrices to design light but effective thermal management systems. The combination of the material and polymer presents novel designs that can be tailored to individual needs and constraints in electronic devices. With the fast evolution of material science, it is increasingly possible to improve passive cooling techniques, which, subsequently, will result in the production of energy-efficient and stable electronic devices. Significantly, application of semipassive systems or heat pipes can maximize thermal regulation through the effective employment of these high-tech materials (Fritz et al., 2025).

7.2.1. Heat Sink Materials

The selection of heat sink materials plays a crucial role in the effective use of passive cooling techniques in electronic use. The use of aluminum and copper has been frequent due to their high thermal conductivity, which guarantees effective dissipation of heat. Aluminum is frequently used in consumer products due to its light nature and low cost, while copper, because it has improved thermal efficiency, is employed in high-performance systems. But with the development of new composite materials and nanomaterials, the heat sink technology landscape is beginning to get redefined. These new materials could enhance thermal performance at lower weight and size. Deployment of new materials for heat sinks is a significant addition to enhancing the cooling capability of electronic equipment in order to remain within safe temperature levels. Recent years have seen new materials show tremendous promise in passive cooling applications and hold great potential, competing favorably with the established aluminum and copper. Graphene, for instance, with its excellent thermal performance, has the potential to outperform conventional materials by a significant amount. Its high thermal conductivity ensures quick heat transfer, enhancing system reliability and efficiency. Another significant contender is ceramic materials, which provide high-temperature stability as well as corrosion resistance, making them applicable to extreme environments. These materials not only improve the passive cooling processes, but they also promote design flexibility that can be tailored to the continuously evolving requirements of electronic device miniaturization.

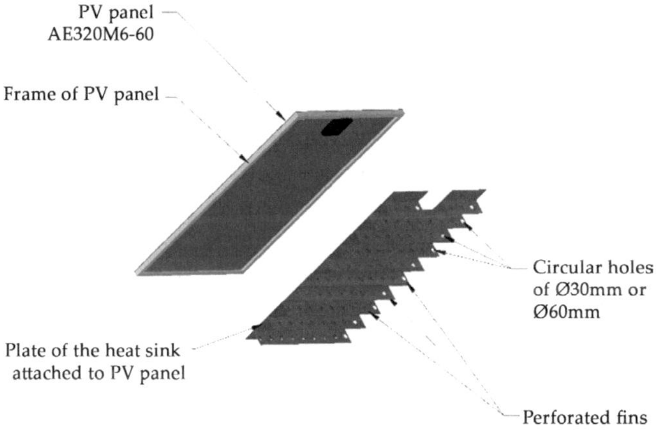

Figure 7.5. Overview of the components of a heat sink attached to a PV panel for cooling.

Source: https://www.mdpi.com/2076-3417/11/23/11323.

Employing these advanced materials can result in tremendous improvements in device life and performance, making them essential in today's thermal management approaches. Also, the integration of heat sink materials in passive cooling systems is a critical aspect of overcoming temperature management issues in electronic devices. As electronic components become more advanced, thermal management emerges as an increasingly pressing issue. Recent research demonstrates the inefficiencies of conventional systems, such as those being used in spacecraft, where significant amounts of energy are devoted to pumping and fan systems for rejecting heat. Other alternatives, such as air-cooled semipassive systems and heat pipes, have also emerged as viable solutions to enhance thermal control. For example, the use of heat pipes can better guide heat flow and distribute it evenly, thereby augmenting surface area heat rejection, as explained in recent research of advanced concepts in thermal management (Fritz et al., 2025). By choosing appropriate materials and configurations, manufacturers can maximize the efficiency of passive cooling solutions and render electronic devices sustainable for utilization in various applications (Hoell et al., 2019) (Figure 7.5).

7.2.2. Phase Change Materials (PCMs)

The application of phase change materials (PCMs) in passive cooling techniques for electronics is now more prevalent because PCMs can effectively manage thermal energy. PCMs do this by triggering latent heat and releasing it when they change phases, which allows for regulation of temperature within electronic systems. This is particularly beneficial where active cooling technology would be in the way or power-consuming. Through the incorporation of materials that undergo solid-to-liquid phase change, the excess heat generated by electronic devices can be suppressed, enhancing device performance and lifetime. The thermal management function of PCMs enables a more constant operating condition, which is the backbone of the dependability of sensitive electronic devices. Phase Change Materials can similarly be engineered to exhibit well-defined thermal properties appropriate to meet the demands of various applications. Recent developments in PCM technology have resulted in composite materials in which the benefits of traditional PCMs and high thermal conductivity are integrated. The result is better heat transfer and greater cooling efficiency. Research has established that the incorporation of high-performance PCMs can lower peak temperatures in electronic devices by a substantial amount, which facilitates extended operation times without overheating. With heightened growth of electronic devices, generating greater heat loads, specifically designed PCMs become vital in maintaining reliable thermal management systems. The ranges of the applications of PCMs may extend beyond traditional electronics to advanced fields such as optomechanical devices. The newly emerging field of optomechanical oscillators (OMOs) emphasizes the need for advanced thermal management techniques as these devices need precision in cooling for proper functioning. By making use of the special character of PCMs in optomechanical systems, systems can be designed not only to operate at appropriate thermal regimes but also to benefit from the latent heat storage character native to PCMs. This collaboration between new materials and sophisticated technology underscores the central contribution of

PCMs to the evolving field of passive cooling, marking an important milestone on the road to more sustainable electronic device design (Gerhard et al., 2025; Hossein-Zadeh et al., 2010) (Figure 7.6).

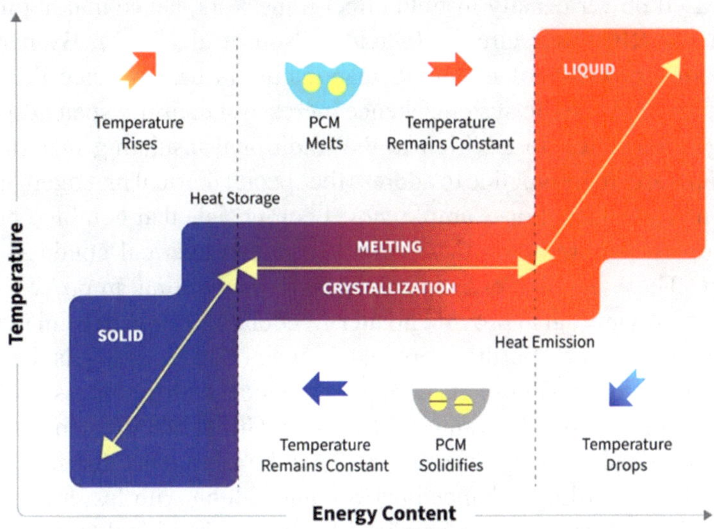

Figure 7.6. Working of phase change materials (PCMs).

Source: https://thermtest.com/phase-change-material-pcm.

7.2.3. Insulating Materials

Among passive cooling technologies employed for electronic systems, the choice of insulating materials plays an important role in managing thermal performance and the longevity of such systems. Insulating materials are applied to minimize heat transfer, maintaining sensitive electronic components at their optimal operating temperatures. Some examples are ceramics, polymers, and composites, which have been popular due to their high thermal resistance and low weight. Their presence in device design not only assists in managing thermal problems but also assists in improving overall efficiency by preventing energy losses. With electronics moving smaller and stronger, insulating material efficiency in heat dissipation without additional weight from active cooling technologies becomes vital to enhancing device performance and reliability. The most promising line for future insulating materials is the research on diamond-based composites, particularly applications in high-frequency electronics. As diamonds have better thermal conductivity and electrical insulation, it has unmatched potential in reducing thermal buildup in high-end devices. Diamond materials have been proven by existing research to be highly applicable in radio frequency (RF) devices, where there is an urgent necessity for efficient thermal management. Diamond's unique characteristics offer low signal loss and increased power density in field effect transistors, hence making it a good candidate for RF power circuits (Davidavidson et al., 2005). By harnessing the insulating property of diamond, manufacturers can enhance the thermal conductivity of electronic systems, hence increasing their life span despite high operating frequencies. In addition to the traditional insulating materials, new formulations are in the pipeline to address the special thermal management needs of modern electronics. For example, novel composites that combine polymers with nanoparticles are being developed to enhance thermal conductivity and insulation. These innovations, in addition to facilitating improved passive cooling performance, also provide greater freedom of design in miniaturization of electronic devices. Furthermore, such materials as aerogels have been reported to exhibit exceptional thermal resistance with lightness, and hence they are ideal for many electronic applications. Utilization of such exceptional insulating materials can significantly counter the heat generated by electronic elements, thus recording enhanced performance along with lesser dependency on the cooling systems that are heavier in nature. It is critical to explore these new frontiers in insulating materials so as to promote efficient passive cooling methods in the rapidly developing world of electronics (Gerhard et al., 2025).

7.3. DESIGN CONSIDERATIONS

In creating passive cooling solutions for electronic devices, one of the most important considerations is how heat that is produced through use may be dissipated efficiently. Effective thermal management is essential to maximize the lifespan and performance of electronic components. As electronic devices are miniaturized, dissipation of heat has led designers to look for new approaches. For instance, the use of the outside surface of a device to dissipate heat, as highlighted in thermal control systems research, can realize significant reduction in reliance on traditional active cooling. This conserves energy but also lowers mechanical failure points. Such aspects matter, particularly in high-performance applications where slight temperature swings impact significantly on performance (Fritz et al., 2025). The application of high thermal conductivity materials is a second critical design need. Graphite, aluminum, and yet more advanced composites are employed to make thermal transfer more efficient within electronic devices. The selection of these materials necessitates that a balance be established with weight and cost issues, which further complicates the design process. Also, ideal thermal contact among different components may be a key to avoiding hotspots that lead to device failure. Contemporary surface treatments and coatings are able to deliver enhanced thermal interface among parts, thus having a role in the general thermal management strategy. As with a detailed technical analysis of sensor technologies, the behavior of these types of materials and their arrangements is very crucial to advanced instrument and system design (Clark et al., 2025). Lastly, the ambient operating conditions are a significant contributing factor to passive cooling system design. The designers must analyze the anticipated heat loads and external temperatures where the device will be operated for its lifespan. Passive cooling systems that are implemented for extreme temperature or varied conditions might need to utilize hybrid systems that can be regulated to work in different situations. This entails consideration of airflow patterns, likely obstructions, and environmental interactions that could affect thermal performance. Focused consideration of such environmental factors makes it feasible to apply passive cooling mechanisms efficiently, as well as consistently reliable and efficient under different operating conditions. Thus, this multi-aspect design approach not only serves to enhance the lifetime of devices but also broadens the boundaries of electronic application in hostile environments (Fritz et al., 2025) (Figure 7.7).

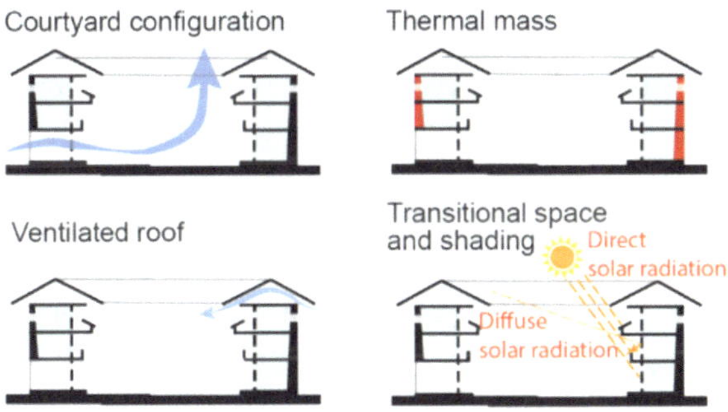

Figure 7.7. Passive cooling strategies.

Source: https://www.researchgate.net/figure/passive-cooling-strategies_fig3_344617668.

7.3.1. Device Layout and Orientation

Device orientation and layout are critical to maximizing the efficiency of passive cooling techniques. Proper device layout occurs when elements are placed judiciously to allow natural air flow to be enhanced, as well as heat dissipation. By placing heat-generating elements away from sensitive components, designers can leverage the use of convection currents to remove heat without wholly relying on active cooling techniques. This spatial strategy further becomes critical in complex systems where thermal management affects performance and reliability. For instance, in microfluidic devices discussed in modern literature, layout optimization has been shown to reduce the fluid channel length and improve area utilization, enabling better thermal management by removing hot spots (Brisk et al., 2019). These advancements not only add to cooling but also add to effective device operation, thus increasing the life of electronic systems. Direction, the other critical aspect of device design, has significant impacts on thermal dynamics. Devices must be placed such that they favor upward airflow and enable heated air to flow away easily. Where electronic components operate under high-stress conditions, i.e., high-rate data processing in particle detection systems, orientation becomes a very important issue. For example, in hybrid pixel detector fabrication in environments like the Large Hadron Collider, strong orientation is coupled with advanced cooling technologies to enable high-throughput data with radiation hardness (Garcia-Sciveres et al., 2018). This intentional focus on orientation, alongside designed component placement, promotes conditions that are conducive to successful passive cooling, highlighting the importance of comprehensive design approaches in thermal control. Additionally, associated layout and orientation are important to the ease of maintenance actions, which indirectly affect thermal performance. Well-planned access schemes for components minimize servicing work but can also ensure that cooling systems work efficiently in the long term. The need for practicality in passive cooling techniques becomes evident since devices are subject to wear and tear due to use stresses. If cooling paths are blocked, or access to significant components is complex, potential overheating risks increase significantly. Thus, the inclusion of the synergistic design and customer-oriented approach forms a design philosophy with an overemphasis on long-term performance, making optimal use of passive cooling methods while minimizing thermal management failure risks for electronic systems.

7.3.2. Surface Area Optimization

In the quest for more efficient passive cooling techniques in electronic devices, surface area optimization is a critical factor. The concept of this optimization is to maximize the area available for heat dissipation, which allows devices to maintain themselves at operating temperatures without active cooling mechanisms. As electronic components generate heat every time they are utilized, their ability to eliminate such heat successfully is directly proportional to their surface area. Thermal management can be improved by maximizing the surface area using various designs, such as the use of fins, heat sinks, or rough surfaces, all of which have higher effectiveness for thermal convection and radiation. Research at present suggests that optimizing surface area not only improves thermal management but also positively impacts electronic system longevity and reliability, contributing to better performance and efficiency in passive cooling solutions. Moreover, high-tech materials are also essential to optimizing surface area for passive cooling. Through the utilization of nanostructured materials, researchers are able to create surfaces with notoriously high heat transfer properties. These materials can be designed to possess personalized geometries and properties that enhance their ability to dissipate heat through conductions, convections, and radiations. For example, the development of hybrid structures that combine metals and polymers has been promising in improving thermal conductivity as well as surface area without adding significant weight. Such a marriage of innovative designs and leading-edge materials is an example of how surface area optimization can meet the rising thermal demands of today's electronic devices. By studying such materials, manufacturers can subsequently effectively address the issues presented by thin electronic devices and high-performance computation and reestablish the role of surface area in passive cooling strategies. The optimization of surface area in passive cooling systems of the future relies on integrated design mechanisms with thermal performance from the very outset of device development. Designers and engineers must collaborate to create not only devices that efficiently manage heat but also devices that optimize complete device architecture to control temperature. The combination of thermal modeling methods and simulation software yields precise estimates of how the various surface geometries affect heat dissipation. Additionally, emerging technologies such as additive manufacturing create opportunities to create surface structures that are impossible to manufacture using traditional methods. By embracing these innovations and continually enhancing surface area optimization methods, the electronics sector can produce devices that operate well at lower temperatures, thus guaranteeing user longevity and device sustainability in a temperature-sensitive technological world.

7.3.3. Integration with Existing Systems

Implementation of passive cooling techniques for electronic components in already installed systems is most essential in the alleviation of heat management issues. As electronic components evolve, the thermal management system must adapt to remain compatible with these developments without incurring costs or downtime. A case in point is incorporating ambient air-based systems into installed cooling systems by utilizing better airflow channels and thermal interface materials (TIMs). This synergy not only increases efficient heat dissipation but also enables the existing infrastructure to be utilized without heavy retrofitting. Further, when passive cooling methods are combined with active cooling measures, they can promote total performance by the virtue of complementing each other, leading to an optimal thermal condition that can support device longevity and reliability. Furthermore, materials science and technology advancements enable new strategies of passive cooling to be seamlessly incorporated into established systems. The optomechanical oscillator, as presented in recent works, emphasizes the potential for new thermal management methods to be infused into existing electronic systems (Hossein-Zadeh et al., 2010). These developments hold the promise of reducing energy consumption while enhancing cooling efficiencies, lowering the environmental impact of electronic equipment. More significantly, the integration of these new approaches with traditional systems yields a more flexible design that aligns with the evolving requirements of contemporary electronics. In this respect, the inclusion of passive cooling techniques is not simply an addition but a well-prepared improvement that enhances sustainability in electronic design. Lastly, the problem is one of maintaining compatibility upon integration while safeguarding system performance. These systems have had thermal management protocols set down, and any integration of new passive cooling must be conducted with these in consideration to avoid system failure. Proper analysis of the existing systems, their thermal loads, and material properties is crucial for successful integration. Thermal conductivity, expansion rates, and mechanical strengths of the parts used should be taken into account to achieve maximum functionality. The study of historical performance and operating statistics of earlier implementations can also have input into integration processes of today and further enhance applications tomorrow, as was demonstrated in earlier predictions of sensor systems emphasizing basic R&D cooperation with component performance parameters (Clark et al., 2025). Hence, the integration of passive cooling techniques with existing systems can be a source of creative yet practical solutions for the ever-evolving world of electronic devices.

7.4. ADVANTAGES OF PASSIVE COOLING

The advantages of passive cooling techniques for electronic hardware are extensive and are essential to the efficiency of today's technology. One of the primary advantages is energy conservation. As compared to active cooling techniques, which consume power in fans or pumps, passive cooling mechanisms take advantage of natural factors such as convection and conduction to dissipate heat. This elimination of electrical input not only reduces energy consumption but also reduces operating expenses, thus serving as an economically practical solution in domestic and commercial applications. Furthermore, with the emphasis on energy efficiency on the rise, passive cooling heralds an adherence to green practices, consistent with global efforts at reducing carbon emissions.

In addition to savings through energy, passive cooling enhances the reliability and lifespan of electronics. Active cooling part-based systems are likely to be mechanically failed due to the presence of moving parts, thereby leading to downtime and expensive repairs. Passive cooling structures like heat sinks and thermal mass strategies do not involve such delicate parts, hence lowering maintenance needs and extending device operating life.

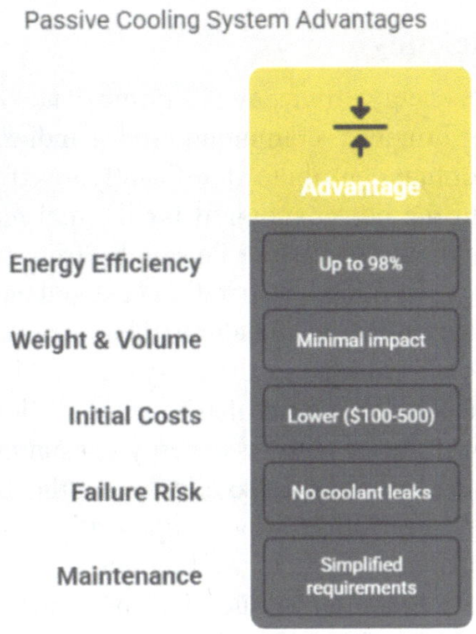

Figure 7.8. The advantages of passive cooling.

Source: https://media.licdn.com/dms/image/v2/D5612AQGctdOerOTOuQ/article-inline_image-shrink_1500_2232/B56ZfOzxmPHoAY-/0/1751521369064?e=2147483647&v=beta&t=ShPPobYaOzjH9V6_z_09L2hG-4xV9pk48N00CXz0Gvo.

Such reliability is particularly vital in high-performance applications such as space travel, where thermal control is vital for functionality. The study of advanced systems, such as air-cooled semipassive designs, is another show case of the viability of passive cooling, as highlighted through research on the efficiency of heat rejection systems in various environments (Fritz et al., 2025). Furthermore, passive cooling techniques lead to overall system weight and size minimization, an overall major consideration when designing electronic devices of small size. By utilizing naturally dissipating heat materials such as copper or aluminum, producers can reduce the amount of bulky cooling equipment that contributes to heavy weight. This is particularly beneficial in uses such as aerospace and hand-held electronics where lightening the device can enhance functionality and mobility. The design of systems incorporating solid-state image sensors is likely to be enhanced by passive cooling, availing themselves of their ability to attain working temperatures at no major cooling equipment requirements (Gerhard et al., 2025). Such developments point to the significance of passive cooling in advancing technology as it serves to overcome inherent heat dissipation issues in electronic devices (Figure 7.8).

7.4.1. Energy Efficiency

In electronic products, energy efficiency is a prime concern due to the growing demand for high-performance computing and handheld devices. Various methods of passive cooling contribute significantly towards enhancing energy efficiency by reducing the power required for thermal management systems. Traditional active cooling systems, like fans and refrigerants, would consume high amounts of energy, with higher operation cost and environmental impact. Passive cooling systems capitalize on natural phenomena such as radiation and convection to dissipate heat without extra energy usage. With the use of materials to release maximum heat or geometries that enhance air flow, these systems can maintain working temperatures with less energy consumption. Use of energy-efficient passive cooling techniques also aligns with the growing emphasis on sustainable technology innovation. Since organizations and companies want to meet environmental regulations as well as corporate social responsibility goals, incorporation of these techniques becomes necessary. For instance, employing thermally conductive materials can provide even heat distribution, which enables greater device life and performance. This technique not only reduces energy consumption but also decreases replacement frequency, creating a more environmentally friendly lifecycle for electronic devices. Additionally, advancements in material science, including phase change materials and metamaterials, have offered new opportunities to optimize thermal performance. Using these technologies, engineers are capable of creating devices that not only conserve energy but also improve the electronic ecosystem's strength and eco-friendliness. In addition, the financial advantages of energy-efficient passive cooling technologies are considerable. Organizations can realize considerable cost savings through decreased energy consumption and lower cooling system expenses. The initial cost of advanced materials or new design can be recovered by these long-term savings, making a more financially sustainable product. For example, Clark et al. (2025) is highlighting the implementation of next-generation sensor technologies based on efficient energy use, in the long term, favoring the producers and consumers themselves. In addition, as the operating cost decreases, the competitive edge of companies that use passive cooling may grow in the market while enhancing their profitability as a whole. Finally, the pursuit for efficiency in energy through passive cooling not only advances the aim for the environment, but it also spurred economic gains, propelling electronics development towards sustainability.

7.4.2. Cost-Effectiveness

An evaluation of possible cost-effectiveness of passive cooling techniques reveals substantial advantage over traditional cooling systems. Using methods like heat pipes and free cooling, electronic components can be made more energy-efficient with reduced reliance on energy-hungry mechanical systems. For instance, incorporation of heat pipes has been seen to enable the substantial reduction of energy consumption, as shown by a study indicating 75% possible energy saving through such systems in data centers. Apart from saving on energy costs, the passive technologies also extend the lifetime of electronic components by insuring optimal operating temperatures, lowering thermal stress, and enhancing overall reliability. In tandem with the growing emphasis on sustainability and environmental responsibility, the economic benefits of passive cooling are all the more compelling, making it a forward-thinking technology for future electronic device uses. Passive cooling methods are particularly well-suited to use in those industries that are burgeoning at their fastest rates, such as data storage and telecommunications. The industry must contend with a growing necessity for cooling methods without sacrificing too much in the way of cost or environmental response. Application of heat-pipe based systems, as suggested in industry analysis, assists institutions in overcoming the aforementioned challenges. These systems handle heat rejection better, significantly reducing operation overhead from traditional cooling. The low energy consumption focus is not only translated into lower utility costs but also extended financial gains. Consequently, the transition towards passive cooling is not only enhanced device performance, but also a significant direction towards economic sustainability in the fast-evolving tech climate, justifying the call for cooling technology innovation. The juxtaposition of cutting-edge cooling methods also speaks to stark differences in working efficiency and economic costs. Experiments with semipassive heat transfer systems, such as those used in spacecraft, show that traditional methods have high energy costs for thermal management. For instance, the deployment of pumped-loop systems by the shuttle space laboratory is one of the most inefficient methods, whereby 1 kW of pumps and fans is needed for every 5 kW of heat rejection. New technology that employs natural dissipation processes of heat, streamlining concepts like air cooling and heat pipes, offer better models of efficiency. Drying up the need for active components, such alternatives offer a cheaper and sustainable model for electronic system thermal management.

With the advancement of the technology ecosystem, pursuing increasingly cost-reducing and energy-conserving cooling strategies will remain the main driver in shaping the future of electronics.

7.4.3. Environmental Impact

The environmental impact of electronic equipment is becoming a concern with the need for computing power continuing to escalate. Traditional cooling methods are primarily energy-consuming systems, and they pose a significant threat to sustainability in terms of enhanced energy consumption and greenhouse gas emissions. Passive cooling provides a new approach that has the capacity to effectively stem these threats. Employing natural convection and thermal radiation to dissipate heat, such methods eliminate the usage of mechanical cooling systems, thereby reducing the energy profile of electronic devices. Passive cooling solutions further assist in promoting circular economies by restricting the waste caused by high-energy cooling systems, contributing towards the development of more sustainable technological environments in the long term. In-depth analyses of advanced passive cooling methods elucidate that their strength resides in their potential to preserve the environment. For instance, one such study in space application research established that conventional spacecraft heat rejection systems are inefficient, under enormous energy inputs to control temperatures (Fritz et al., 2025). However, innovative configurations such as air-cooled semipassive designs can effectively reject heat from devices through the outer casing, with the use of external environments to maintain optimal operating temperatures without the use of much energy consumption. By applying the same tactics to business electronics, manufacturers can reduce carbon footprints by quite a substantial amount without sacrificing performance. Such a change not only is in line with global goals for sustainability but also creates the catalyst for more environmentally friendly technology in other industries. Moreover, incorporating new materials and technologies can further optimize the efficiency of passive cooling techniques, thereby further boosting their environmentally positive aspects. For instance, aerospace applications of fiber optic technologies can potentially improve thermal management processes (Glasheen et al., 2025). Passive cooling methods have the potential to generate impressive weight savings and efficiency improvements over conventional techniques, ultimately translating into reduced energy demands for electronic systems. Advanced materials and designs have the potential to make passive cooling methods propel the electronics industry towards an eco-friendlier future. It is therefore important as we go deeper into cooling technologies that solutions come first that seek to be environmentally friendly, meaning that the advancement of electronic devices comes not at the expense of our environment.

7.5. LIMITATIONS OF PASSIVE COOLING

The law of passive cooling is fundamentally limited by the reliance on ambient conditions, which will be highly variable and unpredictable. Unlike active cooling systems, which provide uniform temperature control regardless of environmental conditions, passive cooling systems are highly dependent on natural processes such as convection, conduction, and radiation. Such methods perform optimally with naturally cooled systems, a definite setback in situations where ambient temperatures are high to begin with. Thus, electronic devices may heat up during high usage or harsh climatic conditions and consequently degrade in performance and longevity. This vulnerability to external variable conditions attests to a major disadvantage of passive cooling systems in the area of offering uniform thermal management to sensitive electronic components. The other significant limitation of passive cooling methods is their design and material constraints.

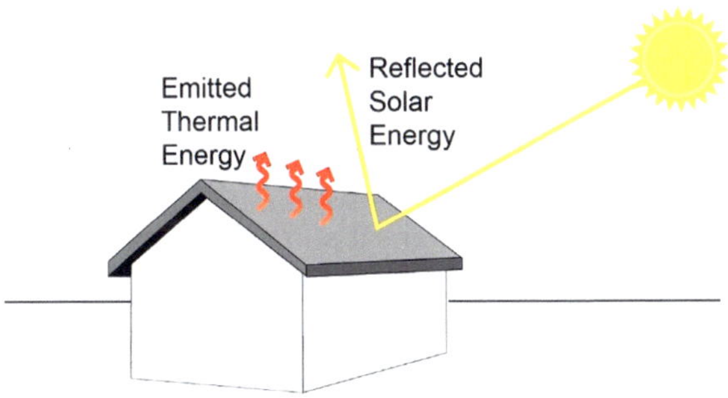

Figure 7.9. The concept of passive cooling is fundamentally limited by its reliance on environmental conditions.

Source: https://encrypted-tbn0.gstatic.com/images?q =tbn: ANd9GcTqcxl0VS2 Mhq-Qgq-YRoCvNJfhcf_Ebu2khw&s.

For instance, the effectiveness of passive cooling methods could be dependent on the utilization of materials with proper thermal conductivity and emissivity. In the majority of cases, the integration of such materials into small electronic designs might pose size and weight compromises, which could influence the overall functionality and handheld portability of the device. Moreover, certain passive techniques such as heat spreaders or heat sinks may require huge space allowances, which cannot be controlled in miniaturized systems. This is an issue that points towards the need for ongoing advancement in material science and engineering to achieve more efficient passive cooling without compromising design rules. Thus, optimization of passive cooling technology is a multi-dimensioned problem that needs interdisciplinarity. Finally, passive cooling's time constraints also make them more sophisticated in heat control. While active cooling mechanisms could directly react to temperature changes through mechanical action, passive systems lag behind by nature of physical law-based operation. Such latency could result in devastating overheating for short bursts of activity or high-density work durations. In addition, some of these emerging technologies, including those developed under cryogenics and sensors on spacecraft, emphasize the disparity in cooling demands that passive systems are unable to provide uniformly. As argued by Abbé et al. (2013) and Clark et al. (2025), such technologies would require more sophisticated thermal management systems than passive cooling allows, eventually suggesting that a hybrid system can be utilized to efficiently meet the aggressive thermal management needs of today's electronic systems (Figure 7.9).

7.5.1. Performance Constraints

In creating and applying passive cooling structures to electronic equipment, performance limits are highly critical to know to provide efficiency and durability. These limits are typically based on material characteristics, environmental conditions, and the specific operating demands of the devices themselves. Thermal conductivity of materials for applications within passive cooling structures, for instance, like heat sinks or thermal fins, is a significant consideration in terms of efficient heat dissipation. This issue is intensified in miniature electronic devices where size constraints are limited. Engineers are then compelled to operate under such constraints very keenly, matching the requirements for effective cooling mechanisms within limit constraints of size. Material and technological decisions thus have direct implications on performance outcomes, which calls for thorough investigation of the applicable technologies and methodologies at hand to bridge these gaps. Besides, application of innovative technologies is paramount in opposing performance constraints in passive cooling systems. For example, research in fields like cryogenics and optical design is key to creating more efficient cooling systems. According to stipulations in previous studies, through the provision of information, they facilitate the creation of innovative designs that render thermal management systems more effective (Clark et al., 2025). By means of application of such sophisticated strategies, passive cooling mechanisms can be constructed by engineers that not only meet current standards of performance but also anticipate future demands from developing electronic devices. Hence, ongoing research on emerging technologies and methodologies is essential in overcoming current performance barriers that hinder the effectiveness of passive cooling techniques. Last but not least, environmental conditions involved in passive cooling techniques cannot be neglected while defining performance barriers. Ambient temperature fluctuations, humidity, and airflow conditions directly influence thermal management of electronics. Practically speaking, climatic conditions may increase or decrease heat in electronic systems, making innovative cooling design relative to specific environments (Appleby et al., 2025). Engineers must design adaptive passive cooling techniques that can maximize performance in different conditions. This demands intimate familiarity with thermodynamic theory and the ability to implement such principles effectively in terms of the context of heterogeneous working environments. Minimization of these limitations guarantees device dependability as well as contributes to the overall prospect of sustainable electronic design and hence makes the investigation of environmental impacts vital to modern research paradigms.

7.5.2. Dependence on Environmental Factors

The environmental sensitivity of electronic devices cannot be stressed enough, particularly when considering passive cooling mechanisms. Optimum thermal management is necessary to provide maximum performance and prolong the life of electronic systems. Variations in environmental temperature, humidity, and airflow play a significant role in affecting the efficiency of passive cooling systems, which rely on natural heat dissipation compared to active cooling systems. As an example, in the regions with temperatures that are very high or very low, the passive cooling techniques may not be able to maintain the proper thermal thresholds, and the electronic components get heated up and destroyed. Thus, the local environment must be known in order to design and realize effective passive cooling techniques for electronic components. One of the interesting aspects of passive cooling in electronics is that it relies on certain material characteristics that can be modulated by surrounding environmental conditions. Recent research puts the advantages of utilizing materials whose thermal conductivity is low into the foreground, particularly where active cooling is not feasible. Bulk acoustic wave (BAW) resonators are among promising solutions because their operation could be significantly improved when these are optimized for low temperature. Under cryogenic conditions, such as those achieved at 20 mK, these resonators are characterized by very low loss rates, resulting in excellent Q-factors for maximizing their performance (Bourquin et al., 2012). This shows a direct correspondence between environmental conditions and operational efficiency of electronic systems, illustrating how cooling techniques can be maximized to leverage environmental conditions towards better thermal management. Furthermore, electronic equipment spatial location also shows dependency on climatological conditions in passive cooling systems. Warmer ambient areas might need novel architectural systems, for instance, the integration of heat sinks or cool surface thermal reflection coatings. Passive cooling of electronics in warm climates is an active field that requires continuous research to overcome challenges posed by increased heat production in dense electronics. Insightful knowledge derived from studies on materials and structures tailored to certain environmental conditions can lead to adaptive cooling devices. The example of the work with acoustic wave oscillators demonstrates how boosting performance parameters through interaction with the environment not only optimizes thermal management but may also pave the way for advancements in quantum technology (Abbé et al., 2013). The integration of such data into the design process will ensure that passive cooling techniques still perform optimally under a broad spectrum of environmental conditions.

7.5.3. Size and Space Limitations

For passive cooling schemes for electronic systems, there are space and size limitations that are big issues to be addressed with care. As electronic devices grow smaller in size, heat generated while they are put to use also gets exacerbated. Hence, effective cooling strategies must be conceived to fit these constrained spaces without compromising on performance. Engineers are challenged to create cooling technologies not only that address thermal management requirements but also meet strict spatial constraints. The pressure for miniaturization has the tendency to lead to increases in temperatures above tolerance levels, and incorporating passive cooling methods has become increasingly necessary in the design of devices in current times. Moreover, the development of passive cooling systems calls for a proper understanding of the thermodynamic characteristics of materials in limited spatial areas. When applying them, thermal conductivity, surface area, and airflow are key factors to consider. Overlooking these factors can render a cooling method ineffective. Research indicates that having access to information on the thermal performance of materials plays an important part in the efficiency of passive cooling systems, particularly in small spaces where the use of conventional methods may not be feasible. This importance is evidenced by findings that depict the influence of terrain surface temperatures in global thermal management of electronics, as presented by Braithwaite et al. (2025). Also, the efficiency of passive cooling is impacted by design approaches employing modules that are applied in electronic devices for reducing size complications. Modular design provides the flexibility to change spatial compositions of the thermal management systems in spatial constraints of a confined space. This method not only reduces the inventory needs for equipment but also enhances the adaptability of systems to diverse operating conditions, as attested by parallel research in low-gravity experiment arrangements where space is even more constrained, as aptly noted by Gorham et al. (2025). By using modular designs and considering the ramifications of size and spatial limitation, engineers can create passive cooling solutions that offer the robustness and dependability of electronic gear in a harsher technological landscape.

7.6. COMPARISON WITH ACTIVE COOLING METHODS

In electronics thermal management, passive cooling techniques present an appealing alternative to active cooling technologies. Active cooling technologies typically involve mechanical components such as fans and pumps that form the center of optimizing heat dissipation. However, such an approach is its own inefficiency, as has been the case with thermal control issues encountered in such complicated systems as shuttle spacecraft, where the process does lose a lot of energy during heat rejection and not effective heat dissipation (Fritz et al., 2025). Passive cooling, conversely, exploits natural phenomena such as convection, radiation, and conduction in rejecting heat without augmenting energy supply. This built-in efficiency makes passive cooling not only a greener alternative but also reduces the overall operational cost of maintaining electronic systems. The dependability and reduced maintenance are two of the key advantages that passive cooling has over its active counterparts. Active cooling mechanisms are subjected to mechanical degradation and wear, leading to system failure and more downtime as well as potential damage to electronic components. The likelihood of mechanical failure reduces substantially with passive cooling approaches, which employ stationary materials and geometries to facilitate heat transport. In addition, as these systems do not employ moving parts, they tend to be quieter, a valuable property in noise-polluted environments like offices or hospitals. Therefore, the use of passive cooling enhances operational dependability as well as provides a more quiet and friendly user environment for electronic devices. Furthermore, the efficiency of passive cooling methods can also be further illustrated by the increasing trend towards advanced design systems, particularly within high-risk environments such as space technology. Research on next-generation passive cooling concepts, like air-cooled semipassive and heat pipe, characterizes the potential for improved thermal performance with the absence of the heavy energy weight and energy needs of active systems (Clark et al., 2025). Sophisticated design utilizes exterior surface areas to the vehicle to reject heat and employs high-tech materials to facilitate thermal conduction. As technology advances, there is on-going transition of balance between active and passive cooling paradigms, which suggests an imminent future where passive cooling will become the major force, particularly in matters of energy efficiency and system lifetime. In general, relative benefits highlight the categorical nature of passive cooling to dictate next-generation electronic device thermal management paradigms (Figure 7.10).

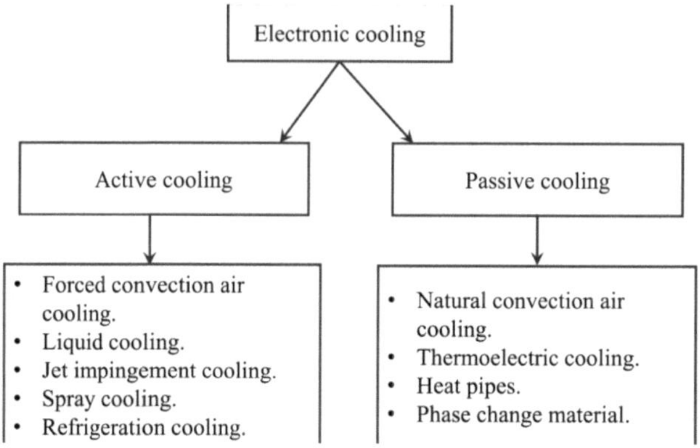

Figure 7.10. Comparison with active cooling methods and passive cooling.

Source: https://www.researchgate.net/profile/Nitin-Ambhore-2/publication/ 375685927/figure/fig1/AS:11431281205160470@1700141457220/Different-techniques-available-for-electronics-cooling.png.

7.6.1. Efficiency and Effectiveness

In cooling electronics, effectiveness and efficiency are the buzzwords. Passive cooling techniques, involving those utilizing heat pipes, have turned out to be reliable solutions for managing heat dissipation in micro- and nano-scale devices. Heat pipes remove heat from the source by using liquid/vapor phase-change mechanisms to a point where it can be dissipated more efficiently. The application of micro-fabricated heat pipes has significantly enhanced the thermal managing capability of such devices with excellent efficiency. Studies indicate that nanoscale heat pipe performance can be optimized with optimal regard for variables such as filling ratio and heat load, revealing an important design versus operating effectiveness interplay within electronic devices (Akkus et al., 2018). This efficient heat transfer not only extends the life of devices but also contributes to overall energy conservation in electronic systems. Additionally, the efficiency of passive cooling can be better illustrated in the data center environment where thermal management is crucial in ensuring operational stability. As the data storage environment is evolving, the growing demand for energy to cool these facilities is posing increasing challenges. The use of heat-pipe-based heat exchangers has been a promising alternative in this area, offering an environmentally friendly method of cooling. Another recent research found the potential for maximizing 75% in energy savings through free cooling from heat pipe technologies (Jouhara et al., 2014). In a demanding way, the intersection of energy use minimization and maximum cooling performance emphasizes the twin goals of efficiency and effectiveness to be realized in these passive approaches. But achieving levels of efficiency and effectiveness is by no means simple. Passive cooling system design demands a detailed understanding of thermal dynamics, material properties, and environmental considerations. For instance, the geometrical parameters of nanoscale heat pipes, including cross-sectional dimensions and aspect ratios, are vital determinants of their efficiency. Optimization of these parameters maximizes both thermal conductivity and fluid mechanics, resulting in enhanced heat management outcomes. As electronic devices get smaller but more powerful with each day that passes, the demand for innovative passive cooling technologies will keep increasing. By addressing these design problems, future passive cooling innovations will extend the efficiency and efficacy, ultimately establishing a sustainable operation paradigm for electronic devices on various applications (Akkus et al., 2018).

7.6.2. Maintenance and Reliability

Reliability of electronics and maintenance are key to making such devices last longer and perform well, particularly if passive cooling techniques are utilized. Adequate thermal management does not only enhance the life of the devices but also lowers potential risk associated with overheating, which can lead to costly equipment damage. The use of passive cooling components, such as heat sinks and natural convection elements, contributes to the saving of energy and the generation of a greater mean time between failures (MTBF). Lastly, a good cooling solution contributes to the overall systems performance by maintaining working temperatures within safe margins, thus ensuring the integrity of sensitive electronic components. Furthermore, the implementation of strict maintenance processes is fundamental to ensuring the efficiency of passive cooling systems over the long term. Regular inspections and analysis of cooling systems allow for wear or clogging that can intrude upon thermal dissipation to be detected early. This proactive approach not only enhances reliability but also minimizes the likelihood of unplanned downtime — a critical consideration in high-risk operation environments. By the application of advanced monitoring tools and the integration of systems knowledge such as those outlined by Glasheen et al. (2025), companies are able to develop an integrated maintenance plan that aims at maximizing cooling efficiencies as well as the life cycle of electronic equipment. Research points out that cooling methodology reliability has a direct impact on the overall performance of electronic systems. In addition, compliance with pre-established protocols during the design and implementation phases of cooling systems should be observed, as illustrated in projects such as those described by Kaushagen et al. (2025). Application of new cooling methods should be balanced with careful maintenance planning to ensure reliability throughout the entire life cycle of electronic equipment. By prioritizing maintenance as well as development in passive cooling technology, not just are organizations' investments conserved but they also enhance thermal stability in their electronics, thereby minimizing the likelihood of thermal-related failures. Implementing effective cooling practices and rigorous maintenance procedures can be instrumental in inducing an enormous change in operational efficiency and reliability.

7.6.3. Cost Analysis

For electronic device passive cooling technologies, the overall cost analysis is critical in determining the feasibility of implementation. Passive cooling technologies such as the use of heat sinks, thermal management coatings, and natural ventilation normally have relatively lower upfront costs compared to traditional active cooling systems. However, operation over the long term is critical when estimating the overall lifecycle costs of such technologies. An analysis of the upfront installation, maintenance, and potential energy saving must elucidate the economic advantages offered by passive cooling measures. Design complexity and material considerations will influence upfront costs, which would need to be weighed meticulously against active cooling options, which tend to be linked to higher energy consumption and maintenance costs. In addition, understanding the broader implications of passive cooling technologies for efficiency in operation explains their cost-effectiveness for the electronic device market. For instance, natural air-flow systems to cool systems could significantly lower energy consumption, leading to long-term savings. There is evidence that innovations such as heat-pipe technologies can realize significant declines in energy consumption, recording up to 75% savings on energy while sustaining maximum cooling efficiency (Jouhara et al., 2014). The declines result in reduced electricity bills and environmental effects, thereby further increasing the general economic feasibility of adopting passive cooling. By quantifying these possible savings, organizations can better counteract the shift away from traditional cooling technologies and towards passive ones, thus encouraging a more environmentally friendly practice for electronic device maintenance. Furthermore, examining the cost drivers of passive cooling in conjunction with global trends indicates a shift in paradigm for electronic device manufacturing and data handling. As the industry is increasingly subjected to legislative pressure for the reduction of carbon footprints, green cooling practices become not only a moral necessity but also an economic solution. The widespread adoption of heat-exchanger systems can show the trend whereby companies can steer clear of hefty bills associated with conventional air conditioning by utilizing passive solutions (Clark et al., 2025). This strategic realignment puts focus on the double benefit of cost reduction and environmental compliance, commenting on the potential of passive cooling strategies revolutionizing cost paradigms in the technology sector. As businesses weather this period, a particular cost segmentation becomes critical in formulating the most appropriate cooling strategies that are in alignment with economic and environmental requirements.

SUMMARY

This research into passive cooling techniques for electronic equipment is meant to highlight the imperative for innovative solutions in thermal management in an expanding high-performance technology context. Such techniques, based on natural processes of heat dissipation, are significant improvements on traditional active cooling techniques. Passive cooling not only conserves energy but also reduces the necessity for mechanical parts, thereby optimizing electronic system reliability and longevity. As demand for more efficient thermal management methods increases, their integration into design models will be fundamental. Therefore, their thorough evaluation and integration into design models will be critical in electronic device performance enhancement. Future research must continue exploring different passive cooling methods to achieve optimal thermal control while promoting sustainability and efficiency. The effectiveness of passive cooling techniques can also be established via their application in other hazardous environments, including the thermal control systems of spacecraft. For instance, studies regarding heat rejection techniques in spacecraft show that existing systems have a tendency to be inefficient in terms of discharging huge thermal loads. Research has indicated that new passive cooling systems, employing air-cooled designs that capitalize on exterior skin rejection and heat pipes designs, can dramatically increase thermal efficiency without maximizing energy needs. Adopting such methodologies can empower future designs to achieve a better balance between heat extraction and energy efficiency. This advance in cooling technologies underscores the potential of passive approaches as viable solutions to high-performance electronics, especially where reliability and sustainability are key issues. Last but not least, the adoption of passive cooling techniques must consider the new advancements in allied areas, such as optomechanics and microresonators, that can impact thermal management strategies. Recent work in optical microresonators demonstrates that advancements in photonic devices, including optomechanical oscillators (OMOs), have new functionalities that can potentially support passive cooling ideas. The interaction between thermal and photonic technologies is an exciting frontier to be explored and utilized, paving the way towards more efficient and responsive thermal control mechanisms. By promoting interdisciplinary interaction, the ongoing development of passive cooling systems can benefit from insights evolved from sophisticated technology and ultimately contribute toward better solutions for modern electronics. The future of thermal management is dependent on embracing these new methods toward high-performance, effective, and sustainable electronic devices.

Effective cooling techniques are vital in the prolongation of lifespan and performance of electrical appliances. Passive cooling systems, using natural occurrences to conduct heat without working parts like fans or pumps, provide an eco-friendly alternative to conventional cooling systems. Convection, conduction, and radiation are the primary methods of passive cooling. All of these exploit inherent physical laws in managing thermal energy. For instance, convection employs air or fluid flow to transfer heat away from an appliance, while conduction employs the passage of heat through materials of varying thermal conductance. Radiation employs surface radiation of heat in a bid to regulate temperature. It is crucial to understand these processes when designing electronics for automation that can function efficiently under various environments, translating further into sustainability and energy conservation in the technology world. Material selection is the key to passive cooling performance. Some TIMs have been found to improve heat transfer capabilities significantly to make thermal management possible.

The material must be highly thermally conductive but very light in order not to compromise the overall design of the devices for passive cooling to be effective. Developments in the field of materials science have led to the development of composites and phase change materials that efficiently absorb and dissipate heat under various operating conditions. Besides this, the use of heat sinks and fins can augment the surface area and even improve heat dissipation. The use of such materials in electronic device design maximizes performance and minimizes energy consumption, thereby meeting today's sustainability goals in electronics manufacturing, according to Kaushagen et al. (2025).

Furthermore, the use of smart design approaches is fundamental in optimizing passive cooling systems. This involves an integrated approach to device placement, where devices are placed to ensure a maximum amount of airflow and heat distribution. For instance, keeping heat-generating components at a distance from sensitive electronics minimizes the likelihood of overheating and enhances overall system reliability. Additionally, making enclosures natural-flow compatible can effectively lower thermal buildup within the device. Applying thermal simulation software to the design process can also help engineers make predictions about thermal performance and locate components accordingly. Through an effective intermingling of design concepts and material physics, passive cooling techniques can be optimized for a variety of applications, leading to improved operational reliability, as indicated by Hoell et al. (2019). Such holistic appreciation of passive cooling not just maintains device longevity but also maintains ultimate goals of energy efficiency and environmental responsibility throughout the electronic sector.

REFERENCES

1. Abbé, P., Galliou, S., Goryachev, M., Imbaud, G., Bourquin, R., & Creedon, D. L. (2013). Advances in development of quartz crystal oscillators at liquid helium temperatures. http://arxiv.org/abs/1309.5445 (accessed on 23rd October 2025).
2. Akkus, Y., Beskok, A., Celebi, A. T., Nguyen, N., Wang, E., & Acar, M. (2018). A first look at the performance of nano-grooved heat pipes. http://arxiv.org/abs/1807.07491 (accessed on 23rd October 2025).
3. Appleby, J. F. (2025). Workshop on Advanced Technologies for Planetary Instruments, Part 1. https://core.ac.uk/download/pdf/42806195.pdf (accessed on 23rd October 2025).
4. Bourquin, R., Creedon, D. L., Galliou, S., Goryachev, M., & Tobar, M. E. (2012). Extremely low-loss acoustic phonons in a quartz bulk acoustic wave resonator at millikelvin temperature. http://arxiv.org/abs/1202.4556 (accessed on 23rd October 2025).
5. Braithwaite, J., Brown, W., & Cook, J. J. (2025). Infrared System Studies for the Earth Resource Program: Final Report. https://core.ac.uk/download/pdf/85242914.pdf (accessed on 23rd October 2025).
6. Brisk, P., Crites, B., & Kong, K. (2019). Directed Placement for mVLSI Devices. https://core.ac.uk/download/323311285.pdf (accessed on 23rd October 2025).
7. Clark, B. P., & Kostiuk, T. (2025). Spaceborne Sensors (1983–2000 AD): A Forecast of Technology. https://core.ac.uk/download/pdf/42848800.pdf (accessed on 23rd October 2025).
8. Davidson, J. L., Esame, İ. O., Gürbüz, Y., Kang, S., & Acar, M. (2005). Diamond semiconductor technology for RF device applications. https://core.ac.uk/download/11738207.pdf (accessed on 23rd October 2025).
9. Fayaz, M., Lajis, A., Nasir, H., Shah, N. S. M., Wahab, Y., & Rahim, M. F. A. (2019). A review on energy consumption optimization techniques in IoT-based smart building environments. https://core.ac.uk/download/300473894.pdf (accessed on 23rd October 2025).
10. Feng, W., Hao, B., Ji, H., Lau, A. K., & Li, D. H. W. (2019). A review of net zero energy buildings in hot and humid climates: Experience learned from 34 case study buildings. https://core.ac.uk/download/323310631.pdf (accessed on 23rd October 2025).
11. Fisher, D., Hand, J., Liesen, R., Wilson, E., & Taylor, Z. (2011). Screening of energy efficient technologies for industrial buildings' retrofit. https://core.ac.uk/download/9033399.pdf (accessed on 23rd October 2025).

12. Fritz, C. H. (2025). Advanced Spacecraft Thermal Control Techniques. https://core.ac.uk/download/pdf/42876319.pdf (accessed on 23rd October 2025).
13. Garcia-Sciveres, M., & Wermes, N. (2018). A review of advances in pixel detectors for experiments with high rate and radiation. http://arxiv.org/abs/1705.10150 (accessed on 23rd October 2025).
14. Gerhard, G. C., Glusick, R. E., Razi, S., Stewart, D., & King, L. (2025). Solid State Image Sensor Research, Phase I. https://core.ac.uk/download/pdf/85248506.pdf (accessed on 23rd October 2025).
15. Hoell, K., McKay, C., Migneau, M., Mullen, C., et al. (2019). Development of innovative heat sinks for power electronics cooling within the more electrical aircraft. https://core.ac.uk/download/294830160.pdf (accessed on 23rd October 2025).
16. Hossein-Zadeh, M., & Vahala, K. J. (2010). An optomechanical oscillator on a silicon chip. https://core.ac.uk/download/4884077.pdf (accessed on 23rd October 2025).
17. Jouhara, H., & Meskimmon, R. (2014). Heat pipe based thermal management systems for energy-efficient data centers. https://core.ac.uk/download/30339678.pdf (accessed on 23rd October 2025).
18. Kaushagen, H. N., & Skinner, K. L. (2025). Implementations Guidelines, Airborne Evaluation Equipment, Advanced System Checkout Design, Phase B: Final Report, 29 Jun. 1965–29 Jul. 1966. https://core.ac.uk/download/pdf/85249966.pdf (accessed on 23rd October 2025).
19. Lopes, A., MacDonald, E., Medina, F., Navarrete, M., et al. (2006). Expanding rapid prototyping for electronic systems integration of arbitrary form. https://core.ac.uk/download/287648527.pdf (accessed on 23rd October 2025).
20. Moy, D. (2025). VEIL Ballarat Center Structural Plan. https://core.ac.uk/download/pdf/30678365.pdf (accessed on 23rd October 2025).
21. Shaker, S. F. (2014). Thermal and flow analysis of piezoelectric fans for cooling LEDs packages. https://core.ac.uk/download/32600178.pdf (accessed on 23rd October 2025).
22. Swann, M. K. (2025). Magnetic Bearings: Fifty Years of Progress. https://core.ac.uk/download/pdf/42806500.pdf (accessed on 23rd October 2025).

CHAPTER 8

HEAT SINKS AND THERMAL INTERFACE MATERIALS (TIMS)

Contents

INTRODUCTION

The value of thermal interface materials (TIMs) and heat sinks in modern electronic devices cannot be overemphasized. With electronics decreasing in size but increasing in power, generated heat needs to be dissipated efficiently so that performance and reliability are assured. Heat sinks are the primary thermal dissipation components, conducting and dispersing heat away from sensitive electronic devices. These are complemented with TIMs, the materials that improve maximum thermal conduction between surfaces. These components guarantee maximum efficient heat transfer. Without the incorporation of such technologies, excessive heat leads to thermal throttling, reduced performance, and ultimately failure of the device. This is the reason why the realization of their significance makes the foundation for driving electronic thermal management solutions. The recent advances in materials science have introduced new technologies for improving the performance of heat sinks and TIMs. Research reports the potential of using new materials such as metal foams and metal matrix composites (MMCs) that possess matured properties to use in heat exchangers. These materials show great promise for improved thermal conductivity and strength for use in heat sinks and alternative heat exchanger configurations. As a case in point, metal foams, as light and porous materials, can greatly increase surface area, improving heat dissipation rates. As demonstrated in current literature, judicious application of such engineered materials can significantly improve heat management practices (Han et al., 2012). This breakthrough serves to underscore the importance of ongoing research and development in this essential field of electronics. Furthermore, the application of advanced technologies in the design and production of heat sinks and TIMs is crucial to meet the demands of current applications. Developments in microfabrication technologies and nanostructured materials such as carbon nanotubes and conductive nanowires could enhance thermal transport efficiency (Emerson et al., 2009). The materials possess enhanced heat conduction characteristics that can reduce thermal resistance by several orders of magnitude at interfaces. With continued development of devices beyond the boundaries of miniaturization and power density, traditional materials may struggle to keep up with performance requirements. Therefore, ongoing exploration of functional nanostructures as well as absolute familiarity with energy transfer phenomena at the microscale is an area of utmost concern for scientists endeavoring to revolutionize thermal management in electronics. Through these innovations, the underlying issues of heat dissipation and thermal interface engineering are resolved successfully, paving way to a new class of reliable electronic devices.

Definition of Heat Sinks

Heat sinks are at the forefront of electronic device thermal management, especially as demand for high-performance systems continues to escalate. As passive heat exchangers, heat sinks are designed to transfer the heat generated by electronic components to the surrounding ambient environment and thereby prevent overheating and allow maximum operation. The devices normally consist of materials with high thermal conductivity, such as aluminum or copper, from which there is efficient heat transfer. Structure of heat sink can also be drastically different, with some bearing intricate designs such as fins that maximize surface area and enhance heat dissipative features. At high-performance applications, utilization of innovative materials such as carbon allotropes, as currently studied, has introduced new dimensions for thermal management enhancement in heat sink design, expanding the range of conductive properties by an extensive margin (Balandin et al., 2011). The efficiency of heat sinks is largely a function of their interaction with TIMs, utilized to bridge gaps between electronic devices and heat sinks. Maximizing the thermal contact, TIMs significantly reduce thermal resistance, hence improving efficiency in overall heat transfer. With the present era of microelectronics miniaturization, where thermal management innovation is needed, selecting the correct TIMs becomes even more critical. There are various types of TIMs, ranging from thermal pastes to phase-change materials, each having inherent properties suited for particular applications. However, traditional tests assuming flat surfaces do not reflect real-world scenarios where surfaces are not parallel (N/A, 2011) (Figure 8.1).

Figure 8.1. Heat sink.

Source: https://www.trentonsystems.com/hs-fs/hubfs/Heat%20Sink%20Computers%20.jpeg?width=6000&name=Heat%20Sink%20Computers%20.jpeg.

Thus, it is imperative to create robust methodologies for TIM characterization under various conditions to enhance the efficiency of heat sinks. Moreover, heat sink design and usage need to accommodate ever more dynamic microprocessor and electronics technologies. As advancements lead to greater thermal output, efficient heat dissipation becomes an issue that requires a sophisticated understanding of both heat sink materials as well as their configuration in order to achieve optimum performance. Emerging studies continue to be targeted towards the promise of nanomaterials, such as graphene and carbon nanotubes, which are more thermally conductive than conventional metals (Balandin et al., 2011). This has opened up new potential for innovation in the heating sink design, giving way to the possibility of lighter, more efficient solutions that can tackle increasingly demanding requirements on thermal management systems. As processors become increasingly powerful and faster, the use of effective heat sinks and TIMs will remain a crucial aspect of system design effectiveness.

Importance of Thermal Management in Electronics

In electronics, efficient heat management is crucial to maximizing the performance and life of devices. Because electronic parts generate heat as they function, this heat can build up and trigger thermal runaway conditions, which consequently lead to failure or inefficiency of operation. Heat sinks are applied for heat dissipation so that components remain in their ideal temperature levels. The employment of TIMs serves to further improve the process by enhancing heat transfer between heat sinks and devices. By lowering thermal resistance, properly selected TIMs have a dramatic ability to improve the effectiveness of heat sinks, thereby ensuring maximum operational lifespan of electronic devices and avoiding the possibility of overheating. As technology's evolution turns towards miniaturization and increased power densities, the demand for smart thermal management solutions is even greater. Emerging technologies have unearthed the potential of applying high-end materials in thermal management, such as for use in heat sinks and TIMs. Metal foams and MMCs have emerged for their tailored material properties, which favor heat exchanger application. The materials possess certain properties that enhance heat conductivity while maintaining lightweight structures, hence favoring performance as well as efficiency in thermal management. With these advancements, industries are beginning to explore the practical applications of metal foams and MMCs for heat sinks with a view to promoting the dissipation of heat in miniature-sized electronic devices. However, large-scale commercialization is faced with some challenges and hence calls for sustained research and development in resolving manufacturing complexities and integration issues (Han et al., 2012). Further, thermoelectric device architecture is one excellent example of how thermal management innovations can revolutionize energy harvesting techniques. New folding schemes have recently been used to optimize thin-film thermoelectric generators for large-area devices through boosting efficiency. The architectures allow for maximum thermal gradients with little leakage of heat, a requirement that is important for optimizing power output. When compared with traditional bulk thermoelectric modules, progress in thin-film technologies draws a focus on thermal interface problems, particularly thermal contact area optimization between heat sources and sinks. The results in this research clearly indicate that thermal efficiency is crucial not only for performance but also for cost-efficient solutions for energy in electronics (Tappura et al., 2018). Thus, the integration of effective thermal management methods remains a cornerstone in the ongoing advancement of electronic devices.

Overview of Thermal Interface Materials (TIMs)

Thermal interface materials (TIMs) play a pivotal role in effective thermal management of electronic devices. With increasingly complex and smaller-sized microelectronic devices, thermal conductivity must increase to ensure system reliability and operation. TIMs provide the critical link between heat-generating components—e.g., processors—and heat sinks, allowing for effective heat dissipation from sensitive areas. Their primary function is minimizing thermal resistance and therefore guaranteeing enhanced thermal coupling. With diminishing transistor size and increasing packing densities, optimizing TIMs cannot be overstressed; they have become front and center in the development of thermal management technologies, guaranteeing not only device lifetime but also overall efficiency (Naghibi et al., 2020).

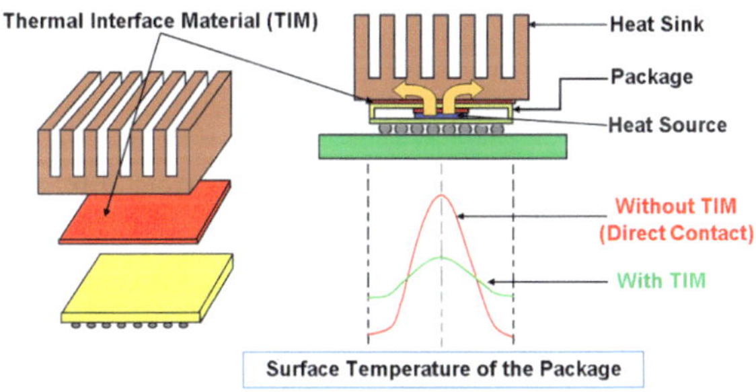

Figure 8.2. Thermal interface material (TIM).

Source: https://tts-grp.com/jp/news-interests/corporate-social-responsibility/news-events/477-introduction-of-thermal-interface-materials-tims-in-thermal-management-solution.

With so many TIMs on the market, selecting the appropriate material must be carefully assessed. Various TIMs, including thermal pastes, pads, and phase-change materials, will have varying uses and performance characteristics. Thermal pastes, for instance, may be superior heat conductors but require certain methods of application, while gap fillers accommodate irregular surfaces but may be less conductive. Recent advancements have seen a significant research focus on the fabrication of high-performance TIMs, particularly through the utilization of materials such as graphene. The superior thermal conductivity and mechanical strength of graphene make available new blends that could surpass traditional thermal paste performance. As such, information regarding the properties and behavior of different TIMs in various conditions is necessary in order to maximize thermal management systems used in modern electronics (N/A, 2011) (Figure 8.2).

TIM performance characterization is also an important aspect in the evaluation of their performance in actual applications. Traditional testing protocols generally utilize perfectly flat surfaces, which are rarely the case in actual electronic assemblies. Filling this void, sophisticated characterization tools have emerged, such as the Angular TIM Tester and Bare Die Tilting Metrology. These methods allow the performance of TIMs to be measured for an array of tilt angles, providing valuable information on how materials of this type will perform in real environments. Testing has shown that even small amounts of deviation from parallel can dramatically impact thermal resistance, emphasizing the requirement for stringent testing protocols (N/A, 2011). As technologies within TIM evolve further, research and innovation will continue to refine selection criteria and application techniques, eventually enhancing heat dissipation performance required for today's electronic components.

8.1. TYPES OF HEAT SINKS

The efficient dissipation of heat in electronic devices largely relies on the nature of the heat sink used in the system. One such type is the passive heat sink, where natural convection is employed to conduct heat without any fan or power requirement. Most commonly built with materials having high thermal conductance, i.e., copper or aluminum, passive heat sinks employ their increased surface area for enhanced dissipation. They are highly effective in low-power applications where the amount of heat generated is low, thereby making compact designs feasible without losing device efficiency. However, passive heat sinks can deteriorate in efficiency as the power levels increase, necessitating the utilization of more active solutions that can keep up with the heat loads found in modern, miniature electronic components. Active heat sinks, unlike their passive counterparts, use mechanical fans or pumps in an effort to dissipate heat more effectively. These systems enhance airflow over the surface of the heat sink, significantly increasing the rate at which heat is able to be removed from key devices. Although active heat sinks tend to be better at cooling, they introduce additional complexity into the thermal system, including higher noise and power consumption. Furthermore, the lifespan of active systems can be compromised by moving components that tend to deteriorate with age. As a result, active versus passive heat sinks are often chosen on the basis of a very broad array of factors including the specific heat dissipation requirements of the application, the operating climate, and overall system design priorities, which could influence performance as well as reliability in electronic components. One newer type of heat sink is the vapor chamber, which functions on the principle of phase change to assist in dissipating heat. In a vapor chamber, the working fluid inside boils off rapidly due to the application of heat and forms vapor that spreads evenly across the devices surface (Figure 8.3).

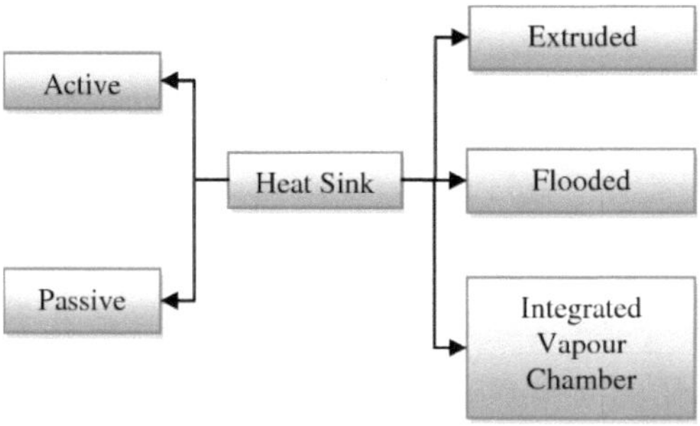

Figure 8.3. Heat sink classification.

Source: https://external-content.duckduckgo.com/iu/?u=https%3A%2F%2Fpick3dprinter.com%2Fwp-content%2Fuploads%2F2020%2F05%2Ftypes-of-heat-sinks.jpeg&f=1&nofb=1&ipt=aeec5b449576bd3115430325d91789efd4321732d2ce733debad4158d864b6d0.

This vapor then forms liquid form as it moves further away from the heat source, thus transferring heat away from electronic components. Consequently, vapor chambers may reduce thermal resistance significantly and improve the thermal performance of heat sinks. They are particularly beneficial in high-performance computing where excessive heat may degenerate overall system performance. Recent advances in TIMs, especially those highlighted in recent reviews, have emphasized the need to boost the thermal conductivity of such systems against the inadequacies such as aging and thermal degradation that have been long recognized to compromise the long-term performance of TIMs in applications involving vapor chambers (Artiaga et al., 2024; Khuu et al., 2009).

8.1.1. Passive versus Active Heat Sinks

The difference between active and passive heat sinks is fundamental to appreciating their function in thermal management systems. Passive heat sinks utilize natural convection to transfer heat away from components by utilizing high-thermal-conductivity materials like copper or aluminum. They typically have finned configurations that increase surface area so heat can be effectively dissipated into the ambient air without any need for external energy input. These systems are typically appreciated for simplicity and reliability in usage where power consumption and noise are important concerns, i.e., in portable electronics. They can, nonetheless, be limited by performance in high-density applications where heat production is higher than the capability to dissipate heat by passive systems, leading to a requirement for an exploration of alternative thermal management strategies (Holloway et al., 2017) (Figure 8.4).

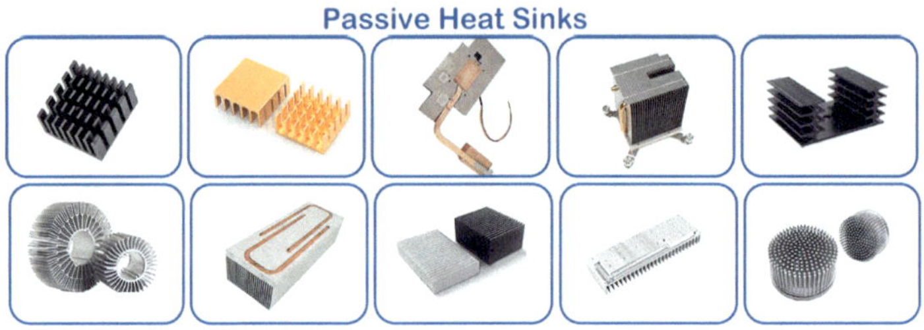

Figure 8.4. Passive heat sinks.

Source: https://www.heatell.com/wp-content/uploads/2023/10/Passive-heat-sink.jpg.

Active heat sinks, on the other hand, utilize mechanical devices such as pumps or fans in order to enhance dissipation, thereby greatly enhancing their thermal efficiency. The forced convection associated with these devices allows for greater removal of heat, particularly in the situation where electronic components create high thermal loads, e.g., in servers or high-performance computing. This increased efficiency is achieved at the expense of higher energy consumption and potential noise generation, conditions which engineers must balance optically in the design of thermal management devices. The choice between active and passive heat sinks is thus dependent on a holistic assessment of working conditions, anticipated thermal load, and acceptable levels of noise and power consumption (Garimella et al., 2017). Furthermore, the use of high-performance materials and engineering has also increased the range of functionality for passive and active heat sinks. Recent advances have included the use of thermally conductive silicone in prosthetic liners, which holds promise for improved thermal comfort in terms of dissipation function. While passive solutions are easier to install and service, they may not suffice to address widespread thermal distress in high-end applications (Holloway et al., 2017). TIMs are crucial to this ongoing evolution because the interfaces between components and their heat sinks can significantly impact overall thermal efficiency if optimized. As researchers and engineers pursue solutions that are application-specific, ongoing debate between passive and active thermal management solutions remains central to achieving optimal performance in different electronic systems (Garimella et al., 2017) (Figure 8.5).

Figure 8.5. Active heat sinks.

Source: https://www.heatscape.com/wp-content/uploads/2017/10/COVER-Active-Cooled-Heatsinks.jpg.

8.1.2. Material Choices for Heat Sinks

The choice of materials to construct heat sinks is very important in deciding their ultimate thermal efficiency and performance. Traditional materials such as copper and aluminum are preferred due to their high thermal conductivity, and copper is usually preferred in high-end applications due to its superior characteristics of heat transfer. But all material has its shortcomings; for instance, aluminum is lighter and less expensive but typically has lower thermal conductivity than copper. Considerations of design transcend raw thermal properties to include such features as weight, expense, manufacturability, and compatibility with TIMs. Therefore, the selection process must be through careful investigation of the operating environment, for example, temperature ranges and potential exposure to corrosive elements, which may influence significantly the lifespan and effectiveness of the heat sink in practical uses. Recent advances in material sciences introduced new options with innovative attributes that enhance the thermal properties of electronic components. For instance, carbon nanomaterials, particularly carbon nanotubes and graphene, have been of interest due to their high thermal stability at high temperatures and good heat transfer properties. They possess the ability to increase heat dissipation efficiency due to their high aspect ratios and surface areas, and the potential to cool more efficiently when coupled with traditional metals. Carbon nanomaterial incorporation as fillers into TIMs or composite structures in heat sinks can improve thermal resistance to a significant extent. As research progresses, characterization of such ensembles of nanomaterials is aimed at optimizing their thermal properties to meet high-temperature applications, to promote overall utilization of improving heat sink efficiencies for small-scale electronic devices (Hao et al., 2016) (Figure 8.6).

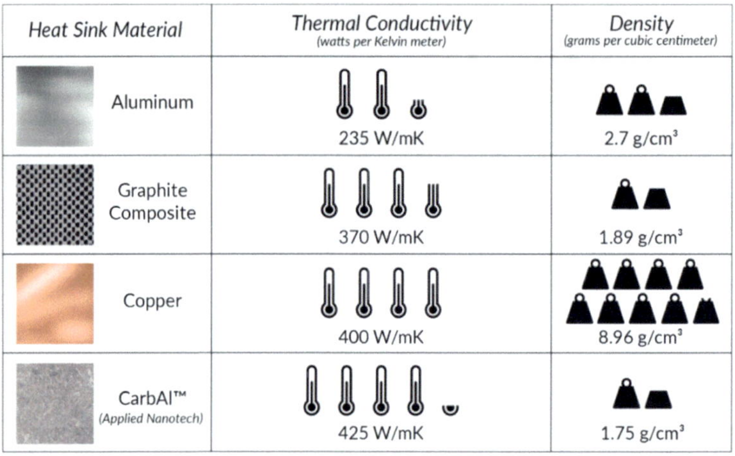

Figure 8.6. Heat sinks materials.

Source: https://northslopechillers.com/wp-content/uploads/2020/03/Heat-Sink-01-1024x704.png.

In addition, advanced TIMs in new designs further enhance heat sink performance. Standard TIMs, being silicone or epoxy-based, are limited by their poor thermal conductivity or long-term degradation. The new thinking has rather concentrated on high-conductivity TIMs, which can withstand long-term high-pressure conditions without degrading their heat transfer capabilities. For example, the application of high-conductivity TIMs in a torsional spring-based device has been shown to facilitate controlled contact between heat sinks and heating elements so that significant heat transfer is improved by closing air gaps and allowing uniform pressure. These designs minimize the common issues with thermal resistance at joints and are especially useful for high-power applications, like in optical transceivers, where they may be using high amounts of power and require effective cooling by way of creative material choice and design (Anonymous et al., 2025).

8.1.3. Design Variations and Their Applications

The primary contribution of heat sink design variations comes in the way of their extensive uses in numerous electronic devices. Fins, plates, and pin designs are a few geometric configurations that could influence the overall thermal performance. By changing these design parameters, the convective heat transfer can be improved, eventually leading to the improved efficiency of the heat sinks. For instance, finned heat sinks are commonly applied where there is air flow available, basically offering greater surface area to conduct the heat. Such design changes are similarly crucial for maximizing space optimization in compact electronic systems to deliver superior cooling without compromising on performance.

This flexibility in design makes it possible for heat management products to be tailored precisely to the specific requirements of each application, ultimately leading to increased device reliability and life. New materials applied in heat sink design and their TIMs have created new possibilities for thermal management. Recent research shows the potential of carbon allotropes such as graphene and carbon nanotubes due to their high thermal conductivity properties. Specifically, thermal conductivity of the materials can actually vary extensively, up to five orders of magnitude, thereby exhibiting the importance of the choice of materials in the design process. The unusual size-dependent thermal conduction properties observed in two-dimensional materials like graphene also strongly suggest that well-directed design modifications can lead to monumental improvement in heat dissipation. Such material engineering insights not only facilitate improved thermal interface performance, but they also bring into play novel applications in modern electronics, where efficient thermal management is growing progressively vital to device performance (Balandin et al., 2011). Moreover, if the complexity of thermal interface resistances and how different materials interact with one another is grasped, then design variations can be further perfected. For example, the orientation of carbon nanotubes in a polymer matrix can create gigantic variations in thermal conductivity due to the complex relationships between the respective interfaces.

Variations in the attachment and functional groups can cause significant thermal resistance problems that must be addressed by designers to provide maximum performance. The study indicates that exploring vertically aligned carbon nanotubes as TIMs can result in new solutions through the elimination of these resistive barriers. Future work in the materials and design variations relationship will form the basis for future TIM development in response to emerging technology demands, offering efficient thermal dissipation across a variety of electronic applications (Louie et al., 2011).

8.2. THERMAL INTERFACE MATERIALS (TIMS)

Thermal interface materials (TIMs) are among the most prominent characteristics of modern thermal management systems, allowing efficient heat dissipation from electronic components to heat sinks. The reduction in the size of electronic devices has seen the power densities and resulting heat evolved rise tremendously, necessitating efficient thermal management as a vogue necessity. The primary function of TIMs is to fill the microscopic gaps between surfaces, thereby enhancing thermal conduction and preventing heat build-up resulting in a loss in performance or failure. The addition of next-generation materials such as graphene in TIMs has been reported to improve thermal conductivity substantially, presenting a possible avenue for heat control in future electronics, as highlighted in the research on TIMs properties (Naghibi et al., 2020). Recent technological advancements in TIM technologies emphasize the importance of fillers in improving the thermal conductance of polymer composites, which are widely favored due to the fact that they are flexible as well as simple to apply. Employing various fillers, researchers have been able to significantly improve the thermal performance of the composites. For instance, the use of graphene and other high-conductivity materials has been established to enhance polymers' effective thermal conductivity while maintaining their integrity of structure. The findings are important, especially as demand for small and efficient electronics continues to rise, necessitating materials that can efficiently dissipate heat without influencing the overall design parameters (Artiaga et al., 2024). Moreover, state-of-the-art research in TIMs focuses on optimizing the trade-off between conductivity and mechanical properties in order to achieve long-term stable and durable thermal management solutions. According to investigations, filler selection and dispersion in the polymer matrix influence the thermal behavior and applicability of TIMs in specified operating environments. Understanding the heat conduction mechanisms in these materials is crucial for the manufacture of the future generation of TIMs capable of meeting the stringent demands of high-end electronics. Ongoing research on the discovery of novel composites and novel processing techniques promises to lead to TIMs that not only enhance heat dissipation but also enhance efficiency and the life span of electronic devices (Naghibi et al., 2020).

8.2.1. Definition and Purpose of TIMs

Thermal interface materials (TIMs) are crucial components of thermal dissipation in electronic systems, particularly with heat sinks. They function as effective interfaces between thermal-emitting devices, such as processors, and thermal dissipators, such as heat sinks, to enable effective thermal transfer. The primary role of TIMs is to minimize thermal resistance at interfaces by containing microscopic surface irregularities. This reduction in thermal resistance facilitates enhanced heat transfer from electronic components into the heat sink and prevents overheating, enhancing the performance and reliability of electronic components. Advances in TIMs have given rise to a broad range of materials, such as thermal greases, pads, and phase-change materials, which possess varying thermal conductivities and are applied for particular purposes. The performance of TIMs relies on various parameters including their thermal conductivity, viscosity, and long-term stability. High thermal conductivity is required to allow for rapid heat transfer, and appropriate viscosity allows for the material to wet the surfaces well in order to create an even bond. Long-term stability is also important because any deterioration with time would lead to higher thermal resistance, which would impair the performance of the system. In the last several years, research has aimed more often at the enhancement of TIMs in a bid to balance thermal efficiency against mechanical properties suitable for real-world applications. The matrices have been filled and added with additives to improve performance while not compromising the needed properties to make the accessible TIMs a broader range of options for numerous thermal management scenarios (Boglietti et al., 2009) (Figure 8.7).

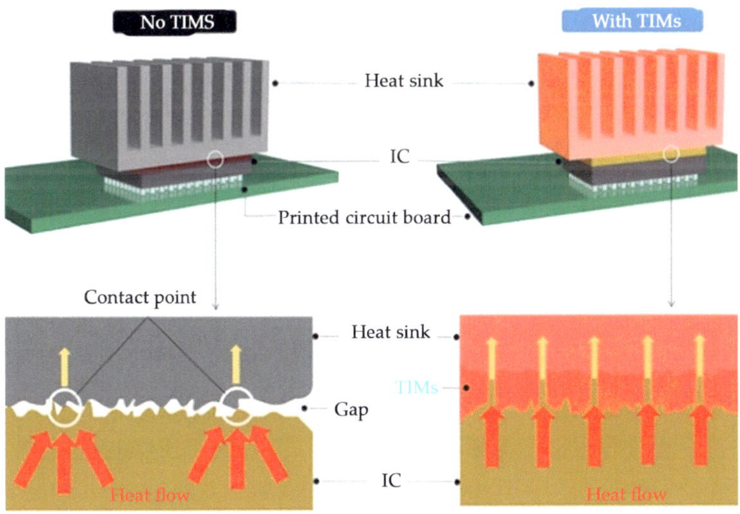

Figure 8.7. Schematic illustration of working mechanism of a thermal interface material (TIM).

Source: https://www.researchgate.net/figure/Schematic-illustration-of-working-mechanism-of-a-Thermal-interface-material-TIM_fig1_371608028.

In actual applications, the selection of an appropriate TIM can have a great influence on electronic system thermal performance. Selection is typically based on the specific needs of the application, including size, weight, and operating temperature range. For instance, TIMs for use in high-end computing may require materials that possess excellent thermal conduction, whereas those in environments with less demanding specifications may stress simplicity of use and cost-effectiveness. Furthermore, recent advancements like spin-wave thermal transport in future materials like yttrium iron garnet (YIG) show ongoing research in this domain, suggesting that the microscopic electric and thermal interactions can be of great help to TIM design (Ando et al., 2018). As technology advances further, understanding the fundamental purposes and nature of TIMs will continue to remain an important part of making improvements in the heat management of smaller and smaller electronic devices.

8.2.2. Common Types of TIMs (e.g., Thermal Pastes, Pads)

Thermal interface materials (TIMs) play a critical role in thermal management, particularly in electronic devices wherein heat dissipation must be effective. Thermal pastes are among the most commonly applied forms of TIMs and are viscous substances created for filling microscopic air pockets between heat-generating components and heat sinks.

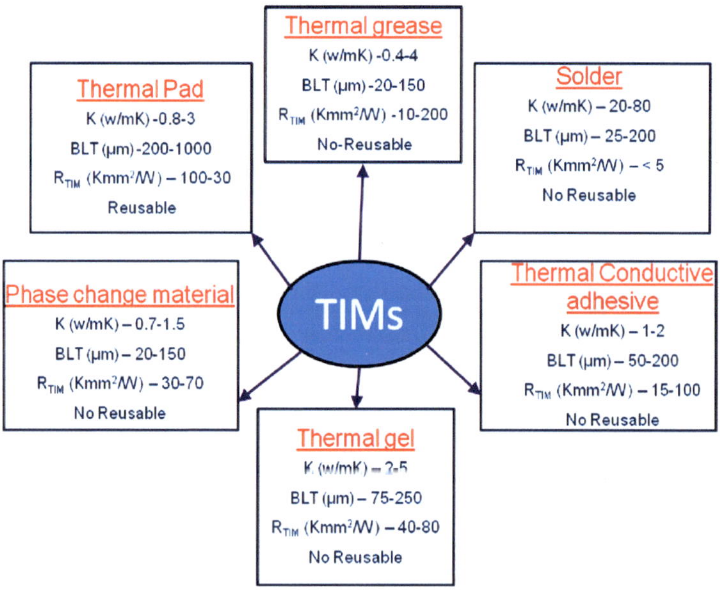

Figure 8.8. Technical aspects of different types of TIMs.

Source: https://www.researchgate.net/figure/Technical-aspects-of-different-types-of-TIMs-Viswanath-et-al-2000-Tong-2011-Bischak_fig2_336906996.

They are typically made from metal oxides or silicone that enhance thermal conductivity. Recent advancement has witnessed the emergence of graphene-based thermal pastes, which have better thermal conductivity due to the properties of the material. Scientists have demonstrated that the non-curing pastes used in thermal applications perform significantly better than traditional ones, as they reach high conductivity at lower filler levels, representing a practical approach for modern electronic devices (Naghibi et al., 2020). The continuing evolution of TIMs is an indication of more attention being placed on hybrid materials that can balance the best thermal performance with compatibility and usability. The other popular category of TIMs is thermal interface pads, which are soft or rigid materials typically used in applications where a consistent thickness and mechanical stability are required. Unlike thermal pastes, pads eliminate the issue of varying application and can be pre-cut for a given component size. Made of materials like silicone rubber with thermally conductive fillers, the pads are effective in thermal coupling surfaces and are readily installable. Through their own natural compressibility, thermal interface pads are capable of conforming to uneven surfaces more effectively, thus enhancing their thermal performance. Furthermore, research has shown hybrid pads by magnetic arrangement of fillers for improving directional thermal conductivity, contributing other enhancements towards the heat transfer mechanisms used (N/A, 2019) (Figure 8.8).

Finally, the choice of TIM—paste or pad—depends on some application needs, surface compatibility, and necessary thermal management characteristics. Apart from pastes and pads, other advanced TIM solutions are entering the thermal management front. One such case in point is liquid metal TIMs, which have been a serious contender due to their high thermal conductivity and low thermal boundary resistance. These materials find application of liquid gallium or gallium alloys, which significantly enhance heat spread capabilities in high-performance devices. Application of liquid metals, however, comes with challenges such as filler material stability and susceptibility to corrosion. Such challenges have been mitigated through the development of protective coatings that enhance particle-particle contact within the liquid matrix (N/A, 2019). Furthermore, advancements in particle engineering and interfacial resistance reduction are leading to increasingly effective TIMs that respond to the evolving requirements of electronic components. This rapid expansion and diversification of potential TIMs underscore the critical role played by effective heat management technologies in contemporary technology, further solidifying the position of TIMs as irreplaceable in electronic function and design.

8.2.3. Performance Metrics for TIMs

To accurately measure the performance of TIMs, many measurements are employed to express their thermal conductivity, thermal resistance, and general effectiveness at heat dissipation. Thermal conductivity is an important property since it has a direct influence on the ability of a material to transfer heat from a heat-generating component, such as a microprocessor, to a heat sink. TIMs' performance can be measured quantitatively with thermal resistance measurements, which describe how well a TIM withstands heat conduction. Lower and more desirable values indicate superior performance, and this requires selecting TIMs that have the best thermal attributes reacting to the special demands of modern electronics. These requirements are the foundation for evaluations so that engineers can pick the right TIMs for applications. In recent studies, the development of improved TIM technology has led to the exploration of novel materials, such as graphene-based composites, which have much enhanced thermal properties. As noted by Naghibi et al. (2020), traditional TIMs prove unsuitable for meeting the cooling requirements of dense and compact electronics. Graphene-based fillers have been proven to cause enhanced thermal conductivity in TIMs compared to conventional materials. The performance attributes of the new TIMs highlight their potency in generating warmth dissipation, thereby avoiding thermal limitation in advanced electronic technology. By comparing the graphene-based TIMs with current commercial counterparts, researchers have fostered greater understanding of their promise in real-world application, thus potentially leading to new avenues for industrial applications (Figure 8.9).

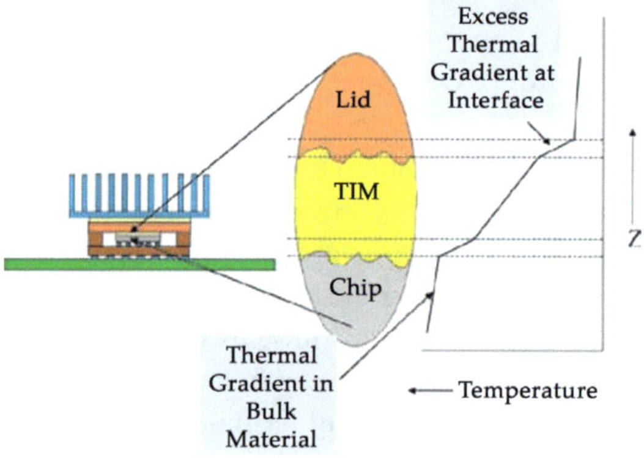

Figure 8.9. Effect of bulk and interfacial thermal properties of TIM on thermal performance.

Source: https://www.electronics-cooling.com/2003/08/calculations-for-thermal-interface-materials/

Furthermore, the introduction of phase change materials (PCMs) to the TIM concept introduces additional performance factors that are worth considering. As elucidated by Ganatra et al. (2016), PCMs possess the unique ability to consume and release heat in the form of phase transitions, which aids in maintaining stability of operating temperatures within electronic systems. Their incorporation into TIMs is capable of enhancing the operating time of electronic equipment under heavy loads significantly, as evidenced by the extreme extension in operating time that some PCMs are capable of. When evaluating these materials, it becomes very important to consider not just their thermal conductivity and resistance but also their latent heat characteristics and transition temperatures. This cross-dimensional performance measure captures the complexity and adaptability of TIMs, and it enables one to design more effective thermal management products that compromise between performance, size, and cost in modern electronics.

8.3. SELECTION CRITERIA FOR HEAT SINKS

In electronic component thermal management, the criteria for heat sink selection are essential in achieving successful thermal dissipation. Thermal conductivity, surface area, and material selection are of fundamental consideration. The thermal conductivity of the heat sink material itself dictates the ability of the heat sink in conducting heat away from source hot components to lower ambient temperatures. Aluminum and copper are typical choices due to their valued thermal properties. However, design of the heat sink should also take into consideration the surface area since a larger surface area allows for more heat transfer through convection and radiation mechanisms. Moreover, design complexities such as fins or geometric arrangements can be used to highly enhance efficiency by maximizing airflow around the heat sink, allowing for more efficient cooling.

This organized analysis of thermal behavior in terms of design choice is critical to optimize the use of heat sinks in any given electronic use. Another major factor that comes into play with heat sink selection is the environment it will be working in, which can influence material performance and life dramatically. Ambient temperature, humidity, and corrosive substance exposure should govern material choice as well as design. In a moist environment, as an example, aluminum might oxidize, causing compromised performance in the long term. Or treatments or coatings may be applied to enhance surface durability at the expense of thermal conductivity retention. The thermal needs for specific heat of the electronic application must also be taken into account because greater packing densities found in newer electronics have a tendency to cause localized overheating. Therefore, understanding both the thermal needs and environmental stresses is crucial when selecting the most appropriate heat sink to ensure reliability and durability of performance in electronic devices.

Neither can the role of the addition of TIMs in the selection process for a heat sink be overlooked. TIMs are employed to reduce thermal resistance between heat-emitting devices and the heat sink, thereby maximizing overall heat transfer efficiency.

The choice of TIM is crucial; thermal pastes or pads in materials vary in thermal conductivity, and developments in materials such as graphene-based fillers promise improvements in performance (Naghibi et al., 2020). These developments portend that future heat sink designs might increasingly rely on high-end TIMs to optimize thermal management. With electronics further evolving into compact yet stronger devices, addressing the interdependence of heat sinks and TIMs will be essential to designing efficient thermal solutions for guaranteeing working stability and extending the lifespan of electronic components (Holloway et al., 2017) (Figure 8.10).

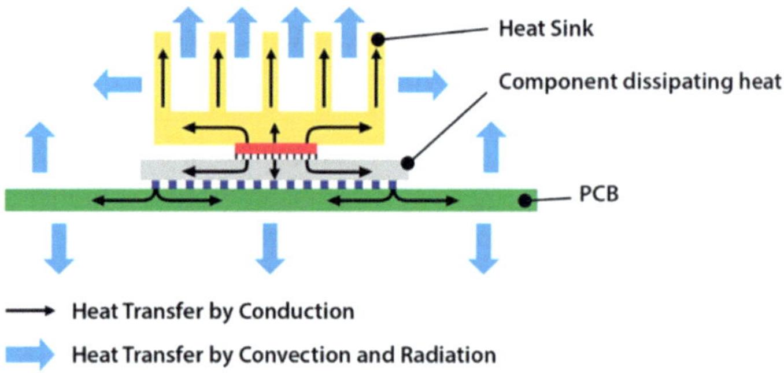

Figure 8.10. Select a heat sink.

Source: https://www.hnhcart.com/blogs/learn/heat-sink?srsltid=AfmBOorJHptc-vxzKmj7GYKEYCIcD-N-Yf7CvSQlVxBKXAUAWy33DzNm.

8.3.1. Thermal Resistance Considerations

In electronic cooling, thermal resistance is of fundamental importance in both TIMs and heat sink design and performance. Thermal resistance is a measure of the capacity of material or assembly to oppose the flow of heat, and its impact on electronic device performance is a direct result. As high-performance systems become increasingly demanded, understanding the various types of thermal resistance is essential. Common contributors to thermal resistance include conduction, convection, and radiation, which can hinder effective heat dissipation when not appropriately managed. The overall thermal resistance then goes on to determine the interface temperature gradient and consequently the reliability and life of electronic devices. Thermal management solutions therefore have to be optimized through an integrated analysis of these resistance components such that devices can stay within operating temperature safety margins. The performance of a heat sink not only depends on its construction and material characteristics but also significantly on the thermal interface between the electronic device and the heat sink. TIMs are a crucial link within the thermal path, lowering interfacial thermal resistance. The selection of TIM should therefore take into account thermal conductivity, thickness, and working environment. For instance, application of novel TIMs can successfully reduce thermal contact resistance such that enhanced thermal transfer from the device to the heat sink is enabled. Particularly in high-temperature applications, as identified in existing research, TIM performance can have a direct influence on the total thermal management strategy (Figure 8.11).

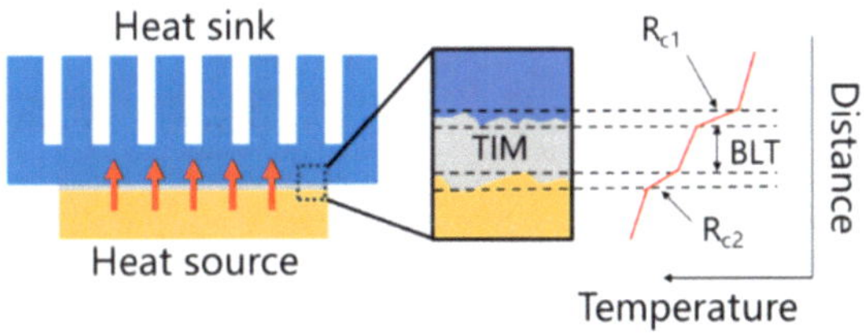

Figure 8.11. Principle of reducing thermal resistance.

Source: https://www.mdpi.com/2079-9292/13/21/4287.

The design must thus be thoughtfully combined with the characteristics of TIMs in an effort to optimize thermal performance and system reliability. Other applications, such as thin-film thermoelectric generators, further highlight the importance of resolving thermal resistance problems comprehensively. The new folding style of these generators enables effective heat handling through attenuated leakage and improved temperature gradient, as described by Tappura et al. (2018).

These developments demonstrate that thermal resistance must be a high-priority field in the design and test phases of thermal systems. In addition, the significance of resistance to thermal radiation, particularly for use at a larger scale like industrial motors, is an indication of the influence environmental conditions play in thermal management. As discussed by Boglietti et al. (2006), calculation of radiation thermal resistances can further improve calculations of thermal performance such that systems are well designed to accommodate varying operating conditions. Such a multi-disciplinary approach to analyzing and treating thermal resistance is critical for enhancing the efficiency and reliability of modern electronic devices (Figure 8.12).

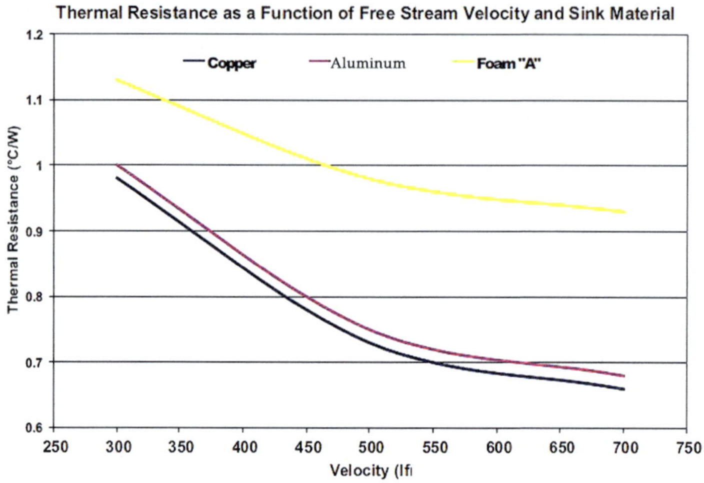

Figure 8.12. Heat sink thermal resistance as a function of velocity.

Source: https://www.qats.com/cms/category/heat-sink-material/

8.3.2. Size and Weight Constraints

Size and weight constraints have become increasingly important in contemporary electronic device design because of the need for miniaturization and portability. This is most evident in portable devices where consumer demand forces companies to innovate with regard to performance while maintaining performance. With smaller components, power density in devices becomes higher, leading to potential thermal bottlenecks since the thermal performance of components has not kept up with the miniaturization. The challenge posed by these constraints necessitates effective thermal management designs that can cope with the constraints put in place by smaller form factors. For instance, traditional active cooling methods like fans are found to be impractical due to their weight and size, which compel engineers to consider more viable alternatives acceptable to the constraints of modern design considerations, thereby triggering the demand for innovative interventions within heat sink designs and TIMs (Ganatra et al., 2016). In addition, TIMs inclusion in electronic component downsizing is also largely size and weight dependent. With ultra-thin design becoming increasingly popular, TIMs are not only expected to be light in weight but also effective in thermal conductivity properties. Researchers have found that new materials, such as graphene, have shown promise as fillers for TIMs and can significantly enhance thermal conductivity with minimal increase in weight. But the optimum performance is reached with the balance of possessing good thermal management combined with the size of the TIMs. As technology continues to advance, the challenge comes in trying to select materials that enhance thermal performance without disproportionately increasing size. The delicate balance is essential in permitting devices to function at acceptable thermal response levels while meeting consumer requirements for thin form factors (Naghibi et al., 2020). Because of the critical need for high-performance thermal management, passive cooling methods such as the use of Phase Change Materials (PCMs) are a far cry from conventional techniques. PCMs have the ability to absorb thermal energy and release it through phase changes, effectively regulating temperatures within sealed environments. Out of the study of utilizing PCMs in compact devices are also opportunities and challenges. Though their use can dramatically extend device life before temperature limits are reached, detailed planning of design parameters is needed to make them fit in the constrained size of contemporary electronics. Thus, it is necessary to identify major parameters that affect the performance of PCM-based systems because engineers are always looking for ways to design thermal solutions not only compliant with the requirements of operation but also within the size and weight constraints dictated by modern consumer electronics (Ganatra et al., 2016).

8.3.3. Cost-Effectiveness and Material Availability

The controversy between heat sinks and TIMs generally orbits around the problem of cost-effectiveness and material availability. With rising demand for effective cooling due to electronics miniaturization, the selection of TIMs becomes a critical factor. Cost-effectiveness is top on the list of manufacturers, who seek materials that are not only more thermally conductive but economical. Their integration within polymer matrices has proved promising to enhance thermal properties without a prohibitive cost. It is thus important to identify the best material combination to efficiently optimize the production process and apply scalable strategies for extensive industrial use such that TIM technology breakthroughs are made accessible without an unrealistic cost. Ease of access to materials is equally crucial in designing efficient TIMs. The rapid progress in electronics needs reliable sources of high-performance materials, particularly with the emergence of novel fillers in the marketplace. For instance, while polymer composites offer an economical alternative for traditional metal-based TIMs, access to innovative fillers like graphene takes center stage in achieving projected thermal characteristics. New manufacturing methods, such as liquid phase exfoliation, have the potential of making these fillers universally accessible so that they can be mass-produced by manufacturers. Therefore, for the successful application of TIMs in heat sink technology, future studies should focus on establishing environmentally friendly ways of sourcing and integrating these materials into existing manufacturing systems. Moreover, cost-effectiveness relative to performance cannot be overstated in evaluating TIMs. TIMs need not only meet technical specifications but also be produced to a degree that is economically feasible for manufacturers. Experiments have shown that non-curing graphene-based thermal pastes achieve improved conductivity at lower filler levels, with such products carrying significant improvements compared to typical products (Naghibi et al., 2020). This not only addresses performance problems but also shows how better thermal management can yield longer lifetimes for electronic components, hence reducing long-term expenses to manufacturers and users alike. In general, the utilization of financially sustainable and readily available materials will undoubtedly characterize the future of heat sinks and TIMs, leading to more efficient thermal management tactics in rapidly advancing electronic technologies (Artiaga et al., 2024).

8.4. PERFORMANCE TESTING OF HEAT SINKS AND TIMS

Efficacious heat sink and TIMs performance testing is essential in mitigating the growing thermal management challenges prevalent in modern electronics. As devices miniaturize, operating temperatures can compromise their performance and longevity. Efficient heat transfer among components and heat sinks is highly reliant on the thermal conductance of the TIMs in question. Current studies highlight the importance of selecting proper TIMs, particularly polymer composites, which have gained increased popularity due to their ability to promote thermal conductivity. However, enhancing their thermal performance while not sacrificing manufacturability is a focus of continued research. In addition, methods such as incorporating varying fillers in polymer matrices exhibit significant thermal conductivity enhancement, validating the double approach to increase the TIM performance and heat sink cooling performance (Artiaga et al., 2024). Test methods for performance must be developed to effectively measure TIMs' thermal interface properties and heat sinks' dissipation effectiveness. Standardized testing methods evaluate key parameters like thermal resistance, conductivity, and cyclic thermal durability. The question of prime importance, namely how the materials really perform under practical conditions, shifts focus towards understanding their operation over a variety of situations from large-scale setups like data centers to portable handheld devices. Doing so also looks into newer trends such as liquid cooling technology solutions and solid-state thermoelectric devices, which reveal an industry that is rapidly evolving to meet growing thermal management needs. Such diverse test environments, as indicated in workshops of thermal management problems, provide invaluable data for future design innovations in TIMs and heat sink efficacies (Garimella et al., 2017). In addition, the future of heat sink and TIM performance testing also rides on embracing state-of-the-art practices in material science and thermal measurement. Developments in nanotechnology, for instance, could lead to novel TIM formulations that outperform incumbent materials, enabling lighter and more efficient thermal solutions. In addition, the intersection of machine learning and data analytics can help enhance predictive modeling of thermal management to enable more precise simulations and analyses of heat sink design and TIM choices. Through ongoing investigation of these frontier methods, care must be exercised in keeping track of an expansive consideration of parameters including cost, manufacturability, and environmental impacts associated with novel materials. Lastly, continued investment in performance testing methods will facilitate the development of effective thermal management solutions required in next-generation electronic products (Artiaga et al., 2024) (Figure 8.13).

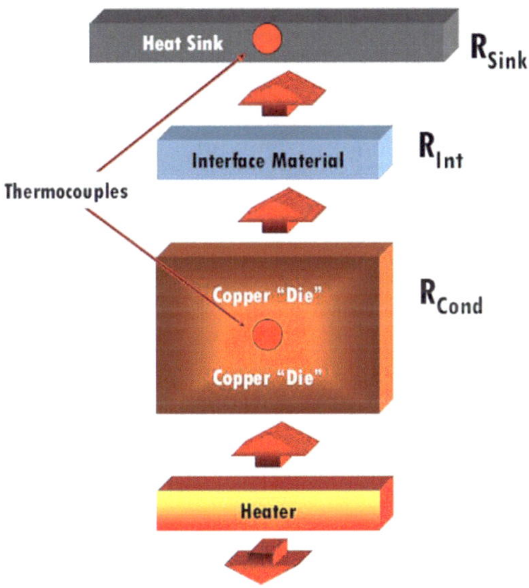

Figure 8.13. Experimental heater and measurement setup.

Source: https://www.qats.com/cms/category/heat-sink-material/

8.4.1. Methods for Measuring Thermal Performance

One of the most widespread approaches in measuring the thermal performance of heat sinks as well as TIMs involves the use of infrared thermal imaging. What this method makes it possible for researchers to capture is real-time temperature maps on a material's surface, thus providing an indication of the dissipation of heat under operational conditions. By measuring changes in temperature with time, infrared cameras generate contour maps highlighting areas of high and low thermal efficiency, allowing for the detection of possible hotspots. The images are vital because they highlight non-uniform thermal distribution, which is harmful to device performance. Like in experiments on materials undergoing phase transformations, as in the Cu–Zn–Al single crystal, infrared thermal imaging has been established to record significant kinetics in the process, placing its significance as a tool in thermal performance measurement at the forefront (Burrows et al., 2011). Another effective means of thermal performance measurement is through the measurement of thermal conductivity by transient methods, such as the laser flash analysis (LFA) technique. This method involves illumination of one surface of a sample with a short laser pulse and detection of the temperature response on the opposing side as a function of time. The thermal trace is then analyzed quantitatively to determine the materials thermal diffusivity, from which thermal conductivity can be approximated when supplemented with density and specific heat data. The merit of the technique lies in its capacity to handle small samples and the fact that it is capable of analyzing materials under varying conditions, hence aiding the testing of TIMs in conjunction with heat sinks (Figure 8.14).

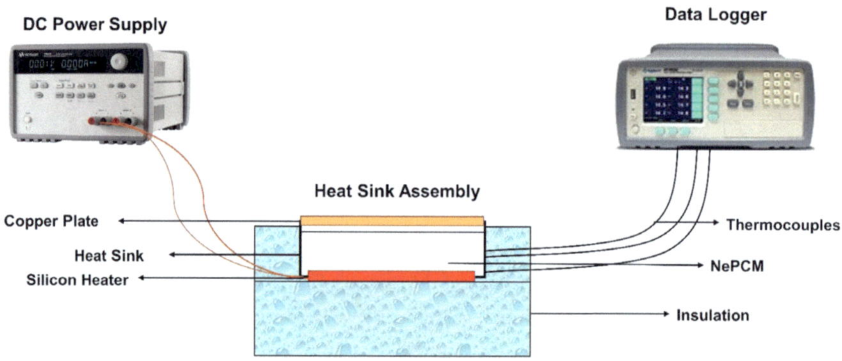

Figure 8.14. Test setup for evaluating the thermal performance of NePCM-based heat sinks.

Source: https://www.mdpi.com/1996-1073/15/22/8416.

Last, LFA provides a measurable value with direct correlation to the quality of thermal management solutions employed within electronic packaging. Second, introduction of new materials to overcome traditional thermal performance typically involves in-depth analysis of their thermal expansion characteristics. An example of how one can condition compositions to control thermal expansion is by using fillers like graphitic fibers in metallic matrices. More advanced measurement methods, such as neutron scattering, can contribute significantly to gaining insight into the interaction between fillers and matrix. According to an investigation into sintered metal matrix composites, interactions result in unanticipated thermal expansion characteristics, which can be critical while designing efficient TIMs (Oddone et al., 2018). By combining these cutting-edge techniques, researchers can do a great deal to improve the thermal performance of heat sinks and TIMs and, in turn, improved heat management in modern electronic devices.

8.4.2. Standard Testing Protocols

Standard test methods provide the basis on which the performance and efficiency of heat sinks and TIMs are determined. The methods provide a systematic approach to investigating a variety of thermal management solutions such that the data obtained are reproducible and also valid under varying test conditions. For instance, standardized practices such as steady-state or transient thermal analyses can be employed to calculate the thermal resistance of a heat sink in conjunction with a specified TIM. Through these practices, engineers not only can calculate the thermal conductivity characteristic of the TIM but also quantify the heat dissipation capability of the heat sink for various load conditions. The results achieved through these protocols are a basic groundwork for further enhancement of thermal design to address issues of efficiency as well as reliability in electronic application. Utilization of standardized test procedures helps remove variability that may be brought about by environmental factors, material variabilities, and measurement techniques. With repeatable test environments, scientists are able to conduct a comparable study between different heat sinks and TIMs with confidence. For example, protocols can specify ambient temperatures and air flow conditions as required for simulating actual uses. Also, the use of exotic composites and materials delineated by N/A (2025) can significantly affect the outcome of these tests. Adherence to standards as intended can eventually enhance the predictive power of thermal models, enabling engineers to pre-optimize design. It allows for the establishment of performance limits, which can breed innovations in material choices and system configurations. Moreover, the creation of test methodologies for heat management solutions is an expression of advancements in theoretical knowledge and practical application. Since the thermodynamic properties engaged in heat conveyance are becoming progressively more complicated, so will the standardized methodologies utilized to test them. New information on technology and materials can enhance test procedures to better reflect new industry needs and new science discoveries. Atmospheric potential oxygen research by Andres et al. (2006), while not specifically relevant to thermal management, demonstrates the subtleties and variations that rigorous scientific examination can uncover. Thus, an adaptive standard testing strategy will not only ensure the ongoing relevance of existing protocols but also enable future innovation in thermal management solutions to keep up with rapidly evolving technological needs.

8.4.3. Interpretation of Test Results

Thermal management test result interpretation, particularly for heat sinks and TIMs, is crucial in determining performance parameters. Test measurement is typically thermal resistance, utilized as a measure of the effectiveness of heat dissipation by important components. There are a variety of factors involved, such as the material makeup of heat sinks and TIMs, ambient condition, and working load, which may affect these results to a significant degree. It is important for designers and engineers to understand how to interpret test results because they indicate the effectiveness of different configurations under actual conditions.

Proper interpretation can also guide follow-up experiments and thermal management innovations in devices to achieve maximum thermal efficiency in electronic devices. Precision in thermal measurements is also important when comparing different heat sinks and TIMs. Tests should have more than one metric, such as steady-state and transient thermal response, to have a full understanding of performance under different conditions. Ambient temperature and airflow variation, for instance, can influence thermal responses, and tests must be run in controlled environments. The use of advanced data handling methods, as discussed by Crane et al. (2025), provides greater capability to evaluate complex datasets with improved understanding of thermal behavior. These methods not only provide greater accuracy but also aid in driving redesigns, so they are an indispensable area of thermal management research. In addition, interpreting the implications of the data as a whole is significant because it connects thermal behavior to system reliability and life. In considering analysis of results, it is important to consider not only temporary thermal response but also long-term effectiveness over a variety of operating conditions. For instance, factors such as long-term material degradation can have a significant impact on the value of TIMs, which then affect the strategy for thermal management. As more designs become driven by environmental factors, analysis must be extended to account for potential impediments such as thermal cycling and mechanical stress. Environmental interactions and particle flux-related problems, noted by Goldberg et al. (2025), reflect the complexity of thermal management and necessitate a holistic examination of test results to develop innovation in heat sink and TIM technology.

8.5. INNOVATIONS IN HEAT SINK AND TIM TECHNOLOGY

The continually evolving technologies of heat sink and TIMs have transformed electronic device efficiency and reliability over the years. Advances in the material science have led to the development of heat sinks that employ advanced alloys and composites, improving thermal conduction while minimizing weight and space. These advanced heat sinks, engineered to enhance efficiency, typically come with optimized airflow designs to ensure that the convective heat exchange is maximized during operation. In addition, integration of active cooling technologies, such as thermoelectric devices, with passive thermal management solutions is a significant improvement. These new technologies facilitate enhanced thermal management in high-performance systems, which meet the increasing demands of current electronic systems and do not allow them to go beyond safe thermal operating conditions, lengthening the life and reliability of the systems (N/A, 2025). One of the most significant areas of innovation in TIM technology is the development of new materials with improved thermal conductivity and improved adhesion properties. New materials such as graphene and carbon nanotubes have been very promising, with unmatched capability to perform much better than traditional silicone-based TIMs. These new materials reduce the thermal resistance at the interface by a significant amount, enabling improved heat transfer from essential components to heat sinks.

Furthermore, developments in the manufacturing process of TIMs, such as using nanomaterials and microstructured surfaces, also enhance their performance by maximizing contact surface area and reducing air gaps. All this plays a significant role in the search for high-performance computing systems in which effective heat dissipation is critical in maintaining performance and preventing thermal throttling (Craig et al., 2007). The incorporation of new materials and new designs in heat sink and TIM technology not only optimizes thermal management but also enables the development of smaller electronic components. Miniaturization has generated the demand for high thermal performance solutions with smaller footprints, and efficient thermal management becomes all the more challenging. In exchange, designers are increasingly looking towards multi-material designs that utilize the unique properties of various materials, such as combining metals and ceramics or polymers. Additionally, greater use of additive manufacturing techniques allows for the creation of complex geometries previously unavailable, with improved thermal performance in confined areas. These advances in technology and ongoing research into new materials are all testaments to the commitment of the discipline to push the envelope of thermal management technology so that future devices will be capable of meeting both performance and size needs without compromise in reliability (N/A, 2025) (Figure 8.15).

Figure 8.15. Thermal interface materials are used both inside and outside of chip packages.

Source: https://www.qats.com/cms/2018/01/24/industry-developments-advances-thermal-interface-materials-electronics-cooling/

8.5.1. Emerging Materials and Their Benefits

Increasing need for efficient thermal management in electronic systems has driven immense growth in future materials. Among them, carbon allotropes in the form of graphene and carbon nanotubes have gained immense popularity because of their superior thermal conductivity properties. These materials have higher thermal conductivities than traditional metals, facilitating more efficient heat dissipation within smaller electronic systems. The studies show that room temperature thermal conductivity of carbon materials spans an impressive range, emphasizing the singular ability of carbon allotropes to meet the thermal requirements of next-generation electronics (Balandin et al., 2011). By exploiting the superior thermal properties of these materials, heat sinks and TIMs can be created with enhanced performance, enabling next-generation devices to operate efficiently and reliably under intense thermal loads. Hybrid composites that include blends of various future materials have been shown to be a promising method for further thermal improvement. For instance, studies on epoxy-based composites with graphene and copper nanoparticles in recent times exhibit a synergy which enhances thermal conductivity by a large margin. This effect is most pronounced in medium fillers, whose combination creates a highly conductive percolation network that is very efficient at sucking heat away from critical components (Balandin et al., 2019). These developments in composite material are not incremental steps; they constitute a paradigm shift in the design and application of TIMs. By adding these composites to thermal sinks, engineers can offer more effective cooling systems while potentially reducing thermal management systems' size and weight (Figure 8.16).

Figure 8.16. Increasing demand for efficient thermal management in electronic devices has catalyzed significant advancements in emerging materials.

Source: https://www.trumonytechs.com/innovations-in-tims-for-enhanced-thermal-management-of-electronics/

The benefits of these novel materials are not limited to thermal performance; they also offer advantages in electrical insulation and mechanical stability. Novel materials like graphene possess inherent better mechanical strengths, and thus they would be the ideal material for robust TIMs that can withstand operating stresses of electronic devices. Moreover, employing such materials in TIMs allows thermal interface resistance to be minimized, which would directly contribute to better device reliability and lifetime. The material design breakthrough—filler concentration variability and exploration of other hybrid mixtures—enables engineers to tailor thermal properties to specific application and environment requirements and usher in a new thermal management solution era (Balandin et al., 2011, 2019). As the technology evolves, application of these advanced materials promises to deliver better performance characteristics across a variety of electronic applications.

8.5.2. Advances in Manufacturing Techniques

Process advances have significantly influenced the production of TIMs and heat sinks, mainly through enhanced fabrication processes to enable enhanced thermal performance. With shrinking electronic components and their increasing power efficiency, there is greater demand for advanced TIMs. Contemporary manufacturing technologies such as additive manufacturing and precision engineering enable the development of increasingly complex geometries and materials optimized for effective heat transfer. The blend of polymer composites with advanced fillers has emerged as a major area of growth, based on scalable manufacturing processes. With improved thermal conductivity via novel design and material technology, the manufacturers can supply TIMs that facilitate improved heat dissipation, leading to prolonged life and performance of electronic components. Moreover, these innovations are important for the purpose of overcoming the constraints posed by high thermal outputs in modern electronics to ensure that users get long-lasting and trustworthy solutions. The need for effective thermal management solutions has inspired detailed research into the use of polymer composites as TIMs, particularly the processes of their manufacture. With the use of sophisticated filler integration in a polymer matrix, researchers have proven the potential of enhancing thermal conductivity in such composites through material means. The use of various fillers, for instance, metals and carbon materials, allows for improved through-plane thermal conductivity, as realized in research into newer production methods. This increasing significance of polymer composites in the electronic industry is further accentuated by their flexibility and scalability, which accommodate a range of applications from big data centers to portable devices. The ongoing research on manufacturing technologies for TIMs not only meets the urgent need for improved heat management solutions but also introduces compliance with economic and environmental standards to manufacturing processes, as is noted in industry trends. To harness the full range of potential for advanced manufacturing techniques for thermal management, cross-disciplinary collaboration is imperative. Workshops and conferences, such as the one that occurred in the course of the 3rd Workshop on Thermal Management in Telecommunication Systems and Data Centers, are meeting grounds where industry meets academia to discuss innovations and emerging challenges. Such conferences emphasize the importance of interdisciplinary perspectives in solving thermal management issues in all facets, ranging from data centers to vehicular technology. By dialogue among cooling technology and acoustic management experts, the thermal management community as a whole can determine common drivers for advancement and establish strategic plans for the future. Through this collaborative effort, ultimately, the refinement and evolution of manufacturing processes continue so that TIMs and heat sinks continue to meet the evolving needs of the electronic world, along with such critical considerations as energy efficiency and sustainability.

8.5.3. Future Trends in Thermal Management Solutions

As the demand for efficient thermal management technology keeps increasing, particularly in electronic applications, researchers are exploring new materials and approaches to assist in the transfer of heat in more efficient ways. One promising avenue is the integration of metal foams and metal matrix composites (MMCs) into the structures of heat sinks and TIMs. These materials provide a unique balance of high thermal conductivity and lightweight structures, making them suitable for applications where space and weight are critical factors. Advances in fabrication methods have enabled the development of materials engineered to meet specific thermal management needs. With the help of MMCs and metal foams, designers can optimize heat sinks for applications in various types, including liquid-liquid and gas-gas heat exchangers, ultimately leading to improved overall system efficiency and reduction of major challenges pertaining to traditional materials (Han et al., 2012). Another prominent trend in thermal management solutions involves the integration of nanotechnology, which has offered a few alternatives to greater thermal performance of TIMs. Nanomaterials such as graphene and carbon nanotubes possess high thermal conductivity and can be made to enhance thermal interfaces by large margins. In TIMs, they are employed to lower thermal resistance considerably, thereby increasing heat transfer efficiency between heat sources and heat sinks. The ability of thermal dissipation could also be further enhanced by applying nanostructured coatings on the surfaces of heat sinks. These advancements offer a path towards the development of smaller and more efficient cooling systems that are most essential in high-density electronic applications where spaces are limited, and thermal management is crucial (Han et al., 2012). In the future, thermal management solutions also offer prospects for smarter and adaptive systems with real-time monitoring and control capabilities. The use of Internet of Things (IoT) technology in thermal management systems allows it to change dynamically with temperature fluctuation and the demand for workload, thus providing enhanced performance. Adaptive thermal management systems can evaluate various parameters, such as ambient temperature and device usage, in a bid to regulate cooling attempts accordingly. This adaptability is particularly beneficial in uses like data centers, where unnecessary production of heat can lead to operational inefficiencies. By tapping into advances in sensor technology and data analysis, these smart thermal management solutions have the potential to lengthen significantly the life of electronic components and enhance long-term energy efficiency.

SUMMARY

In concluding the topic of heat sinks and TIMs, it is critical to observe the central role these components play in thermal management across different engineering uses. The effective dissipation of heat is integral to increasing the life expectancy and efficiency of electronic devices. As can be observed, advances in materials, specifically the creation of new paradigm-breaking carbon-based materials, have contributed significantly to the advancements of the thermal conductivity attainable. Research shows that graphene and carbon nanotubes possess high thermal characteristics, therefore qualifying to be suitable candidates for future TIMs. Thus, making such material implications possible allows engineers to create more efficient thermal management strategies that can adapt to the continuously imposed demands of electronic devices (Balandin et al., 2011). The other overarching idea that emerges from TIMs research is the imperative of constantly developing innovations in reducing thermal discomfort in actual uses, e.g., in prosthetics for medical use. The compromise between user comfort and material efficiency cannot be overstressed, particularly in the case of prosthetic device users suffering from severe thermal issues. Recent work demonstrates that other conductive materials, though promising with respect to liner design improvement, do not generally surpass baseline performance of standard silicone liners. This means that despite progress, thermal discomfort would have to be tackled by using more active solutions than the passive upgrading of material today. This discovery opens the door to further research to design active systems that could enhance user experience in prosthetic applications (Holloway et al., 2017). Finally, the culmination of research on heat sinks and TIMs underscores the imperative of interdisciplinarity and innovation. Ongoing advancements in materials science, coupled with practical engineering applications, will dictate the degree to which heat management concerns are resolved. While electronics continue to miniaturize but power increase, the thermal interface will remain a critical area of study, ensuring maximum device performance. In addition, thermal dynamic knowledge not only leads to more efficient electronic systems but allows users to enjoy improved quality of life in being able to access thermal management technologies across various environments. Accordingly, future research on TIMs and heat sinks must remain dynamically sensitive to shifting technology environments, forging a convergence between material innovations and utilitarian ends.

The addition of heat sinks and TIMs is essential in managing dissipation of heat in electronic systems. Effective thermal management not only guarantees operational dependability but also affects performance and device longevity. Heat sinks, through their function of soaking up and radiating heat off sensitive components, supplement TIMs, that serve to fill microscopically small air cavities between surfaces in order to achieve maximum thermal conductivity. Material choice is important; for instance, aluminum is extensively employed in heat sinks due to it being light and possessing excellent thermal conductivity. Besides, a number of TIMs, ranging from thermal pastes to phase change materials, are important in lowering the thermal resistance, hence improving component-to-heat sink heat transfer. Such synergy of heat sinks and TIMs underscores inter-disciplinary approaches to electronics design and accelerates the evolution of thermally efficient technologies. Additionally, comprehending heat dissipation principles is fundamental to improving design processes in electronics. Studies into the contributions of TIMs properties to system performance offer seminal concepts towards optimizing heat management solutions. For example, the thermal conductivity of TIMs is particularly germane to their performance with heat sinks. The application of advanced modeling techniques is becoming more and more common for engineers in order to forecast temperature profiles and material flow through thermal management systems. Numerical models, such as those recently under consideration simulating process parameters of material, assist in determining the optimum configurations of heat sinks and TIMs. Such modeling work not only increases our knowledge about thermal behaviors but also makes it easier to design future cooling solutions that can match the increasing needs of high-performance applications (Arbaoui et al., 2014). In addition, past examples demonstrate the dangers of incorrect heat sink and TIM applications. An example of the same can be referred to in the case analysis of the Magellan Radar digital unit where low-quality thermal management resulted in failure of joints in terms of thermal expansions of materials (Ross et al., 2025). The example highlights the requirement of rigorous test and design methodologies in order to ensure that such failures do not occur in electronic hardware in the future. It exposes how the amalgamation of thermal materials can contribute towards severe operational failures if not corrected. Therefore, a rational grasp of the thermal dynamics, material response, and potential failure modes is required. By reconciling these lessons with contemporary design practice, engineers are capable not only of promoting safer and more efficient systems but of ensuring compliance with industry standards in electronic component reliability.

REFERENCES

1. Ando, Y., Dushenko, S., Shigematsu, E., Shinjo, T., et al. (2018). Spin-wave-induced lateral temperature gradient in a YIG thin film/GGG system excited in an ESR cavity. https://arxiv.org/abs/1810.06875 (accessed on 23rd October 2025).
2. Andres, A. C. M., Battle, M., Bender, M., et al. (2006). Atmospheric potential oxygen: New observations and their implications for some atmospheric and oceanic models. https://core.ac.uk/download/2768216.pdf (accessed on 23rd October 2025).
3. Arbaoui, L., Camilleri, D., Cater, S. R., Galloway, A., et al. (2014). Advances in friction stir welding of steel: Project HILDA. https://core.ac.uk/download/17037941.pdf (accessed on 23rd October 2025).
4. Artiaga, R., Farooq, Z., Guo, Q., Li, Y., et al. (2024). Fillers and methods to improve the effective (out-plane) thermal conductivity of polymeric thermal interface materials–A review. https://core.ac.uk/download/614588595.pdf (accessed on 23rd October 2025).
5. Balandin, A. A., Barani, Z., Coleman, D., Geremew, A., et al. (2019). Thermal properties of the binary-filler composites with few-layer graphene and copper nanoparticles. https://arxiv.org/abs/1905.08725 (accessed on 23rd October 2025).
6. Balandin, A., Casher, A., Dhar, A., Khitun, A., Lippi, A., Reina, A., et al. (2011). Thermal properties of graphene, carbon nanotubes and nanostructured carbon materials. https://arxiv.org/abs/1106.3789 (accessed on 23rd October 2025).
7. Boglietti, A., Cavagnino, A., Mejuto, C., Mueller, M., et al. (2009). Evolution and modern approaches for thermal analysis of electrical machines. https://core.ac.uk/download/pdf/11409050.pdf (accessed on 23rd October 2025).
8. Boglietti, A., Cavagnino, A., Parvis, M., Vallan, M., et al. (2006). Evaluation of radiation thermal resistances in industrial motors. https://core.ac.uk/download/pdf/11370380.pdf (accessed on 23rd October 2025).
9. Burrows, S. E., Dixon, S., Edwards, R. S., Mañosa, L., et al. (2011). Temperature contour maps at the strain-induced martensitic transition of a Cu–Zn–Al shape-memory single crystal. https://core.ac.uk/download/1386882.pdf (accessed on 23rd October 2025).
10. Cooper, A. H., Farrant, A. R., & Price, S. J. (2010). Karst, GIS and geological hazard planning and management in Great Britain. https://core.ac.uk/download/56754.pdf (accessed on 23rd October 2025).

11. Craig, J. E., Felicelli, S. D., & Wang, L. (2007). Thermal modeling and experimental validation in the LENS™ process. https://core.ac.uk/download/287648624.pdf (accessed on 23rd October 2025).
12. Crane, R. B., Malila, W. A., & Richardson, W. (2025). Discrimination techniques employing both reflective and thermal multispectral signals. https://core.ac.uk/download/pdf/42898534.pdf (accessed on 23rd October 2025).
13. Ganatra, Y. Y. (2016). Passive thermal management using phase change materials. https://core.ac.uk/download/145190889.pdf (accessed on 23rd October 2025).
14. Garimella, S. V., Gektin, V., Persoons, T., Weibel, J. A., et al. (2017). Electronics thermal management in information and communications technologies: Challenges and future directions. https://core.ac.uk/download/127583165.pdf (accessed on 23rd October 2025).
15. Goldberg, E. D. (2025). Remote sensing and problems of the hydrosphere. https://core.ac.uk/download/pdf/42867044.pdf (accessed on 23rd October 2025).
16. Han, X. H., Jacobi, A., Park, Y. G., Sommers, A., et al. (2012). A review of metal foam and metal matrix composites for heat exchangers and heat sinks. https://core.ac.uk/download/55763987.pdf (accessed on 23rd October 2025).
17. Hao, M. (2016). Carbon-based nanomaterials and their ensembles for high temperature thermal applications. https://core.ac.uk/download/220145977.pdf (accessed on 23rd October 2025).
18. Holloway, C., Washington, E. D., Foort, J., Hoaglund, F. T., Levy, S. W., Miodownik, M., & Williams, R. J. (2017). The effect of liner design and materials selection on prosthesis interface heat dissipation. https://core.ac.uk/download/132195080.pdf (accessed on 23rd October 2025).
19. Khuu, V. (2009). Evaluation of thermal interface materials and the laser flash method. https://core.ac.uk/download/56106444.pdf (accessed on 23rd October 2025).
20. Louie, B. M. (2011). Design of a steady state thermal conductivity measurement device for CNT RET polymer composites. https://core.ac.uk/download/484223014.pdf (accessed on 23rd October 2025).
21. N/A. (2025). Highly efficient mechanism for cooling high-power pluggable optical transceivers via enhanced thermal contact at elevated ambient temperatures. https://core.ac.uk/download/658903378.pdf (accessed on 23rd October 2025).

22. Naghibi, S. (2020). Noncaring graphene thermal interface materials for advanced electronics. https://core.ac.uk/download/323082404.pdf (accessed on 23rd October 2025).
23. Oddone, V., Reich, S., & Wimpory, R. C. (2018). Understanding the negative thermal expansion in planar graphite–metal composites. https://core.ac.uk/download/228064792.pdf (accessed on 23rd October 2025).
24. Poulikakos, D. (2009). On emerging micro- and nanoscale thermofluidic technologies. 2nd Micro and Nano Flows Conference (MNF2009). https://core.ac.uk/download/9633457.pdf (accessed on 23rd October 2025).
25. Ross, R. G. (2025). Magellan/Galileo solder joint failure analysis and recommendations. https://core.ac.uk/download/pdf/42824952.pdf (accessed on 23rd October 2025).
26. Tappura, K. (2018). A numerical study on the design trade-offs of a thin-film thermoelectric generator for large-area applications. http://arxiv.org/abs/1801.06418 (accessed on 23rd October 2025).

CHAPTER 9

RELIABILITY AND THERMAL STRESS IN ELECTRONICS

Contents

INTRODUCTION

Electronic system complexity is on the rise, and accordingly there is an increased requirement for a comprehensive understanding of reliability under thermal stress. Electronic components can degrade due to various reasons, but primarily due to thermal loads generated during normal operation. Thermal effects can lead to failures in critical components, and ultimately the total ability and life of systems are negatively impacted. Therefore, there is a necessity to talk about the impact of thermal stress on electronics' reliability and developing techniques that would be able to forecast future failures. As electronic devices continue to become smaller in size, managing thermal stress is not only critical in maintaining performance but also in order to ensure user safety and satisfaction. Knowledge of the thermal stress-reliability relationship provides the basis for optimizing the design and maintenance of electronic products for both performance and long life. Ongoing monitoring of the health status of electronic systems is an important part of maximizing reliability, especially in application environments where thermal stress is common. Novel approaches such as prognostics make the assessment of an electronic products health possible, which forecasts potential failure before it happens. Monitoring thermal loads is essential due to the fact that factors such as power cycles and temperature gradients have a direct impact on component degradation. With the application of continuous monitoring techniques, now it is possible to construct data that provides predictive algorithms with information regarding the functional condition of an electronic system. Thus, implementation of real-time analytics can add a new dimension to proactive maintenance practices and reduce lifecycle costs. Not only do these prevent unexpected downtime but also improve the support and design infrastructure for future electronic systems, thus leading to improved reliability. Solder joints, being a critical link of electronic assemblies, provide a best example of thermal and mechanical stressors' challenges. These bonds are typically the first failure locations due to their thermal degradation sensitivity, which has the potential to cause severe operational issues. Defining the dynamics of solder joint condition is vital since degradation often starts on the surface, where the high-speed signals converge, a situation referred to as the skin effect. With digital signal characterization techniques, scientists have demonstrated the capability of non-destructive assessment of solder joint integrity. This strategy not only enables the real-time monitoring of health but also as a preventive step for the observation of early signs of failure. By embedding such monitoring systems, electronic equipment can be made more reliable and longer-lived, opening the door to trusted systems that can withstand the ordeal offered by thermal stress (Pecht et al., 2006; Yoon et al., 2016).

Definition of Reliability in Electronics

The electronics reliability principle is essential to providing consistent performance across various applications, particularly those applications that are subjected to thermal stress. Reliability can be defined as a measure of the probability that a device will fulfill its intended function in described conditions over a specified time. In electronics, therefore, not only is the ability to operate without failure involved but also the maintenance of performance during changes in temperature and other environmental factors. A reliable electronic component is therefore one that can withstand thermal cycling, which can introduce stress that could lead to accelerated modes of failure. This focuses on the delicate interplay of reliability and thermal stress management during electronic system design and operation and highlights the necessity of rigorous testing and quality assurance requirements during production. Furthermore, reliability assessment extends beyond mere operating statistics to examine comprehension of the failure mechanisms that can happen under ordinary use, particularly under conditions of thermal stress. These modes can be thermal fatigue, electromigration, and mechanical deformation, all of which have the potential to degrade the integrity of electronic circuits with time. Building a firm reliability basis involves consideration of these aspects within the context of particular applications. For instance, in the aerospace sector in which components experience severe thermal fluctuations, NASAs own study aims to develop materials and systems that not only perform efficiently but also provide hints towards sites and life expectancy of imminent failure (Cazier et al., 2025) (Figure 9.1).

Figure 9.1. Electronics reliability.

Source: https://ketiv.com/wp-content/uploads/2021/05/ansys-electronics-2.jpg.

Thus, by determining thermal stress-related vulnerabilities, engineers can enhance electronic parts and systems' reliability and ensure better efficiency and longevity. Finally, the general implications of reliability extend to the design phase of electronics, where thermal stress is integrated into the component's life cycle analysis. It is a holistic process that involves the selection of materials and engineering practices aimed at avoiding thermal-induced failures from occurring. These prophylactic actions include the provision of thermal control systems and use of stable materials that deliver consistent performance under varied conditions. An example is, in the context of considering next-generation manned launch systems for space exploration, NASA highlighted the fact that effective thermal management is significant for ensuring the longevity and operational efficiency of electronic systems in vehicles designed for hostile environments (Brown et al., 2025). By focusing on reliability at the time of product design, manufacturers are better able to mitigate the risk of thermal stress and ultimately supply more reliable electronic products across a wide range of industries.

Importance of Thermal Management

Good thermal management is essential to electronic device reliability and lifespan, particularly in a world where devices are shrinking and increasing in power. Elevated temperatures lead to thermal stress, which subsequently degrades electronic performance and reliability. Not only do elevated operating temperatures degrade the integrity of the components, but they also are the root cause for mechanisms of failure such as electromigration, thermal cycling, and delamination. By application of thermal load tracking and management techniques, these effects can be reduced, and devices perform optimally. As thermal profiles are under good control, related performance gains not only extend electronic system life in operation but also improve electronic systems as functioning optimally and more enjoyable to use, with tight thermal management, hence, playing a critical role in electronic design. Further, the reliability of modern electronics is highly susceptible to many thermal parameters, such as temperature gradients and thermal cycling. The interplay between these and performance is crucial; under some situations, devices at high temperatures can jeopardize signal integrity, which has a direct effect on overall performance (Gonzalez et al., 2019). Acknowledging and combating the thermal stresses caused by these conditions is central to the development of more resilient electronic systems. By ongoing monitoring and use of novel materials and designs, e.g., low temperature co-fired ceramics, engineers are able to design components that will function properly at higher temperatures without diminished performance. Such a proactive action serves to pre-detect failures and enable timely maintenance and replacement, thereby enhancing the necessity of proper thermal management. The integration of prognostic methods in thermal management strategies is a giant step towards guaranteeing electronic component reliability. By employing thermal load statistics, designers can anticipate failures and make proper design decisions to enhance performance.

As recent studies confirm, prognostics enable manufacturers to establish differences in product performance and predict future reliability through real-time testing (Pecht et al., 2006). As electronic components evolve, proper thermal management becomes increasingly necessary. Combining thermal mechanics with prognostic analysis can lead to comprehensive strategies that not only treat current thermal conditions but also forecast future needs of electronics. This forecasting is essential for high-performance electronic system-based industries, further emphasizing the necessity of thermal management for reliability.

Overview of the Relationship Between Reliability and Thermal Stress

The relationship between reliability and thermal stress of electronics is multifaceted, primarily controlled by the intrinsic thermophysical characteristics of electronic materials and their operational environments. Electronic devices generate heat during operation, leading to thermal cycling that is well documented to influence the system reliability. Thermal stress arises from changes in temperature, which have the potential to induce mechanical stresses into the materials, thus altering their structure. This degradation can come in many forms like solder joint failure and material delamination, particularly under conditions of high-speed temperature cycling or continuous high temperatures. Engineers come to design electronic systems to survive such stress by understanding how thermal stress impact's reliability, extending the lifespan and performance of their devices. Implementation of thermal management techniques should be incorporated to offset adverse impacts of thermal stress and thereby improve overall reliability. Continuous monitoring of thermal loads plays an important role in the enhancement of electronic system reliability. Real-time data acquisition technologies provide the capability of comprehending thermal behaviors, e.g., identifying stress points predisposed to failure. Putting in place prognostics techniques, as demonstrated in other research, helps one calculate when the parts might fail due to thermal stresses, which makes it simpler to have maintenance plans beforehand. For instance, thermal data can be integrated with models of stress-and-damage to assess the wellness of electronic components and estimate their working life. This proactive approach not only optimizes maintenance schemes but also helps redesign parts to address found vulnerabilities, an undertaking that is a priority in equipment where reliability is the determining factor. Such advances demonstrate the underlying relationship between thermal management and electronic system reliability. The role of solder joint integrity in electronics cannot be overstated, given its exposure to thermal and mechanical stress. Solder joints are often critical interfaces for electrical connections in electronic assemblies, and solder joint failure can be catastrophic to system reliability. Under operation of electronics under alternating thermal environments, solder joints can be subjected to thermal cycling that induces consequential issues such as thermal fatigue or fatigue crack growth. Research indicates that signal integrity monitoring through high-speed digital signals can be employed as a non-destructive test technique to diagnose initial signs of solder joint degradation to undertake interventions at a timely manner before catastrophic failure (Yoon et al., 2016). This process deems important continuous health monitoring in avoiding possible failures with thermal stress, thereby strengthening the critical connection between extreme thermal management and electronic system reliability (Pecht et al., 2006).

9.1. FUNDAMENTALS OF THERMAL STRESS

In electronics, thermal stress basics are essential in ensuring reliability and extending the life of devices. Thermal stress happens due to abrupt temperature changes, leading to differential expansion and contraction of materials. Its impact is most significant in solder joints that are necessary for electronic assembly functioning.

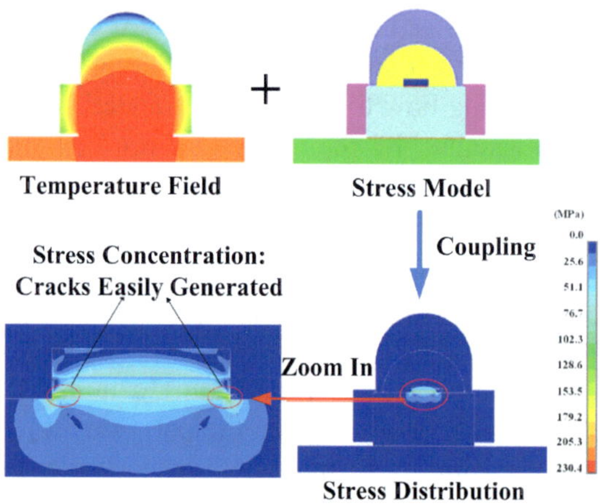

Figure 9.2. Thermal stress analysis of high-power LED.

Source: https://www.researchgate.net/figure/Thermal-stress-analysis-of-high-power-LED_fig8_241288934.

As the temperatures vary with use, varying thermal expansion characteristics of materials can lead to high internal stresses, resulting in mechanical failure. Material behavior under conditions of varying temperatures must therefore be understood to avert the dangers posed by such stresses. Experiments, such as those performed using the Johnson-Cook model, provide valuable insights into the impact of thermal conditions on the mechanical characteristics of solder materials under dynamic load, especially under high strain rates and varying initial temperatures (Wang et al., 2014). One of the most significant factors in the assessment of thermal stress is the impact of thermal stress on solder joint deterioration. Solder joints are usually the weak points of electronic assemblies, exposed to thermal, mechanical, and chemical stress throughout their operating life. Joint degradation can lead to failures that may destroy system integrity (Figure 9.2).

Notably, impacts like the skin effect will at high frequencies exacerbate such issues by causing stress concentrations in solder joints. Continuous monitoring of signal integrity in digital systems is necessary because early degradation processes may be detected by alterations in digital signal characteristics. This assessment not only permits the detection of possible failure before its occurrence but also highlights the importance of understanding thermal stress mechanisms in electronic part design and test phases (Yoon et al., 2016). Furthermore, employing non-destructive testing methods to monitor the integrity of solder joints is a significant advancement in thermal stress effect control. By employing digital signals to gauge the condition of solder joints, engineers are able to ascertain the integrity of electronic assemblies in real time. It encompasses monitoring signal parameters, such as eye parameters and jitter, which bear a direct correlation with solder connection integrity. In high-speed life testing, the progressive failure of stressed solder joints can be accurately described under controlled levels of stress, whereby useful information regarding the involved failure modes is achieved. The forward-looking monitoring technique not only aids in sustaining system performance but also aids in optimizing design requirements, which eventually means more reliable electronic devices able to withstand the inherent challenges of thermal stress conditions (Yoon et al., 2016).

9.1.1. Explanation of Thermal Stress in Materials

Thermal stress in materials takes place because of uneven temperature distributions that result in differential contraction or expansion within the material structure. Materials contract or expand with the variation of temperatures in proportion to their thermal coefficients. But in electrical equipment, the heating might differ because of periodic power use or environmental conditions, leading to the development of localized temperature gradients. Such gradients cause internal stresses because different parts of the material respond differently to temperature changes. This stress can compromise the integrity of parts, particularly in solder joints and other interfaces, where there is the potential for fracture under thermal fatigue conditions. Understanding the physics of thermal stress is crucial in estimating the reliability of electronic systems because materials undergo thermal cycles all day and night throughout their life, with potential failure modes such as cracking or delamination. For thermal stress management in electronic packaging, interfacial bonding is critically enabled. Materials like underfill resin used in flip chip packaging are designed to neutralize thermal fatigue in solder joints, thereby enhancing reliability. Lacking proper evaluation of the interface strength during temperature cycling will lead to significant weakening of the structure. Traditional approaches presume perfect bonding without considering the complexities arising from micro-scale structural imperfections in the interface. Recent studies suggest a bimaterial model that takes into account such defects, modeling stress fields surrounding interfacial cracks through finite element analysis (FEA). Experience shows that the mechanical behavior of the interface is greatly influenced by variations in bonding quality, evidencing the need to take into account realistic conditions in material evaluation to avoid catastrophic failure in electronic devices (Matsumoto et al., 2006) (Figure 9.3).

Figure 9.3. Thermal stress in solder joints.

Source: https://www.chemtronics.com/content/images/thumbs/0002249_how-to-identify-and-solve-thermal-stress-issues-in-solder-joints.jpeg.

In addition, the advent of more sophisticated modeling techniques has facilitated understanding thermal management in new electronic products, e.g., SSL lamps, with much ease. Thermal and thermo-mechanical modeling together play a vital role to ensure maximum reliability in these devices. By applying sophisticated simulation tools like ANSYS and CoventorWare, researchers can create precise three-dimensional thermal distribution models of LED lamps, which are thereafter verified through empirical thermal measurements. These models not only support prediction of regions of interest for heat transfer but also extend to impact better material and configuration designs that optimize light quality and life. As more demand is placed on electronics with high performance, consequences of full-scale thermal stress analysis become more important in safeguarding electronic systems against reliability and lifespan problems in a rapidly changing technological landscape (Formanek et al., 2012) (Figure 9.4).

Figure 9.4. Fractured solder joint.

Source: https://www.chemtronics.com/how-to-identify-and-solve-thermal-stress-issues-in-solder-joints?srsltid=AfmBOoqDNRnWBQS7gjCzn6pVSLTO1U4Swf3B5__ZY4kZe18aaKPNjIvi.

9.1.2. Factors Contributing to Thermal Stress

Electronic devices are controlled by thermal stress, which is characterized by fast temperature fluctuation that occurs during operation. The fluctuations can be induced by changing operational conditions such as power cycling and load changes that create gradients in the material during heating of an electronic device. As different materials heat up at different rates, mechanical stress is thus produced. This stress can lead to microcrack initiation and eventually material failure. Understanding this effect is essential in creating robust devices that are able to withstand such thermal stresses over their lifetime. A practical example includes the utilization of heat-dissipating materials and effective cooling methods to compensate for the adverse effects of thermal cycling, hence enhancing the reliability of electronic devices as a whole. Research indicates that addressing such temperature fluctuations at the early stage of the design process can yield significant long-term performance and life benefits. Material usage is also another reason that causes thermal stress. Different materials possess different coefficients of thermal expansion; thus, when they are subjected to thermal stress, they react differently to changes in temperature. For instance, metals will tend to expand more than ceramics or polymers and thus may cause likely delamination at interfaces where such materials meet. This disparity is likely to initiate component and solder joint assembly failures, which play a crucial role in maintaining electrical contact. In addition, technology advancement, such as the development of composite materials, has proved promising in offering better thermal stability and reduced weight. These materials can accommodate more thermal gradients without enduring significant damage and, thus, are creating new design opportunities with a focus on longevity and dependability. This is an important field of material science when considering the interplay between design and functionality in electronic devices. Additionally, external environmental factors have a significant contribution to thermal stress. Equipment is subject to many environmental conditions, from humidity, temperature extremes, and mechanical vibrations. These contribute to the internal thermal cycling that takes place in electronics. For instance, prolonged exposure to elevated ambient temperatures can trigger thermal degradation mechanisms and enhance thermal expansion mismatch phenomena. In addition, representative test environments for real-world applications, e.g., space missions or high-performance computing, are critical to the analysis of these stresses since in integrated systems like the Integrated Power and Attitude Control System (IPACS) (Eisenhaure et al., 2025). Robust designs must be capable of tolerating these variations in order to be reliably repeatable, which suggests that careful testing and material choice procedures are necessary in reducing device failure rates due to thermal stress.

9.1.3. Impact of Thermal Cycles on Electronic Components

The thermal cycles greatly affect electronic component reliability by engaging materials in expansions and contractions due to variations in temperatures. Most thermal cycles induce mechanical stress that can result in material fatigue, cracking, or even failure of components. It is very significant in solder joints and interconnects, where differential coefficients of thermal expansion

(CTE) between different materials can create strain. Because electronic components are bearing ever increasing thermal loads as they operate, the working relationship between temperature cycling and stress induced is critical to understand. How these thermal cycles operate on a continuous basis can provide insight into component operational life, enabling designers to develop predictive maintenance methods based on the outcomes of Pecht et al. (2006). Thermal cycle influences not only mechanical stress but also the physical properties of the materials, which change significantly. For example, electronic components, while undergoing changing thermal conditions, may transform their microstructure, inducing local plasticity affecting overall performance. An example is research on through-silicon vias (TSVs) because thermal expansion coefficient incompatibility between copper vias and silicon wafers can result in stresses that reduce reliability. Experiments using tools such as wafer curvature and finite element analysis have shown that the thermal cycling can create huge internal stresses that can initiate failure in stacked die structures. These issues foreground the significance of a critical understanding and prognostic models that can predict failure modes due to thermal stressors, as noted by Ho et al. (2014). New methodologies are emerging to address the negative effect of thermal cycles on electronic components. One such approach is the use of prognostics based on continuous temperature data monitoring to inform maintenance practice and design change. By correlating thermal loads to algorithms of stress and damage, engineers can better predict when the components will fail so they can intervene more effectively. Being proactive not only reduces life cycle expenses but also enhances the overall reliability of electronic systems. Therefore, the inclusion of real-time thermal information in engineering practice increases insight into the impact of thermal cycles on material properties and behavior, leading to more resilient electronic designs that withstand the strains of their environmental conditions as referenced by Pecht et al. (2006).

9.2. TYPES OF ELECTRONIC COMPONENTS AFFECTED BY THERMAL STRESS

Thermal stress is a basic problem to most electronic devices, and understanding the complexities of the manner in which these devices react to temperature changes is crucial in a bid to enhance reliability. Semiconductors, particularly transistors and diodes, are particularly susceptible to thermal stress owing to the nature of their materials and the manner in which they operate. Temperatures lead to variations in carrier concentration, which in turn affect their performance, eventually resulting in thermal runaway conditions. The purity of the semiconductor material and packaging are essential to its heat transfer capability; hence, effective thermal management techniques, including heat sinks and TIMs, are an integral part of sustaining functionality when under stress. All these need to be addressed, as compromised silicon performance can result in catastrophic failures that lower electronic system reliability. Another group of electronic components significantly affected by thermal stress are capacitors, especially ceramic and electrolytic capacitors. These also experience physical changes through temperature-driven expansion and contraction, which might result in dielectric breakdown or even total failure at worst. Capacitor performance characteristics vary extensively with temperature, affecting their capacitance, equivalent series resistance, and leakage current. Besides, repeated thermal cycling can accelerate aging effects, degrading capacitance retention and even risking increased failure risk in circuit usage. Thermostable design is required to ensure long-term reliability, e.g., designing circuits to support outlined temperature ranges and choosing capacitors rated for those values. All these are critical in applications where component failure can lead to tremendous system downtime or pose safety risks. Lastly, passive components like resistors and inductors are also vulnerable to the adverse effects of thermal stress. As temperature varies, the value of a resistor's resistance may change depending on the resistance temperature coefficient, which can lead to signal integrity issues. Thermal stress can alter inductance in an inductor due to variations in core material properties and structures of the windings, impacting performance in power management devices. The threat of diminished reliability under thermal stress necessitates correct rating and selection of such components to withstand the working conditions. The engineers must consider these at the designing level because the interaction between thermal variation and electronic component reliability is the key to guaranteeing the vitality and lifespan of electronic systems in general. Mitigation of such problems by new design technologies and material developments is vital to the future development of electronics in extreme environments.

9.2.1. Integrated Circuits (ICs)

Integrated circuits (ICs) are fundamental building blocks of modern electronics, which enable devices to be smaller yet more capable in a wide range of industries. Being small facilitates high performance as transistor density is enhanced, which is a need for advanced applications. More specifically, three-dimensional integrated circuits (3D ICs) are of interest because they enable them to stack silicon dies in layers so that heterogeneous technologies can be combined even further. This stacking process presents unique thermal management challenges, primarily associated with heat generated by closely packed devices. An awareness of these issues and solutions to them are needed to improve the reliability of ICs. Scientists are thus exploring various methodologies that can minimize thermal stress, which can have a very significant effect on integrated circuit performance and lifespan, and this focus has brought new frontiers of innovation in electronic reliability assessment and design (Hopsch et al., 2016). IC reliability is also intrinsically connected with thermal behavior since increased temperatures can accelerate failure mechanisms like thermal fatigue and electromigration. As components operate, they generate heat, and this leads to large temperature gradients in various regions of the chip. This temperature gradient can potentially introduce mechanical stresses that degrade the integrity of interconnects and semiconductor material with the passage of time. Comprehensive models of component lifetime that consider thermal cycling and stress are essential to precisely predict failure. There are two main methods of predicting reliability: the military handbook empirical reliability prediction technique, which is very simple to use, and complex degradation mechanisms modeled in circuit simulators. The latter is easier to use but does not offer any design parameters to incorporate built-in reliability, leaving a great void in IC lifecycle analysis (Hu et al., 1994) (Figure 9.5).

Figure 9.5. Integrated circuits (ICs) are foundational components in modern electronics.

Source: https://cdn.sparkfun.com/assets/7/a/6/9/c/51c0d009ce395feb33000000.jpg.

Thermal management research on ICs is crucial to ensuring their functional viability and reliability in extreme conditions. Of key interest is the development of next-generation materials and design techniques that enhance thermal conductivity and better dissipate heat. In addition, new packaging approaches will improve airflow and thermal coupling, again minimizing the danger of thermal stress. Since the trend is to reduce devices but make them even stronger, the role of ICs in achieving this cannot be emphasized enough. Constant studies on thermal stress effects not only ensures the long-term integrity of individual components but also aids in the overall reliability of electronic systems. This symbiotic relationship between performance and reliability underlines the need for ongoing research in integrated circuit design, leading to increased future performance and reliability in electronics (Hopsch et al., 2016).

9.2.2. Power Electronics

Power electronics play a very critical role in enhancing the performance and efficiency of most electrical systems, particularly in current applications such as electric vehicles and aviation technology. With the increase in the range of such applications, power electronics component reliability has grown to be of concern due to the critical role they play in ensuring reliable operation. It is essential to integrate state-of-the-art materials and smart design strategies to address the challenge of thermal stress.

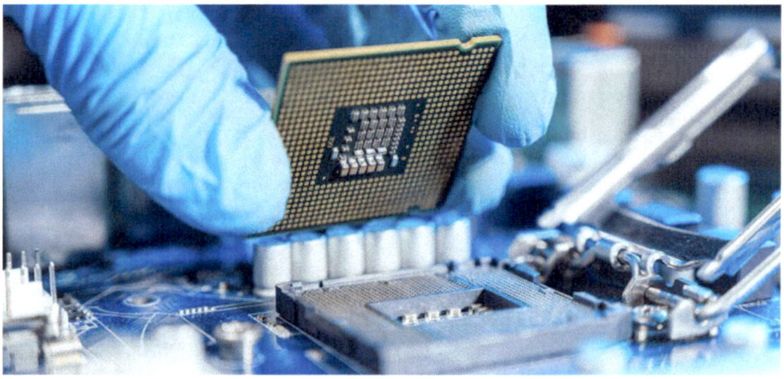

Figure 9.6. Power electronics play a critical role in enhancing the efficiency and performance of various electrical systems.

Source: https://images.ctfassets.net/wp1lcwdav1p1/3cAQ87K4KsHjfXGrQdkzBZ/b47523eae1fb2ac5b12bdf6619d71f2f/GettyImages-1307216100.jpg?w=1500&h=680&q=60&fit=fill&f=faces&fm=jpg&fl=progressive.

High operating temperatures are detrimental to the lifespan of such components, and they lead to performance degradation and ultimate failures. There is a need to understand the thermal dynamics in power electronics systems so that they can be optimized during design and maintenance for long-term reliability. The performance of power electronics relies heavily on thermal loads that result from continuous operation and ambient temperatures. Steady-state temperatures, power cycling, and thermal gradients are some of the factors that contribute significantly to the degradation of power electronics components. With prognostic methods, one can be capable of monitoring these thermal stresses and predict the reliability of power electronics components in the future. For instance, prognostic models based on known thermal monitoring in real-time can offer informative data on when exactly a component is expected to fail, enabling the development of preventive maintenance plans. This proactive measure not only minimizes downtime in operation but also optimizes the total longevity of the electronics, thereby reducing lifecycle costs as demonstrated in various case studies (Pecht et al., 2006) (Figure 9.6).

Moreover, recent work has employed advanced probabilistic modeling techniques, such as Monte-Carlo simulations, to estimate power electronic reliability and remaining useful life. Such models are beneficial for the estimation of the degradation path of components under various thermal and operational conditions, with a statistical model of predicting failure. Probability distributions representing the process of degradation can be employed to simulate various conditions that could influence component reliability. This method enables stakeholders to make informed choices regarding system design and servicing, thus enhancing electric vehicle performance as well as other advanced power-electronic-based systems. Such studies bring to light the importance of including reliability analysis at design levels to ensure that thermal stress is adequately addressed to prevent risks (Alghassi et al., 2014).

9.2.3. Passive Components (Resistors, Capacitors)

The role of passive elements, i.e., resistors and capacitors, is highly significant in the design and reliability of electronic circuitry, particularly thermal stress into account. Resistors generate heat by dissipating power, possibly affecting their operating temperature and, hence, their performance and life. Operating at high temperatures can lead to higher resistance and likely failure modes. Therefore, it becomes necessary to select resistors that will withstand temperature variations with precision and stability. Techniques that incorporate the use of materials with high thermal coefficients can be used to compensate for these effects so that the parts are stable under varying temperatures. In high-requirement applications like power electronics, the design of passive devices takes on a greater importance, requiring advancements in the area of material science to allow continued functionality under conditions of heat stress. Capacitors and resistors likewise experience thermal stress-related issues that can compromise their dependability in electronic circuits. Because capacitors store energy, their dielectric material is at risk of breakdown or leakage due to excessive heat.

Capacitor performance dependability is primarily a function of the choice of dielectric material and the process of making it, e.g., certain ceramic capacitors are more thermally stable than organic capacitors. Consequently, the demand for high-temperature capacitors has prompted R&D in new materials and manufacturing technology.

Recent developments report that the development of passive components for use under such conditions not only facilitates their functionality but also ensures durability in applications subject to thermal cycling, a common scenario in modern electronic systems (Dziedzic et al., 2013) (Figure 9.7).

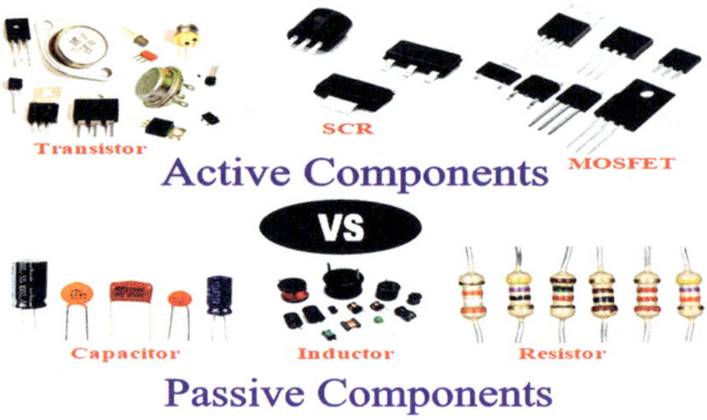

Figure 9.7. Passive components.

Source: https://images.theengineeringprojects.com/image/main/2024/02/difference-between-active-and-passive-components-3.png.

High-temperature passive components' evolution necessitates revision of manufacturing processes for enhanced reliability when under thermal stress. Screen-printing techniques that have been conventionally applied in passive component production, particularly for thick-film or Low-Temperature Co-fired Ceramics (LTCC), may be modified to ensure that micropassives may efficiently operate within the specified temperature ranges. Specialized methods such as comparing standard methods with photolithography or laser cutting can yield resistors and capacitors which meet the stringent demands placed upon them by high-temperature application.

Furthermore, long-time temperature stability is a severe challenge since complex interfacial phenomena related to these micro-fabricated products come into play. However, as exemplified through existing research, precise material selection and deposition procedures can result in the creation of reliable micropassives intended to endure the ruggedness of high-operational temperatures and thus cater to the emerging field of high-temperature electronics (Abbé et al., 2013) (Figure 9.8).

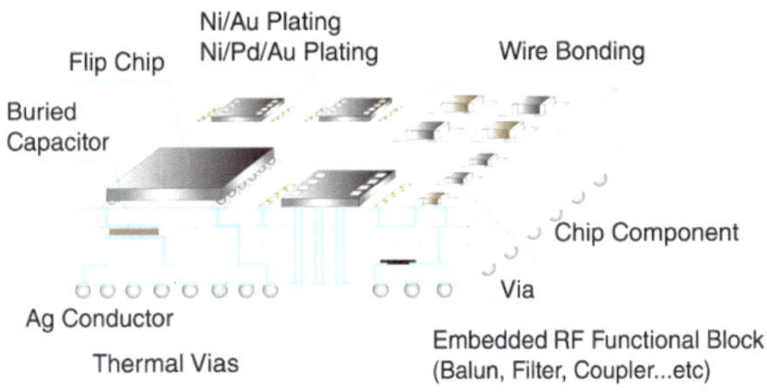

Figure 9.8. Low temperature co-fired ceramics (LTCC).

Source: https://orbray.com/en/product/jewel/product/images/01-14-01-02.png.

9.3. MECHANISMS OF FAILURE DUE TO THERMAL STRESS

Thermal stress is primarily due to differential contraction and expansion of materials in electronic components and can result in serious reliability issues. As the materials heat up, they expand when in operation; they contract when cooled. This repeated cycle can lead to the mechanical stresses, especially when materials of different types are bonded together, like the solder joints of integrated circuits. These elements' operating thermal gradients can result in fatigue, cracking, and eventual failure. It is crucial to comprehend these mechanisms because they enable engineers to design systems with lower thermal stress risks, thereby extending the life of electronic components and making them reliable (Pecht et al., 2006). One of the most important failure mechanisms of thermal stress relates to power cycling, which generates localized heating in electronic components. For instance, in Insulated Gate Bipolar Transistors (IGBTs), on-current flow generates localized thermal loads that impose mechanical stresses in the bond materials. Cycling thermal stress may have a severe impact on the reliability of power electronics and requires advanced monitoring techniques of thermal loads. Recent research supports the virtue of *in-situ* thermal monitoring in an attempt to understand better these stresses, and predictive maintenance can be so designed that it can warn users of impending failures prior to their occurrence, thereby ensuring utmost system reliability (Medjahed et al., 2012) (Figure 9.9).

Focusing on proactive measures, the integration of prognostics in electronic systems has been promising to counteract thermal failure mechanisms. By monitoring thermal performance for periods of time and utilizing algorithms to predict reliability in terms of measured conditions, engineers can extend the operational life of electronic devices. Prognostics, this process can significantly improve maintenance schedules and determine design modifications that enhance thermal management. For electronics subjected to cyclic thermal loads, e.g., power electronics in the automotive sector, such predictive insights are invaluable. Hence, acquisition of knowledge on mechanisms of thermal stress and utilization of effective prognostic methods can lead to dramatic improvement in electronic reliability and, consequently, minimize lifecycle costs as well as enhance system performance (Pecht et al., 2006).

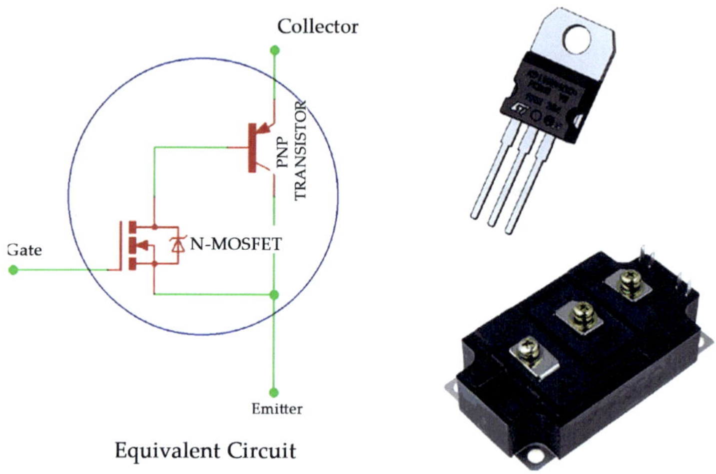

Figure 9.9. Understanding insulated gate bipolar transistors (IGBTs).

Source: https://www.ic-components.com/upfile/images/50/20240524172834464.png.

9.3.1. Mechanical Fatigue and Cracking

Mechanical fatigue poses a significant risk to electronic component reliability, particularly when devices are further miniaturized. Mechanical fatigue processing happens when materials are subjected to cyclic loading and develop a gradual accumulation of damage leading to cracks. Thermal cycling that occurs in electronic packaging, particularly involving solder joints, hastens the process, resulting in interfacial cracking that jeopardizes performance. For instance, underfill resins, which are regularly utilized in flip chip packaging, play a critical role in minimizing thermal fatigue fractures. However, when the size of these electronic devices reduces, the challenge in actually evaluating and estimating interfacial strength rises, where it is challenging to estimate reliability and durability of electronic assemblies, especially in fluctuating thermal conditions (Matsumoto et al., 2006). It is vital to understand the cracking mechanisms to maximize component design. Because the operating environment fluctuates in temperature, thermal stress placed upon the materials will be the reason why micro-cracking occurs in solder joints and interfaces. Such cracking progressing under continuous thermal cycling will ultimately lead to complete failure in the electronic assembly. One of the most interesting examples was when the Magellan spacecraft failed due to high thermal coefficient of expansion material causing open circuit solder joints, thus creating a significant technical problem. The incident highlighted the need for rigorous testing and design to minimize the potential of mechanical fatigue and cracking in future electronic systems (Ross et al., 2025). Resolving such reliability problems not only enhances the working life of electronic devices but also ensures their performance in extreme conditions. To reduce mechanical fatigue and interfacial cracking, innovative modeling techniques have been proposed recently to account for imperfect bonding conditions. Traditional analyses typically relied on idealized conditions under which perfect bonding was assumed; however, actual interfaces typically exhibit structural micro-scale imperfections. The application of finite element analysis (FEA) in mechanical models can be used to forecast stress distribution along interfacial cracks, providing information on crack propagation behavior and process zone properties. With these advanced methodologies integrated together, engineers can maximize the reliability of electronic packages and develop strategies for improving interfacial adhesion and overall material performance against thermal stress. Finally, these methods not only help to establish a better understanding of mechanical fatigue but also open up the door for the development of durable electronic systems with the ability to endure extreme thermal environments (Matsumoto et al., 2006).

9.3.2. Dielectric Breakdown

Dielectric breakdown is a failure mechanism that occurs when an insulating material can no longer resist electric current, leading to sudden and rapid conduction of electricity. The breakdown often has severe consequences for the reliability of electronic components, particularly those subject to high voltage and heat stress. Breakdown strength of dielectric material is important in determining the quality of insulation and device lifetime when used in various applications including power systems and electronic circuits. Dielectric breakdown not only lowers energy efficiency but also the unexpected failure of components, leading to costly repair and downtime. An understanding of dielectric breakdown mechanisms, especially thermal-stress dielectric breakdown, enables one to know how the reliability of electronic systems can be enhanced. Recent work has shown that there is a possibility of having improved breakdown characteristics due to enhancements in the dielectric material. For instance, incorporation of nanostructures such as graphene and carbon nanotubes into insulating oils has been reported to trigger spectacular enhancements of breakdown voltages, enhancing the reliability and performance of stressed dielectric materials. Incorporation of such nanofillers makes the delayed formation of streamers reduce ionized channel establishment rates inside the dielectric. Experimental findings have revealed improvements of approximately 70–80% in breakdown voltage, and it suggests that nano-colloids can potentially offer deep meaning for the economical and safe operation of power systems at high voltage levels (Chattopadhyay et al., 2018) (Figure 9.10).

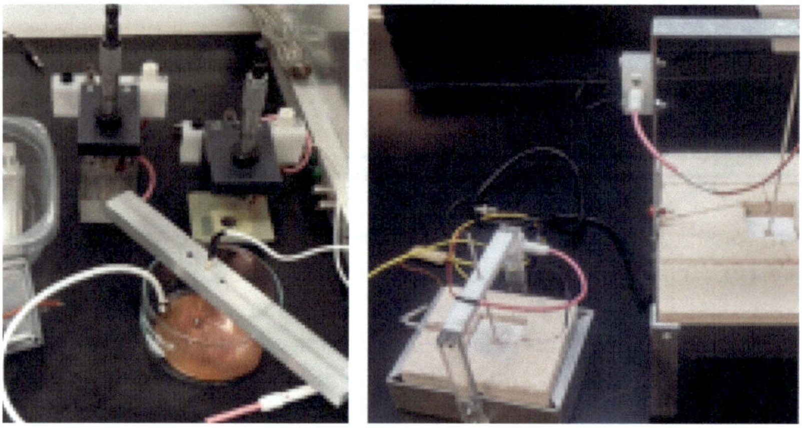

Figure 9.10. Dielectric breakdown.

Source: https://www.mri.psu.edu/sites/default/files/styles/large/public/inline-images/dielectric-2.png?itok=RijVq7o_.

This new method presents a possible avenue for reducing the risk of dielectric breakdown with certainty of long-term performance in electronic devices. In addition to dielectric improvement, the influence of design and material on electronic devices is a significant consideration for reducing the risk of dielectric breakdown. For example, in trench power n-type low-voltage MOSFETs, cell pitch and avalanche current have been found to be related with electrical degradation mechanisms and stability. The power-law relationship between threshold voltage variation and avalanche cycles highlights the sensitive relationship between performance reliability and design parameters. More specifically, decreased cell pitches have been linked to improved stability and reduced hot-hole injection into the gate dielectric, which is necessary for long-term performance maintenance under cyclic stress conditions (Alatise et al., 2010). Thus, understanding the relationship between design, material properties, and environmental stressors helps in developing aggressive electronic devices capable of withstanding the rigors of dielectric breakdown.

9.3.3. Solder Joint Failure

Solder joint failure is a significant concern in the area of electronic reliability, particularly in systems subjected to long-term thermal stress. Solder joint failures have generally been blamed on coupled fatigue and creep mechanisms that cause the integrity of solder junctions to weaken with time. Solder joints undergo mechanical stress under thermal cycling, where temperature variations cause expansion and contraction, which can result in cracking and eventual joining separation.

This is most prevalent in automotive and aerospace electronic industries, where one failed joint can compromise the overall performance of the system. Researchers have emphasized the necessity of establishing robust testing methodologies for predicting and preventing such failures in ensuring long-term performance and reliability in electronic assemblies, as mentioned in the context of IPC-9701 (Goti, 2025). Accelerated shear fatigue testing is one technique that yields crucial information on the degradation mechanism of solder joints under thermal conditions of stress. By simulation of varying stress amplitudes and residence times, researchers can experimentally explore systematically the effects of creep and fatigue on solder joint reliability. For instance, under different conditions of stress, the nature of failure of the solder joint is analyzed quantitatively by new techniques like Coffin–Manson and Morrow Energy models.

Such techniques allow for damage due to cyclic fatigue to be differentiated from damage formed by creep and hence inform design and manufacturing methods for enhancing solder joint strength. Thus, the comprehension of such mechanisms of failure not only helps in making electronic systems more dependable but also works in tandem with evolving industry standards and best practices, eventually resulting in product reliability improvement (Abueed et al., 2022). Machine learning software combined with artificial intelligence also has the capability of revolutionizing how solder joint reliability is assessed and improved. By employing AI-based techniques, manufacturers can automate defect detection schemes and thereby significantly increase the efficiency of reliability test regimens. It is not merely a case of making it faster but also employing data analysis to predict in advance where failures are likely to emerge in solder joints, so that more proactive measures can be taken during design and manufacture. The use of such technologies has the potential to provide a competitive advantage in applications where system reliability is paramount, ensuring electronic components meet strict operational criteria. Furthermore, AI-based methods are being developed to meet the IPC-9701 test criteria, providing a nascent future of increased reliability and reduced failure rates in solder joint applications (Goti, 2025).

9.4. METHODS FOR ASSESSING RELIABILITY

Reliability testing is paramount in electronics, especially given the enormous influence thermal stress exerts on component functionality. Prognostic health management system integration is one of the primaries means of developing reliability. The systems leverage continuous monitoring of operating conditions and using machine learning algorithms to analyze data trends. By tracking the operating parameters such as temperature and power cycles, one can forecast probable points of failure before they occur. These prognostic metrics enable organizations to refine maintenance planning and minimize unplanned downtimes, resulting in overall system reliability. Having information on the convergence of thermal loads and reliability measurements provides engineers with the tools for optimizing product designs and extending lifecycle expenses. Thus, application of real-time monitoring systems is a cornerstone in modern reliability assessment methods (Pecht et al., 2006). Another trustworthy method of reliability assessment relies on non-destructive examination and monitoring of solder joint integrity in electronic assemblies. Solder joints, typically the weak link of electronic circuitry under thermal and mechanical stresses, are a particular challenge to assessing reliability. New methods utilize digital signal characterization to quantify the health of such joints without actually needing to disassemble them. Accelerated life testing can be conducted under controlled thermal stress, simulating actual environments to capture slow degradation. Signal integrity of a device, measured through parameters like jitter and eye patterns, is a reliable measure of solder joint performance. Correlating these signal characteristics to thermal load exposure, engineers can determine early failure markers. Such a preventative measure thus leaves space for interventions at the right time, allowing for higher reliability in electronic systems. As explained above, high-speed digital signals are a valuable tool when analyzing and predicting solder joint reliability (Yoon et al., 2016). Finally, the addition of failure mode and effects analysis (FMEA) adds further reliability analyses by providing a disciplined methodology for identifying possible failure modes in electronic components. The analytical process allows designers to review how many thermal and operational stresses would result in failures, allowing more comprehensive understanding of reliability issues. When performed alongside statistical analysis of historical data, FMEA helps determine priorities where design solutions or engineering improvements would be applicable. With concentration on the most vulnerable components with highest exposure to thermal stress, such as transistors and capacitors, design teams can provide valid solutions that minimize the risk of failure. In doing so, combining FMEA with predictive monitoring technologies enables businesses to construct a holistic platform for reliability enhancement. This multi-faceted approach emphasizes the importance of active measures to reduce vulnerability and ensure long-term sustainability of electronic devices exposed to thermal stress.

9.4.1. Accelerated Life Testing (ALT)

Accelerated life testing (ALT) is an essential procedure used in the electronics industry to validate product reliability against stress conditions that simulate extended use over a shorter time frame. Utilizing increased temperatures, humidity, and voltage, engineers can expedite failure mechanisms quicker than under normal use. Through ALT, engineers are able to locate potential weak points in a design before products are implemented in the field. However, the credibility of ALT as a predictive measure of estimating hardware life can be challenged in the context of an accurate estimation of the hardware life. The evidence indicates that there are no quantitative methods available for the reliable prediction of hardware life for system components, as reflected by the requirement for an improved understanding of failure modes in accelerated testing (N/A, 2025). The uncertainty can challenge the utilization of ALT as a standalone indicator for long-term electronics reliability estimation. ALT is a vital tool employed in the context of thermal stress to understand the impact of over-temperature on electronic devices. Thermal loads may either be steady-state or transient conditions, and both have a critical impact on the reliability and performance of electronic products. Continuous observation of such thermal conditions can prove helpful information for prognostics, which gauges deviation from regular conduct and forecasts future dependability (Pecht et al., 2006) (Figure 9.11).

Figure 9.11. Product reliability is essential for success, especially for electronic products such as printed circuit boards (PCBs).

Source: https://images.ansys.com/is/image/ansys/electronic-accelerated-life-testing-pcb?

Through the utilization of stress-and-damage models together with thermal monitoring, scientists and engineers are better able to predict the initiation of failure of components. The addition of thermal stress data into prognostic models renders ALT more efficient and permits elevated reliability assessments, allowing proactive actions to avoid incipient failures in electronic systems. Ultimately, while ALT is critical to making electronic items more reliable, its limitations must be approached with caution as well. The challenges to extrapolating hardware life from accelerated testing alone necessitate a multi-disciplinary method to reliability engineering. Sophisticated prognostic techniques, which integrate ALT results with real-time environmental monitoring, offer a more detailed insight into the effects of thermal stress on device performance over time. By using these observations, the logistics support and design of products are enhanced by the manufacturer, as well as the optimization of maintenance schedules, to reduce life cycle costs (Pecht et al., 2006). Even though ALT is a useful resource in ensuring electronic reliability, its use must be served with unceasing innovation in monitoring and predictive technology to properly combat the additional complexity brought by thermal stress.

9.4.2. Thermal Cycling Tests

Thermal cycling tests are the focal point of electronic component reliability evaluation under fluctuating temperatures. Thermal cycling testing replicates the thermal stresses resulting from the on/off switching of heating and cooling, which are typical in real-world use. By subjecting electronic products to rapid temperature fluctuations, engineers can quantify how materials and solder joints respond to such stress and identify likely failure points. Repeated cycling through temperatures causes expansion and contraction of materials, which can cause fatigue, leading to mechanical failure in the form of cracking or delamination. Understanding these failure mechanisms is necessary to enhance the longevity and performance of electronic devices, and thermal cycling tests hence become an essential component of reliability testing protocols.

Thermal cycling tests are used for purposes beyond mere failure detection. These assessments provide useful insight into the thermal management methods needed in high-reliability applications, such as power electronics. As suggested by recent studies, cycling thermal stress has a significant influence on component lifetime, impacting parameters such as current flow and localized heating in devices. The necessity for more advanced modeling approaches is emphasized by the need to predict how designs can be optimized for thermal robustness. For instance, numerical 3D finite element models have been employed to simulate the thermal and mechanical stresses developed during cycling. This enables a comprehensive analysis of electrical performance versus thermal response interaction, and hence improves design processes (Medjahed et al., 2012). In addition, the utility of thermal cycling tests is enhanced by the use of monitoring technologies, such as infrared systems that track temperature variation during testing cycles. Through this, engineers are able to visualize temperature distribution in real time, which is necessary for identifying points of thermal stress and failure in a device. Information gained through these tests can be used to guide design changes and material choice to oppose detrimental effects of temperature. With the increasing concern for the dependability of power electronic devices by business organizations, it is essential to develop intensified testing methods, including thermal cycling tests. These proactive steps lead to innovations in packaging designs as well as materials and thus eventually to the total reliability of electronic systems with exposure to thermal stress (Gao et al., 2006).

9.4.3. Reliability Prediction Models

The development of reliability prediction models is becoming increasingly important in the electronics industry, especially with the added complexity and shrinking size of electronics devices. The models will generally estimate the likelihood of a failure occurring over a specific time period and under certain operating conditions. By integrating historical performance data, environmental factors, and operating stressors, reliability prediction models offer designers and engineers insightful information regarding the durability and lifespan of their products. The accuracy of these models relies greatly on the integrity of the input data, methodologies employed, and failure mode assumptions. Thus, accurate prediction models are essential in optimizing product design, optimization of maintenance planning, and, in general, guaranteeing greater consumer confidence in electronic products. Thermal stress is a key factor impacting electronic component reliability, and predictive models that can represent thermal dynamics become imperative. The combination of thermal fluctuations and physical material integrity often leads to thermal fatigue, one of the primary failure modes in electronics. Using measurements of parameters such as steady-state temperature, power cycling, and thermal gradients, predictive models can be enhanced to reflect actual operating conditions. Long-term thermal monitoring not only makes one better understand failure modes but also enables stress-and-damage model incorporation to better anticipate future reliability. Consistent with that, application of such advanced methodologies can offer valuable insights regarding maximum operating limits, enabling manufacturers to take proactive measures to rectify imminent failures and make overall system reliability better (Pecht et al., 2006). Considering the advancements in reliability prediction models, research argues that there is no comprehensive framework for hardware reliability prediction yet. Evaluation of the existing methodologies reveals a large gap between theoretical frameworks and practical utilization, especially for those systems that require long-term reliability over varying conditions. Current methods are usually not precise enough for successful prediction of hardware life, which is further compounded by variability of hardware faults and actual application complexity. The implication of this situation is the fact that there must be ongoing research and development in this field for the sake of enhancing prediction techniques and mitigating imbalances in the available literature. In order to address these problems, more advanced testing methods and familiarity with failure modes are needed to make the reliability models for electronics more predictive (N/A, 2025) (Figure 9.12).

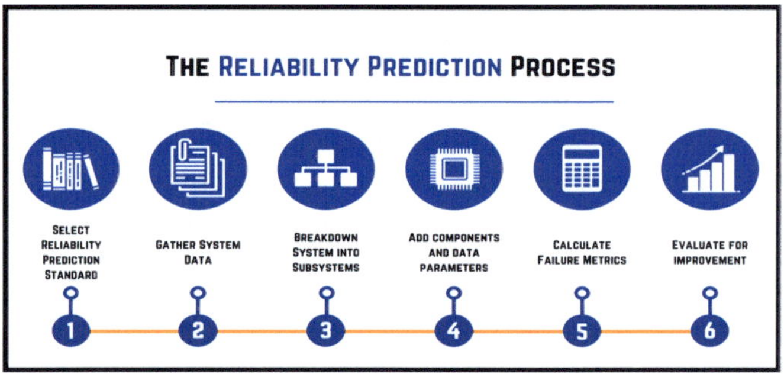

Figure 9.12. Reliability prediction process.

Source: https://relyence.com/2022/05/31/reliability-prediction-best-practices/

9.5. CASE STUDIES ON THERMAL STRESS IN ELECTRONICS

The investigation of thermal stress in electronics is necessary to enhance reliability, particularly in power electronic devices such as insulated-gate bipolar transistors (IGBTs). Thermal cycling is the dominant failure mechanism in these devices, where temperature changes induce local heating and concomitant mechanical stress. Case studies of this situation utilized advanced experimental setups and numerical simulations to analyze the electro-thermal performance under various working conditions. For instance, through a three-dimensional finite element model, simulating thermal stress enables researchers to visualize how temperature gradients impact bonded wire integrity through active current cycles. The overall aim is to develop an in-depth understanding that will lead to the design and packaging of electronic devices in such a manner that thermal stress-induced failure is prevented in future application, as explained by Medjahed et al. (2012). Apart from mere experimentation, prognostics also play a key function in sensing and anticipating thermal stress impacts. By monitoring a component's thermal loads permanently, researchers are able to gather valuable information that highlights deviations from the typical behavior. Permanent examination allows precursor reasoning algorithms to be integrated, which generate predictive output on potential faults due to thermal anomalies. For example, as abstracted by recent studies, the correlation of temperature gradients, power cycling, and steady-state temperatures with reliability outcomes assists engineers in designing and planning for maintenance to achieve maximum product and maintenance design. The prognostic methods have been imperative in achieving maximum electronics reliability under varying thermal loads, basically extending component life, and maximizing operational efficiency, as reported by Pecht et al. (2006). Moreover, the thermal stress consequences extend beyond specific components to whole systems and operational dependability. Thermal loads and their effects on electronic systems have been investigated to determine the interconnectedness of thermal dynamics, operation parameters, and failure mechanisms. It has been demonstrated through case studies that aspects subject to severe thermal gradients tend to be prone to mechanical delamination and failure, indicating the need for radical design methods dealing with thermal robustness. By studying real-life cases where thermal stress is a major driver, scientists are able to identify prevailing failure modes and paths for improved designing. Thus, focused research on thermal stress not only enhances our understanding but also targets packaging and materials breakthroughs to enhance the overall reliability of electronic systems.

9.5.1. Analysis of Failures in Consumer Electronics

In the world of consumer electronics, failure analyses typically contain valuable information regarding product reliability. One key feature of gauging such failures is the dual mechanisms that contribute to device degradation over time. For instance, while traditional reliability prediction methods provide an easy means of modeling component lifetimes, they tend to assume a constant failure rate. This simplification is not only undesirable, particularly as it also dismisses the phenomenon of infant mortality where first-stage failure affects overall reliability in a disproportionate manner (Hu et al., 1994). With a clear comprehension of these mechanisms, engineers can devise more robust designs that inhibit likely issues, thus leading to consumer electronic products lasting longer and being more reliable. Additionally, the effect of thermal stress on electronic component reliability cannot be overstressed. Severe temperatures and severe temperature fluctuations are capable of inducing serious physical change in materials used in consumer electronics, often leading to premature failures. As electronic devices have a tendency to shrink and get more powerful, heat load management has emerged as one of the main design and manufacturing challenges. Through monitoring systems that track these thermal parameters—i.e., ramp rates and steady-state temperatures—designers can gain knowledge of real-time stressors on operational devices (Pecht et al., 2006). Not only does it give feedback on the wearability of products in service, but it also aids in designing future devices with increased resistance to the stresses of use. The integration of prognosis methods is a viable option to address the limitations of failure analysis in consumer electronics. Through ongoing tracking and damage-and-stress modeling, a predictive strategy can be used in maintenance and design improvement. By identifying the condition of degradation of a product, producers can predict oncoming failures in advance and come up with corresponding preventive measures. This forecasting capability significantly reduces lifecycle cost and increases customer satisfaction through the assurance that products are functional and sound over their estimated lifespan. As the business increasingly embraces these methods, the emphasis on reliability through rigorous thermal management and failure analysis will most likely form the foundation of the next generation of dependable consumer electronic products.

9.5.2. Examination of Automotive Electronics Reliability

The reliability of auto electronics has evolved as a key region of consideration due to the functionality and complexity of modern-day vehicles. As vehicles continue to transform to support advanced driver assistance systems, the electronic components controlling such systems should be highly reliable. Auto electronics are usually exposed to harsh environmental conditions, e.g., long periods of elevated temperatures, humidity, and excessive vibration, which pose the risk of disrupting their functioning integrity. Previous works have primarily focused on understanding the impact of single stressors, such as temperature or humidity, on systems, hence ignoring the simultaneous exposure to multiple stresses under actual driving conditions of vehicles. This ignorance is significant, considering the fact that the manner in which different stress factors interact may heavily influence the performance and longevity of automotive electronics, prompting the need for an integrated reliability analysis methodology within this industry (Pandurangan et al., 2025). Advanced packaging structures are key to enhancing the reliability of car electronics. These structures, like flip-chip ball grid arrays, are thermally optimized and electrically optimized and are structured to minimize the failure modes anticipated for high-power applications. Effective TIMs play a critical role in these structures, as they allow heat to be transferred from electronic devices to heat sinks. However, potential failure at the ChipUF or Substrate-UF interface often occurs prior to that of the chip interconnects and requires targeted reliability screening. The interfaces' reliability under extended environmental stress, i.e., high temperature and humidity, remains inadequately researched and requires additional work in understanding the parameters governing the behavior of these materials in the true automotive environments (Pandurangan et al., 2025) (Figure 9.13).

Figure 9.13. Examination of automotive electronics reliability.

Source: https://www.keysight.com/blogs/en/tech/educ/2024/media_19d0d0228e0e7957334a4f8b7225bcc8063831331.png?width=750&format=png&optimize=medium.

Also, some components used in automotive systems, such as the function block in a fuel rail assembly, are excellent examples of high requirements for reliability testing. The part is exposed to high pressure and heat loads, so it is imperative to test its durability when it is exposed to cyclic stress conditions. By applying Finite Element Analysis (FEA) and advanced high-tech mechanical pulse tests, researchers can quantify the stress experienced by such components and give S-N curves that determine the applied stress against cycles to failure. The findings highlight that such components like the function block can have extremely small failure probabilities, which is a statement of how strong they are when exposed to demanding conditions. This approach not only adheres to stringent automotive reliability requirements but also addresses the importance of ongoing research initiatives to further refine the reliability of crucial electronic parts in the automotive industry.

9.5.3. Lessons Learned from Aerospace Applications

In looking at aerospace uses and what has been discovered from them, worth mentioning are the stringent reliability requirements under which aerospace electronics are developed and utilized. The space flight environment places special demands upon equipment and systems to endure harsh conditions, including intense thermal stress. By examining the current design and quality control practice for low Earth orbit space flight hardware, engineers can ascertain factors of importance that enhance the reliability of electronic systems. Glover et al. (2025) prescribes the primary environmental and design constraints inherent within aerospace applications. The attentive focus on component selection, thermal management, and stress testing procedures not only adds to the longevity of space-traveling electronics but also provides insight into best practices for other industries dependent upon electronic systems. Moreover, the development of a Supportability Technology Development Roadmap for the NASA Constellation program is a stark reminder of the aerospace industry's lessons learned regarding maintenance and sustainability. This roadmap puts great value on logistics autonomy on long-duration missions, including missions to the Moon and Mars. Oeftering et al. (2025) recognizes measures with diagnostic, maintenance, and recycling strategies that maintain spacecraft system longevity and operational capacity in resource-constrained environments. The recommendations of the roadmap offered form a set of principles for relief from thermal and operational stress through anticipatory maintenance and repair tactics. These observations signal a fundamental paradigm change in aerospace engineering from reliance on support to proactive support capability, and electronic reliability benefit in particular. Additionally, the movement away from highly logistic-dependent operations to increasingly self-sustaining systems has profound implications for electronics reliability. The emphasis by Oeftering et al. (2025) on embedded and process technologies affords important insight into how aerospace applications have progressed toward eliminating dependency on continuous external support. This approach not only challenges thermal stress in terms of more efficient designs with lower heat outputs but also guarantees the robustness of electronics via innovative maintenance regimes. With additional industries now experiencing resources in scarcity and operating stresses aside from aerospace, implementing these learned methodologies would promote improved reliability in electronic systems in other industries so that they are adequately prepared to withstand similar conditions for space exploration.

SUMMARY

Investigation of reliability and thermal stress has revealed important questions still affecting modern electronic design and manufacturing. With gadgets growing smaller and smaller, complexity in thermal management systems is on the rise, and extra expertise and innovative solutions are indispensable. Thermal fatigue and interfacial strength coupling, particularly in solder joints, dictates the need to assess them stringently by complex mechanical models. Tests have confirmed that micro-scale interface bonding defects are the primary cause of reliability issues, and the need to shift from traditional tests that are based on ideal bonding assumptions. This delicate understanding triggers further research of the complexities of thermal effects, eventually leading to high-performance device reliability standards, especially with electronic devices. Research on high-temperature electronics stresses the importance of advances in materials to provide stability during thermal stress. The application of new technologies such as silicon carbide and low temperature co-fired ceramic (LTCC) is a promising direction of future research and development. These technologies not only withstand high temperatures but also have increased parameters of performance in conditions that previously were considered hostile for conventional electronics. By addressing devices that are designed to operate in high-temperature applications, significant strides can be taken toward reducing the adverse effects on signal integrity that consistently plague electronics under heat stress. Thus, a paradigm shift toward such revolutionary materials has the potential to culminate in advanced electronic systems capable of meeting the demands of today's applications. In summary, the cyclical malfunctioning of reliability and thermal stress in electronics necessitates a multi-faceted approach that incorporates theoretical and practical solutions. Advances in the knowledge of interfacial mechanics, as well as the identification of new materials, provide a critical door to enhancing electronic device performance and lifespan. As the industry evolves further, integration of findings from recent studies will be crucial. Examples encompass techniques making use of the results of interfacial fracture mechanics and the successful incorporation of high-temperature technologies to guarantee that the reliability issues vital in contemporary electronic systems are addressed. Such an openness to innovation and adjustment based on new thermal challenges naturally guarantees that the emerging electronic designs are reinforced in resilience.

Understanding the reliability of electronic systems is more critical than ever, particularly as they are increasingly subjected to thermal stresses that impair function and life. Prognostics, a central component of reliability assessment, involves determining how changes in operating conditions can be indicative of future failures. Through this real-time monitoring, it becomes feasible to predict the future reliability, enabling optimal maintenance and decreased life cycle cost. Specifically, when *in situ* thermal loads—steady-state temperature and temperature gradients—are quantified, they can provide critical input to forecast models. These enhancements can enhance dramatically the design and logistics support for existing and future electronics systems, such that reliability becomes a multi-dimensional issue that demands strategy planning as well as creative engineering practices (Pecht et al., 2006). Thermal management is pivotal to electronic component performance and longevity. Through thermal and thermo-mechanical simulation, engineers can create solutions to enhance the reliability of an electronic device. As an example, in SSL and LED applications, precise thermal models are necessary to maximize heat dissipation from components into the surrounding environment. Simulation studies utilizing software such as ANSYS have illustrated the importance of simulating the thermal distributions in order to locate critical regions affecting heat transfer. Such models not only predict the thermal behavior of components but also aid in the selection of optimal materials and lamp shapes—a step crucial to improving efficiency, lifetime, and reliability. Such complete knowledge of thermal physics demonstrates the focal interlinking of thermal management and electronics dependability (Formanek et al., 2012). The interrelation between thermal stress and reliability necessitates ongoing research and innovation in material science as well as design engineering. Designing sophisticated models that simulate and predict the effects of thermal loads on electronic components encourages proactive reliability management. As evidenced by LED technology research studies, material selection for electronic devices is decisive in determining thermal performance, ultimately affecting reliability. Predicting latent failures and managing thermal stresses through the use of sophisticated simulations and modeling techniques can directly reduce unplanned downtime and maintenance costs. With advancing electronics, care in the incorporation of thermal management strategies will be critical to ensuring greater reliability in increasingly complex and performance-intensive applications (Formanek et al., 2012; Pecht et al., 2006).

REFERENCES

1. Abbé, P., Galliou, S., Goryachev, M., Imbaud, G., Bourquin, R., & Creedon, D. L. (2013). Advances in development of quartz crystal oscillators at liquid helium temperatures. http://arxiv.org/abs/1309.5445 (accessed on 23rd October 2025).
2. Abueed, M., Al Athamneh, R. A., Tanash, M., & Hamasha, S. (2022). The reliability of SAC305 individual solder joints during creep–fatigue conditions at room temperature. Crystals. https://www.semanticscholar.org/paper/04ee9a65af96019bc4b8353fd489b379c0cf61ff (accessed on 23rd October 2025).
3. Alatise, O. M., Heppenstall, K., Kennedy, I., Khan, A., et al. (2010). The impact of repetitive unclamped inductive switching on the electrical parameters of low-voltage trench power nMOSFETs. https://core.ac.uk/download/1384352.pdf (accessed on 23rd October 2025).
4. Alghassi, A., Perinpanayagam, S., Sreenuch, T., Xie, Y., et al. (2014). Probabilistic Monte-Carlo method for modelling and prediction of electronics component life. https://core.ac.uk/download/42144212.pdf (accessed on 23rd October 2025).
5. Bauer, C., Bernhard, S., Capasso, M., Diebold, S., et al. (2015). FlashCam: A fully-digital camera for the medium-sized telescopes of the Cherenkov Telescope Array. http://arxiv.org/abs/1509.02434 (accessed on 23rd October 2025).
6. Brown, J., Chen, I., Earhart, R., Ehrlich, D., et al. (2025). Advanced Manned Launch System (AMLS) study. https://core.ac.uk/download/pdf/42811271.pdf (accessed on 23rd October 2025).
7. Cazier, F. W., Jr., & Gardner, J. E. (2025). Space Transportation Materials and Structures Technology Workshop. https://core.ac.uk/download/pdf/42808223.pdf (accessed on 23rd October 2025).
8. Chattopadhyay, A., Dhar, P., Harikrishnan, A. R., Maganti, S., et al. (2018). Streamer evolution arrest governed amplified AC breakdown strength of graphene and CNT colloids. http://arxiv.org/abs/1807.02698 (accessed on 23rd October 2025).
9. Dziedzic, A., & Nowak, D. (2013). Thick-film and LTCC passive components for high-temperature electronics. https://core.ac.uk/download/30311962.pdf (accessed on 23rd October 2025).
10. Eisenhaure, D. B., & Oglevie, R. E. (2025). Advanced Integrated Power and Attitude Control System (IPACS) Study. https://core.ac.uk/download/pdf/42842428.pdf (accessed on 23rd October 2025).

11. Formanek, J., Husak, M., Jakovenko, J., et al. (2012). High power solid state retrofit lamp thermal characterization and modeling. https://core.ac.uk/download/30311799.pdf (accessed on 23rd October 2025).
12. Gao, S., & Holmes, A. S. (2006). Thermoscopic flip chip interconnection using electroplated copper column arrays.
13. Glover, D. (2025). Design considerations for space flight hardware. https://core.ac.uk/download/pdf/42825339.pdf (accessed on 23rd October 2025).
14. Gonzalez, D. (2019). Characterization of high temperature optocoupler for power electronic systems. https://core.ac.uk/download/215464228.pdf (accessed on 23rd October 2025).
15. Goti, A. B. (2025). AI-driven PCB reliability testing for IPC-9701 compliance. International Journal of Scientific Research and Management (IJSRM). https://www.semanticscholar.org/paper/0cd178a7f42899288fb56316a10a16da31d8a10c (accessed on 23rd October 2025).
16. Ho, P. S., Huang, R., Im, J., et al. (2014). Characterization of thermal stresses and plasticity in through-silicon via structures for three-dimensional integration. https://core.ac.uk/download/211336033.pdf (accessed on 23rd October 2025).
17. Hopsch, F., Khursheed, S., Marinissen, E. J., et al. (2016). Guest editors' introduction: Robust 3-D stacked ICs. https://core.ac.uk/download/80776038.pdf (accessed on 23rd October 2025).
18. Hu, Y., Brombacher, A. C., Claessen, J. H. J., Campbell, P., Peck, R., Dunn, W., Zelenka, J., et al. (1994). Component lifetime modelling. https://core.ac.uk/download/pdf/11474562.pdf (accessed on 23rd October 2025).
19. Kurt, S. (2024). Reliability assessment of the casted steel function block in gasoline fuel rails. Orclever Proceedings of Research and Development. https://www.semanticscholar.org/paper/4d1021577333b17f2296c9e2e431646a6615406b (accessed on 23rd October 2025).
20. Luo, Y., Abidian, M. R., Ahn, J. H., Akinwande, D., Andrews, A. M., Antonietti, M., Bao, Z., et al. (2023). Technology roadmap for flexible sensors. ACS Nano, 17, 5211–5295. https://doi.org/10.1021/acsnano.2c12606 (accessed on 23rd October 2025).
21. Matsumoto, T., Matsuzaki, T., Shibutani, T., et al. (2006). Modeling with structure of resins in electronic components. https://core.ac.uk/download/15488709.pdf (accessed on 23rd October 2025).
22. Medjahed, H., Nogarède, B., & Vidal, P. E. (2012). Thermo-mechanical stress of bonded wires used in high power modules with alternating and direct current modes. https://core.ac.uk/download/12044299.pdf (accessed on 23rd October 2025).

23. N/A. (2025). Evaluation of methods for determining hardware projected life. https://core.ac.uk/download/pdf/85233673.pdf (accessed on 23rd October 2025).

24. Nazir, A., Gokcekaya, O., Billah, K. M. M., Ertuğrul, O., Jiang, J., Sun, J., & Hussain, S. (2023). Multi-material additive manufacturing: A systematic review of design, properties, applications, challenges, and 3D printing of materials and cellular metamaterials. Materials & Design, 226, 111661–111661. https://doi.org/10.1016/j.matdes.2023.111661 (accessed on 23rd October 2025).

25. Oeftering, R. C., Struk, P. M., & Taleghani, B. K. (2025). A lunar surface system supportability technology development roadmap. https://core.ac.uk/download/pdf/10562170.pdf (accessed on 23rd October 2025).

26. Pandurangan, A., Choudhury, P., Kasturi, M. L., Lall, P., & Suhling, J. (2025). Interface reliability of advanced packaging under sequential stresses of high-temperature and temperature-humidity in automotive environments. 2025 IEEE 75th Electronic Components and Technology Conference (ECTC). https://www.semanticscholar.org/paper/407c277bd945e996743a47d5903ecc52b1ce581d (accessed on 23rd October 2025).

27. Pecht, M., & Vchare, N. (2006). Enabling electronic prognostics using thermal data. https://core.ac.uk/download/15488695.pdf (accessed on 23rd October 2025).

28. Ross, R. G. (2025). Magellan/Galileo solder joint failure analysis and recommendations. https://core.ac.uk/download/pdf/42824952.pdf (accessed on 23rd October 2025).

29. Wang, B., Harding, J., Haward, S., Johnson, P., Kappor, A., Kolsky, I., Lim, K., et al. (2014). On the high strain rate behavior of 63–37 Sn-Pb eutectic solders with temperature effects. https://core.ac.uk/download/29140278.pdf (accessed on 23rd October 2025).

30. Yoon, J. A. (2016). An approach to assess solder interconnect degradation using digital signal. https://core.ac.uk/download/79707745.pdf (accessed on 23rd October 2025).

CHAPTER 10

THERMAL MANAGEMENT IN POWER ELECTRONICS

Contents

INTRODUCTION

The advent of power electronics has introduced unprecedented advancements across various sectors, including renewable energy, electric vehicles, and consumer electronics. The technology of thermal management lies at the heart of all these advancements and plays a crucial role in the reliability and efficiency of electronic systems. As power densities increase, the heat generated upon operation needs to be properly managed to prevent performance degradation and failure of components. Current studies indicate that poor thermal performance leads to high temperatures, which are detrimental to device longevity and operational dependability (Hoell et al., 2019). Thus, thermal performance is not a technicality; rather, it is vital in the creation of robust, high-performance power electronic systems. Furthermore, with the global energy demands on the rise, the importance of energy efficiency cannot be overstated. Data center expansion is one of the greatest illustrations wherein power consumption has risen exponentially because of the humongous volume of data processing and storage needs. Thermal management solutions are becoming significant strategies for addressing this situation because they contribute to both power consumption and cooling requirements (Mittal et al., 2014). As more pressure is put on designers to reduce energy consumption without compromising system performance, newer methods directed at thermal management are under research. These include enhancements in material, cooling, and system design, which all benefit the thermal performance of power electronics. Also, the intersection of thermal management and environmental sustainability provides a compelling narrative in the case of power electronics. Due to enhanced awareness of climate change and its impact, there has been a great emphasis towards the development of green technologies that minimize wastage of energy. Thermal management contributes not only to the efficiency of operation of electronic devices but also to the overall mission of reducing carbon footprints for industries relying on power electronic systems. With the world demanding greater sustainable methods, greater R&D in the coming years will go towards merging superior thermal management technologies with energy-efficient designs, ushering in a greener technological future. Both methods will be crucial in shaping the future of electronics while ensuring environmental requirements are met.

Definition of Power Electronics

Power electronics would be the field of electrical engineering concerned with the conversion and control of electric power using semiconductor devices. Power electronics would be an extensive area of applications from power supplies to electric drives and renewable energy systems. The essence of power electronics is its capability of efficient regulation of electrical power by converting it from one form to another, i.e., AC to DC or DC to AC, and switching the voltage and current levels. It employs various components such as transistors, diodes, and capacitors to achieve high efficiency and reliability in power conversion systems. With the advancement of power electronics, its usage has increased extensively, and designs with improved performance that aim at managing heat become necessary because high temperatures can compromise the working and life of electronic devices. Thermal management is a central field in power electronics as it entails the management of heat that arises as a natural byproduct during power conversion processes. Proper dissipation of the heat is required to prevent damage and realize optimal performance of electronic devices. Examples include applications such as electric vehicles and renewable energy harvesting, which are typically coupled with large thermal loads that require careful planning and innovative cooling systems. For instance, the design of components such as cooling fans, heat sinks, and TIMs plays a significant role in guaranteeing performance effectiveness. Furthermore, materials science and engineering advancements have enhanced to create new thermal management systems that can enhance heat dissipation. These innovations make the power electronics systems more reliable by avoiding the potentially destructive effects of overheating (Figure 10.1).

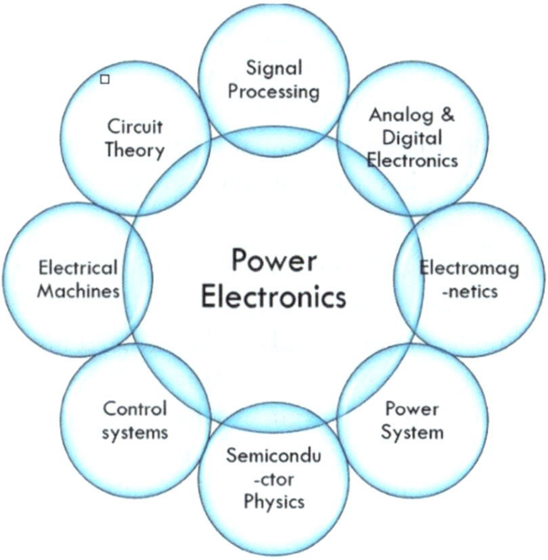

Figure 10.1. Relation of power electronics with other disciplines.

Source: https://electronicscoach.com/power-electronics.html.

Inclusion of thermal management methods not only ensures the safety and longevity of power electronics systems but also enhances their efficiency. Emphasized in dynamic system studies, effective temperature regulation is central in sustaining performance under varying operation conditions, particularly in high-power systems (Rao et al., 2015). Besides, innovative design approaches in the design process, such as utilizing multi-layered TIMs, can significantly reduce thermal resistance and increase heat dissipation. Furthermore, as shown in developing next-generation electronic systems, combining efficient heat management with cutting-edge technology allows for greater energy efficiency and more secure operations. Generally, a general view of power electronics combined with a good thermal management plan will allow the field to progress and address the complexity brought by intricate power systems.

Importance of Thermal Management

Thermal management is of great significance in power electronics, especially with high power dissipation rates and with tiny size present in recent electronic devices. With devices shrinking in size, the thermal output per unit area of area increases, which necessitates a critical demand for efficient heat dissipation techniques. Poor thermal management can cause overheating, which not only deteriorates performance but could also cause catastrophic failure. With the application of various thermal management methods, such as heat sinks, phase change materials, and active cooling systems, engineers can enhance reliability and life. Thermal management also allows for energy efficiency improvement, which is critical as businesses strive to reduce energy consumption as well as costs incurred. Green thermal management methods can thus be revolutionary both in terms of efficiency in operation and also environmental effects and are a part of power electronics.

Thermal management is also noteworthy outside the reliability framework into the general picture of energy usage, especially that of large systems like data centers. With cloud computing and big data accelerating at a very fast rate, the demand for processing of data has risen significantly, leading to massive power usage in such facilities. Recent research has indicated that effective management of heat generation is most important when it comes to improving energy efficiency in such facilities. Using methods like server consolidation and dynamic voltage and frequency scaling (DVFS) can lead to peak performance while cutting down energy usage drastically. As power dissipation continues to remain a key challenge, the implementation of advanced thermal management schemes becomes a necessity to ensure that energy efficiency is maximized in modern data centers, aligning with the necessity of appropriately designed thermal strategies in power electronics (Mittal et al., 2014).

Lastly, the integration of advanced thermal management techniques is crucial for both short-term performance and electronic component long-term lifespan. Effective thermal solutions enable higher operating temperatures and power densities, essential to applications ranging across the board from consumer equipment to industrial equipment. Further, the design principles of thermal management can simplify the design process, facilitate innovation, and enable flexibility in the rapidly changing arena of power electronics. A systems-level thermal management involves taking into account material design and system architecture on a par with how electro-optic sensor systems are designed to share common components and design efficiencies (Barlow et al., 2007). By therefore giving thermal management priority within design and operating practices, engineers are able to guarantee improved performance and reliability, supporting the progress of power electronic systems as a whole.

10.1. FUNDAMENTALS OF THERMAL MANAGEMENT

The significance of thermal management in power electronics is the latter's contribution to enhanced device efficiency and performance operation. Proper temperature control is essential given that devices in power electronics generate high temperatures upon use. The heat, if not managed, will lead to decreased performance or complete device failure. Thermal management techniques can be divided into passive and active methods, and passive techniques aim to provide heat dissipation via conduction, convection, and radiation. These technologies involve the use of heat sinks and TIMs that are intended to promote heat transfer. Active thermal management systems, on the other hand, incorporate mechanisms like forced air or liquid cooling to help control temperatures in a more dynamic fashion.

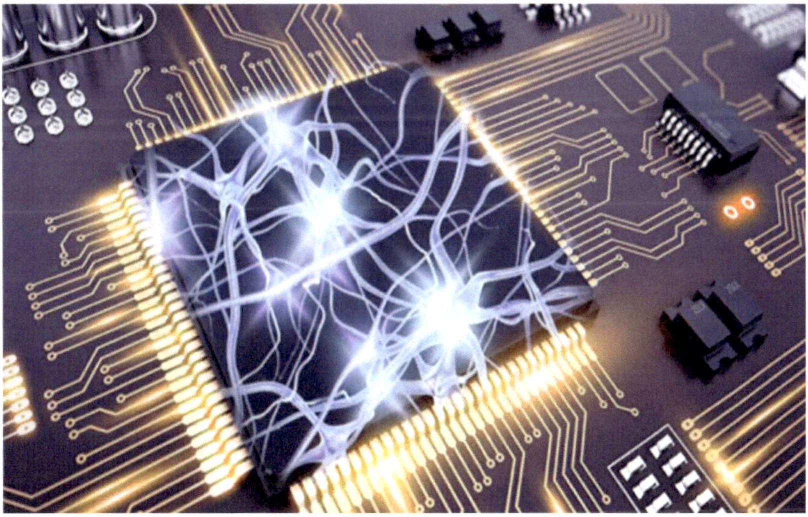

Figure 10.2. The significance of thermal management in power electronics.

Source: https://external-content.duckduckgo.com/iu/?u=https%3A%2F%2Fwww.eetindia.co.in%2Fwp-content%2Fuploads%2Fsites%2F4%2Fimages%2Ffab8c4b5-eb27-4732-97c6-33465add3185.jpg&f=1&nofb=1&ipt=27b09c5fdf73f4e112f9a125549608cce718305d1420d8bfc52b68102ee44fcb.

Through the incorporation of state-variable model mechanisms, designers can predict thermal performance for various operating conditions, allowing for quicker thermal configuration evaluation without compromising accuracy as described by Bingham et al. (2004). Another central aspect of thermal management is materials science for heat dissipation solutions. TIMs evolution has led to the evolution of materials with higher thermal conductivity with better mechanical stability and material compatibility. For example, advances in flexible polymer-based materials may find future applications in microfuel cells because these technologies require good thermal management for increasingly small scales. From nanostructured materials like carbon nanotubes to sophisticated cooling systems that use liquid coolants, the potential for efficiency improvements in the future is vast. New materials can provide for better conduction of heat and cooling of systems, in general, which is particularly critical in those instances where high-density electronics are required. New methods such as microchannel cooling, which delivers high heat extraction rates, show the real-world applications that directly lead to the energy efficiency of real systems in real applications, highlighting the necessity for further research in this field as mentioned by Emerson et al. (2009) (Figure 10.2).

The cross-over with control strategies and thermal management must also be taken into account, specifically within system reliability under fault conditions. Zero-voltage switching (ZVS) devices enable huge switching losses reduction, making thermal management during transient events significantly improved. Improving operation frequencies based on the individual behavior of resonant converters benefits designers in lessening heat generation and achieving superior efficiency. Besides, the application of cutting-edge control techniques enables systems to react effectively to varying load conditions and hence provide maximum thermal performance. High-performance applications call for such innovations in which minor variations in temperature can lead to operational issues. The application of dynamic thermal management methods accommodates power electronic systems' strength, which gives them longevity and reliability in maintaining energy conservation, an advantage that is extremely crucial in high-speed technology environments, as indicated in the findings of Bingham et al. (2004).

10.1.1. Heat Generation in Power Electronic Devices

The generation of heat in power electronic devices is a highly critical concern that has profound impacts on their efficiency and reliability. Due to the increased need for higher performance, these devices are biased to work at higher temperatures due to elevated power dissipation. This can be highly apparent in small devices where miniaturization restricts the heat dissipation process. As power transistors and integrated circuit's function, their conduction loss and switching loss result in a tremendous amount of heat generation that, without being managed, will result in failure. They introduce the necessity for proper thermal management techniques in order to allow these systems to function and operate in the long term. Insufficient thermal energy control also causes thermal runaways wherein excessive heat generates catastrophic device failures, underpinning the need to design with enhanced heat dissipation (Chen et al., 2019). Effective thermal management of power electronics also entails the understanding of the heat generation profiles under various operation conditions. Current research has identified the need to consider, in addition to heat generated by the primary device, the thermal contribution of the surrounding components, such as batteries and capacitors, to total thermal accumulation. As an example, thermal generation in mobile electronics like smartphones from the overcurrent battery management system is typically unnoticed. Unless well managed, failure to consider this could lead to unprecedented thermal events, like device explosions, otherwise. By studying and predicting heat distribution in these devices, engineers can create combined cooling solutions to enhance the thermal performance of power electronic devices while ensuring reliability and safety (Chen et al., 2019) (Figure 10.3).

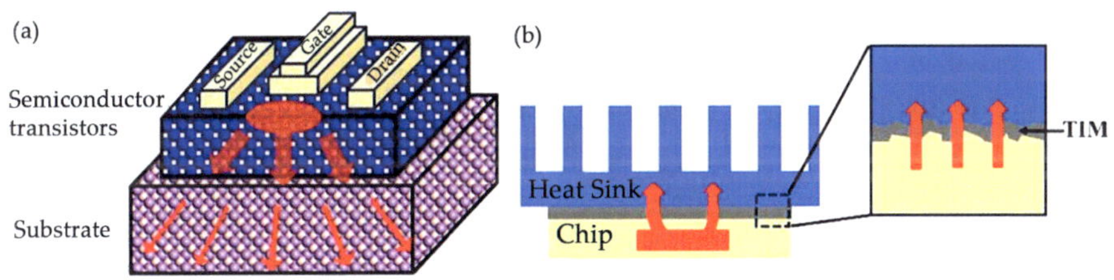

Figure 10.3. (a) Schematic illustration of heat generation and dissipation in a semiconductor device: Heat is produced at the semiconductor transistors and dissipated through the substrate. (b) Schematic of heat dissipation in an electronic chip, where the heat generated in the chip is transferred to a heat sink. The inset shows the role of the thermal interface material (TIM) in enhancing heat transfer efficiency between the chip and the heat sink.

Source: https://www.researchgate.net/figure/a-Schematic-illustration-of-heat-generation-and-dissipation-in-a-semiconductor-device_fig1_383194201.

While alleviating heat generation in power electronics, high-technology cooling technologies play an innovative role in taking the performance of devices to a new level. Among them, heat pipes are proved to have tremendous potential as a mechanism for heat transfer due to their high efficiency. However, their performance is limited by operational conditions and device miniaturization, necessitating further research on their design and use to be improved. In addition to conventional active and passive cooling technologies, integration of advanced materials that have higher thermal conductivity is also a means of maintaining low thermal resistance. Comprehensive analysis of the relationship between device design, material selection, and cooling technologies remains pertinent in formulating integrated thermal management approaches. By mapping heat generation and dissipation interactions in detail, engineers can unlock the way for power electronic devices that are more efficient and safer to use and that will foster their application in environmental-friendly systems.

10.1.2. Thermal Conductivity and Resistance

In power electronics, thermal conductivity is a significant parameter that determines the performance and reliability of electronic devices. Thermal conductivity is the process by which a material transmits heat and influences thermal management strategies employed to shield devices against overheating.

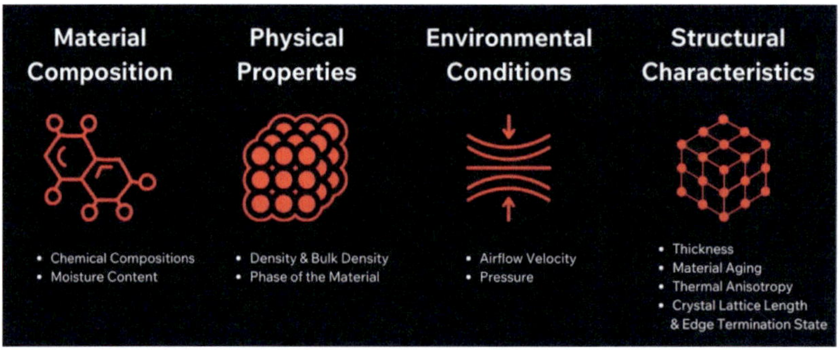

Figure 10.4. Factors that influence thermal conductivity.

Source: https://thermtest.com/wp-content/uploads/thermal-conductivity-measurement-1024x576.png.

As power electronics are becoming smaller and more powerful, it becomes increasingly difficult to handle heat dissipation. Materials with high thermal conductivity provide effective heat extraction from heat sources such as processors and batteries, thereby maintaining optimal operating temperatures. Ineffective heat dissipation might lead to performance degradation or even failure, making advanced materials and designs where thermal conductivity is a primary parameter a priority in power electronics (Chen et al., 2019). Thermal resistance is intrinsically connected with thermal conductivity, as a hindrance to heat conduction. Thermal resistance in power electronics may exist in many interfaces, e.g., at the interface between semiconductor materials and heat sinks. Each interface bears a thermal resistance that should be made as small as possible to enhance heat transfer. Effective thermal resistance management demands careful material selection and engineering techniques, such as heat sink geometry optimization and TIMs that ensure maximum surface contact. Innovations in design and material technology, such as the incorporation of metal foams and metal matrix composites, present the possibility for decreasing thermal resistance and improving heat exchanger efficiency (Han et al., 2012) (Figure 10.4).

Through thermal resistance, engineers can improve power electronic system reliability and efficiency. Moreover, the interdependence of thermal conductivity and thermal resistance plays a significant role in achieving holistic thermal management in power electronics. The synergy of increasing thermal conductivity and reducing thermal resistance leads to a better thermal management system. A thorough understanding of how these two properties interact enables engineers to design systems that can best manage heat when exposed to a multitude of operating conditions. For instance, tactical positioning of components such as application processors and batteries can influence overall thermal behavior to a great extent. Furthermore, incorporation of cutting-edge cooling technologies, such as heat pipes and phase change materials, further enhances thermal management technology to avoid overheating of power electronic devices (Chen et al., 2019). New material discovery and cooling technology highlight the fact that there is always a need for improved thermal conductivity and resistance for achieving more reliable power electronics.

10.2. THERMAL CHALLENGES IN POWER ELECTRONICS

The thermal challenges of power electronics have become increasingly significant with the growing needs for high-performance devices. Since power electronic devices have to operate at increased frequency and power levels, heat generation by them becomes a critical concern. The development of efficient thermal management concepts is crucial in order to ensure optimal performance and longevity of electronic components. Improvements in semiconductor technology, particularly the utilization of wide-bandgap materials such as silicon carbide (SiC) and gallium nitride (GaN), are significant advancements in resolving such thermal issues. These materials not only have better efficiency but also produce less heat compared to traditional silicon devices, thereby mitigating thermal issues that result from high-power applications in power electronics (Hossain et al., 2024). Apart from material developments, integration and design of effective cooling devices also play an imperative part in managing thermal challenges in power electronics systems. Traditional cooling techniques such as air or liquid cooling may not be sufficient for future usage where compactness and efficiency are extremely paramount. New cooling such as microchannel heat sinks and phase-change materials is thus under research and development. These advanced cooling technologies are dominated by greater thermal conductivity and increased heat extraction, aligning them accordingly for power electronic high-density applications. Moreover, adaptive thermal systems may alter thermal management strategies dynamically in response to dynamic thermal data and thus optimize performance and reduce operation-related risks associated with overheating (Mittal et al., 2014). Finally, knowledge of how power electronics interact with their thermal surroundings is essential in effectively addressing thermal issues. As power electronic system complexity increases, predictability of heat and system reliability become tougher. By combining thermal modeling and simulation tools into their design process, designers are better positioned to ascertain thermal behavior under varying operating conditions. In so doing, they can effectively identify potential hotspots and failure sites within the system, facilitating proactive thermal management. Secondly, ongoing research in energy-efficient architectural approaches demonstrates commitment towards maximizing thermal performance of power electronics. Despite the fact that power electronic thermal issues are enormous, additional research on material, cooling, and thermal management techniques will ultimately yield more efficient and better power electronic systems (Hossain et al., 2024).

10.2.1. High Power Density and Its Implications

High power density is a more focal concept within power electronics, particularly with the sheer revolution in semiconductor technology and thermal management processes. High power density allows for electrical systems to be miniaturized with greater efficiency and performance, as is beneficial in use in electric cars and renewable energy systems. As power density increases, so too does the challenge of thermal output management. Proper heat management is required in order to preserve maximum operating efficiency and prevent overheating, which can lead to component failure. One can see that in order to achieve higher power densities; manufacturers must at the same time design advanced cooling systems and components that can withstand increased temperatures without compromise to reliability or efficiency and indicate the fine line between performance and thermal management. This development in high power density systems also aligns with increasing penetration of renewable energy into the grid, and innovative power electronic interfaces are a requirement. The likes of the solid-state transformer (SST) are prime examples, merging renewable energy with higher grid stability and efficiency. SST advantages include bidirectional power exchange and improved system control, which are valuable in terms of managing the variability of renewable energy. However, these benefits come at the cost of increased operating temperatures with significant thermal management issues. Therefore, the development of advanced thermal management techniques is extremely crucial in enabling the stable and efficient operation of high-power density interfaces as well as being capable of meeting stringent efficiency and performance demands in the market (Figure 10.5).

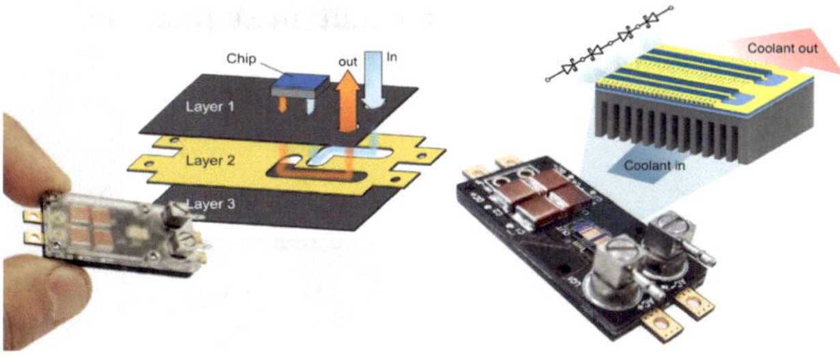

Figure 10.5. High power density converters.

Source: https://www.epfl.ch/labs/powerlab/wp-content/uploads/2023/02/power-converters2.png.

Also, as power electronics improve with greater emphasis on high power density, the implications extend beyond cosmetic performance enhancements. For instance, the possibility of reduced emissions in transportation is substantial with the shift towards hybrid and electric cars. Use of high-power density technologies has the potential to drastically restrict fuel usage and pollutant emissions, thus ensuring environmental sustainability. While urban cities are facing increasing pollution challenges, applying the technologies presents an effective method to mitigate against the adverse effects of transport emissions outlined in existing research, and revealing broader effects on public health and city environment (Capata et al., 2018). Moreover, along the path towards the integration of renewable energy into the power grid, the evolution of power electronic systems with high power density characteristics can potentially enhance the reliability and stability of modern energy systems at large (George et al., 2016). For these implications to be addressed, it will be essential that the advancements in power density and thermal management also continue to foster innovative technologies in power electronics.

10.2.2. Effects of Temperature on Device Performance

The effect of temperature on performance in power electronics is drastic, impacting efficiency as well as reliability. With an increase in temperature, the intrinsic properties of semiconductor materials are influenced, leading to variability in electrical conductivity and mobility. For instance, increased thermal energy can increase charge carrier mobility but may also contribute to increased leakage currents, which can undermine performance. The impact is further pronounced in equipment with high power density, where heat generation and dissipation interact more significantly. Understanding such thermal behavior is essential, since thermal runaway caused by temperature excursions beyond specific levels can result in a measurable reduction in component life and system reliability.

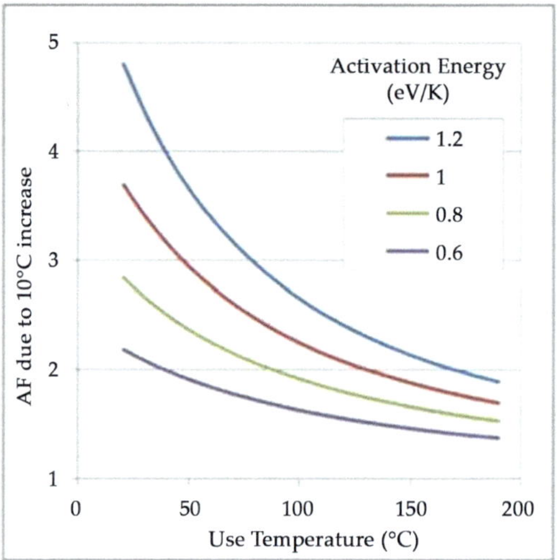

Figure 10.6. Acceleration factor for a 10°C increase in temperature.

Source: https://www.electronics-cooling.com/2017/08/10c-increase-temperature-really-reduce-life-electronics-half/

Adequate thermal management practices must be utilized to counter the adverse influence of temperature on power electronic devices. It has been evidenced through research that the use of materials such as diamond brings a significant boost in applications demanding efficient heat dissipation due to its thermal conductivity as well as resistance to extreme conditions. FET devices are particularly enhanced through this technique since they can achieve high power densities in addition to efficient heat management. In addition, technologies such as liquid-cooled heat sinks, intended for industrial use, take advantage of high heat transfer coefficients to effectively deal with the frequently significant heat fluxes produced by these devices. In turn, through such technologies, engineers are able to reduce temperature impacts and improve device performance in a variety of operating conditions (Davidson et al., 2005). The connection between temperature management and device longevity cannot be overstated, since uncontrolled thermal conditions speed up material degradation towards early failure. Modern thermal management techniques embrace advanced cooling technologies, including microchannel cooling systems, to maintain optimal operating temperatures. Microchannel systems utilize highly optimized internal geometry to promote heat transfer efficiency, thus preventing localized overheating. They were numerically simulated in modeling to predict thermal-fluid distributions, justifying advantages they provide for temperature regulation of electronics. Their use greatly improves thermal performance of electronic devices, reducing the chances of overheating and extension of lifespan of operations (Ding et al., 2011). Thus, ongoing innovation in thermal management not only enhances the efficiency of the device but also enhances the reliability and lifecycle of power electronic systems in high-impact applications (Figure 10.6).

10.2.3. Reliability Issues Related to Thermal Stress

Thermal stress due to temperature cycling in operation significantly impacts the reliability of power electronic devices. Temperature cycling can cause material expansion and shrinkage, generating mechanical stresses that compromise the device's integrity with time. With power electronics being applied to increasingly demanding fields like electric vehicles and renewable energy systems, it is essential to understand how thermal cycling impacts device reliability. Research has proven that thermal stress is among the failure mechanisms like solder joint fatigue and insulation breakdown, which are directly related to temperature fluctuations. Due to the vital nature of such components, resolving thermal management becomes essential in order to improve their lifespan and operation. Both passive and active thermal management methods are being considered to minimize such effects, but the challenge is to balance these solutions with the overall size and efficiency of the electronic devices. The second key component of the reliability of power electronics due to thermal stress is the material selection to be used in device fabrication. For instance, semiconductor materials like silicon carbide (SiC) and gallium nitride (GaN) possess higher thermal conductivity and can operate at high temperatures than traditional silicon devices. However, they possess widely differing thermal expansion coefficients compared to similar packaging and joining materials. This disparity may also increase thermal stress, leading to more rapid device failure. Advanced packaging technologies, such as thermally conducting substrates and heat spreaders, are required to alleviate such stresses and increase reliability. Moreover, as miniaturization and efficiency are more and more needed in electronic systems, being familiar with the interaction between material selection and thermal management is of utmost importance. Material compatibility and methods of optimizing thermal dissipation will dominate the achievement of further functionality in power electronics reliability for high-performance applications. The role of thermal stress in reliability issues extends beyond mere material selection; it is also true for system-level design solutions. The integration of thermal management components such as heat sinks and fans must be designed in understanding the system-wide thermal dynamics of the application. An example would be inefficient airflow within a densely packed design, which can enhance thermal accumulation to lead to localized overheating and burn-out of components. Also, developments such as miniaturization and integration of power electronics alongside the control circuitry into a single package make thermal management even more complicated. To mitigate against such developments, advanced simulations and modeling aids become essential in predicting thermal performance under varying loads conditions. Through such forecasting aids, engineers are able to design more reliable systems that react to failure owing to thermal stress, ensuring that the long-term life and performance of power electronic devices are significantly enhanced. The significance of exploring these interconnected factors is emphasized through various studies, such as the representations of the SBIR programs which suggest a system approach toward ensuring reliability through innovative thermal management technologies (Gilman et al., 2025).

10.3. MATERIALS FOR THERMAL MANAGEMENT

In power electronics, effective thermal management is crucial due to the high-power densities and operating temperatures of modern devices. Thermal management materials play a critical role in enabling safe and trustworthy operation and durability through effective heat transfer established during operation. Emerging advanced materials such as high-conductivity thermally conductive polymers and metal matrix composites are increasingly being utilized to enhance thermal transport properties. These materials not only enable improved heat dissipation but are also linked to weight savings, beneficially applied in size and weight-restricted applications. Studies put such improvements into perspective, indicating that the application of tailored materials can lead to enhanced thermal management systems that improve the performance and reliability of power electronic systems (Hoell et al., 2019). Thermoelectric materials, which can be used to transform temperature differences into electricity, are increasingly key players in power electronics' thermal management arena. New designs, such as thin-film thermoelectric generators from a recently introduced folding structure, with an increased thermal contact area and minimal heat leakage (Tappura et al., 2018). These devices capitalize on efficient heat transfer mechanisms that are present in the plane of the thermoelectric thin-film, thus making it possible to achieve packaging densities required for large-area applications. Being able to harness waste heat for ancillary energy generation not only improves system efficiency but also helps maintain optimal operating temperatures, demonstrating a two-fold energy harvesting and thermal management capability. This synergy demonstrates what can be done with thermoelectric materials to extend thermal management from mere heat dissipation. In addition, material selection of TIMs remains critical in power electronics in bridging heat sources and heatsinks. The efficiency of TIM can significantly influence the system heat transfer rate and thermal resistance as a whole. Costly graphene or carbon nanotube composites offer excellent thermal conductivity and flexibility and conformability required to provide effective heat spreading. In addition, these materials address the issues brought by temperature fluctuation and mechanical stress, providing resilient solutions that can withstand long-term performances. With the growing demand for higher efficiency and longer operating lifetimes in power electronics, innovation and use of new materials in thermal management will become essential to ensuring system reliability and performance in the face of changing technological environments (Hoell et al., 2019).

10.3.1. Thermal Interface Materials (TIMs)

Thermal interface materials (TIMs) contribute to the enhancement of thermal management in power electronics through effective heat transfer from and to components, such as semiconductor devices and heat sinks. The primary function of TIMs is to minimize thermal resistance at hotspots where heat is generated. By seeping in tiny air gaps that form between surfaces, the TIMs facilitate improved conduction and enable increased heat dissipation, which prevents overheating and maintains the reliability of electronic devices. Various formulations of TIM include thermal greases, phase change materials, and conformable pads with varying thermal conductivity properties suited to diverse applications. As the demand for higher-performance and smaller-size power electronic devices is on the rise, the development of new-generation TIMs is required to support maximum thermal performance over a broad range of operating conditions. Investigations into advanced materials for TIM have identified carbon nanotubes (CNTs) as an extremely promising candidate due to its outstanding thermal conductivity properties. CNTs, upon incorporation in TIM formulations, add significant improvement in thermal performance compared to traditional materials. Recent studies have shown that electrophoretic deposition (EPD) can be utilized effectively to the uniform deposition of CNTs on heat spreaders to create a robust interface for improved heat transfer properties. Among the studies, one demonstrated that applying 0.50 mg/mL of CNT concentration led to a deposition thickness of 4.78 μm when an applied voltage of 150 V was used, indicating that deposition parameters control properly can create a significant impact on the thermal effectiveness of TIMs. The resulting innovations enable power electronic devices to perform optimally, even under high loads, thus enhancing their operating life and reliability (Mohamed et al., 2017) (Figure 10.7).

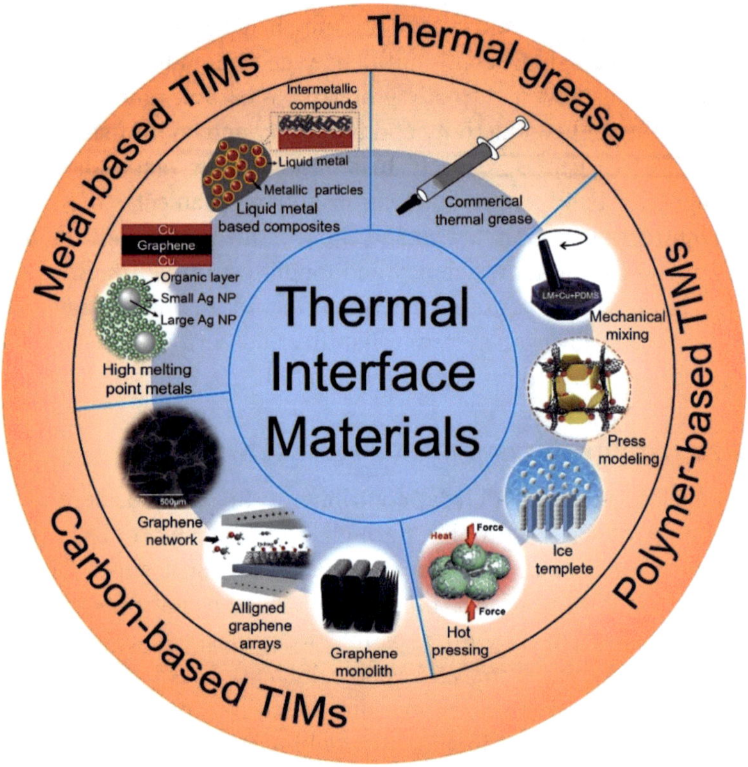

Figure 10.7. Thermal interface materials (TIMs).

Source: https://www.mdpi.com/nanomaterials/nanomaterials-12-03365/article_deploy/html/images/nanomaterials-12-03365-g001.png.

The implications of selecting proper TIMs cannot be overstated, particularly in light of emerging technologies in power electronics. With electronics becoming more integrated and smaller, the ability of TIMs to adapt to varying thermal loads becomes extremely critical. The need for TIM solutions that are thermally efficient and substrate- and device-architecture-compatible offers challenges and opportunities for additional research and development. Additionally, future development of TIMs must balance sustainability and affordability without compromising thermal performance. These innovations are paramount in satisfying industries' future demands for high-performance electronic systems, adding further to the importance of TIMs as an anchor technology in thermal management applications (N/A, 2025).

10.3.2. Heat Sinks and Their Materials

The efficiency of heat sinks in thermal packaging depends significantly on the material utilized for their construction. Traditionally, the most preferred material has been aluminum, as it possesses good thermal conductivity, is physically light, and relatively low-cost. However, with the rise of power electronics and the expanding needs, interest in alternative materials with better thermal properties is growing. Copper, for example, is another metal choice with superior thermal conductivity than aluminum and therefore an excellent one for applications in which maximum heat loss is necessary. Having knowledge of the thermal characteristics of the materials forms the foundation of power electronics heat sink design optimization and performance due to the fact that the efficiency of the heat sinks to transfer heat dictates the components' reliability and longevity. In recent decades, materials technology has discovered new means of heat sink performance improvement. Among these is the use of metal foams and metal matrix composites (MMCs). These possess remarkable properties that play a significant role in thermal management. Metal foams, for instance, possess a remarkable surface area-to-volume ratio, which has the ability to significantly increase the heat transfer rates in heat exchange applications. Furthermore, their properties in light contribute to overall system efficiency without additional weight. The potential for tailored MMC properties to specialize can offer tailor-made solutions capable of solving specific power electronic thermal management problems. With their growing use, it is imperative to analyze their process of fabrication and consider issues related to their broader application to commercial products, particularly to heat sinks (Han et al., 2012). Even with the new materials that provide better thermal management of power electronics, their full potential can be realized only when some of their most essential problems are fixed. One of the biggest problems is the possibility of scaling up production techniques to accommodate large-scale manufacturing without compromising material integrity and performance. In addition, their operating condition durability over the long term needs to be tested and inspected to verify that they will be capable of withstanding thermal cycling and mechanical stress. With these problems resolved by manufacturers, however, the application of new materials such as metal foams and MMCs in heat sinks should lead to more effective heat management solutions, culminating in enhanced performance and longevity of power electronic systems. As the field continues to grow, constant research and development efforts will unlock the full potential of such high-tech materials (Hoell et al., 2019) (Figure 10.8).

Figure 10.8. The effectiveness of heat sinks in thermal management systems.

Source: https://static4.arrow.com/-/media/arrow/images/miscellaneous/h/heat-sink.png.

10.3.3. Advanced Materials for Thermal Management

The development of new high-performance materials for thermal management is increasingly significant in the field of power electronics, where high heat dissipation reliability is crucial to both performance and reliability. Traditional thermal management options typically do not hold up under high power density device loads, leading to overheating and device failure. Materials science breakthroughs, notably involving nanocomposites and phase change materials (PCMs), are creating new opportunities for enhancing thermal conductivity and storage. The incorporation of graphene into polymer matrices, for instance, has shown good prospects with the potential of a lightweight material that enjoys superior thermal transport properties. With power electronics moving towards more efficient and smaller systems, these new materials are critical in maintaining the best operating temperature levels to foster the longevity and reliability of critical electronic components. Furthermore, the function of advanced TIMs in power electronic assemblies is critical in maximizing heat transfer between various layers of power electronic assemblies. Unlike traditional epoxy or silicone-based TIMs, emerging technologies from carbon nanotube and other high-end composite materials are enhancing thermal conductivity while reducing thermal resistance significantly. Apart from enabling the efficiency of heat dissipation, this transition to next-generation TIMs is also easing the solution of the challenging thermal management demands of modern systems such as electric vehicles and renewable energy systems. The application of such materials during implementation guarantees uniform heat distribution across elements, thus preventing hotspots that may lead to premature failure. With the application of the unique characteristics of such advanced materials, designers can assure that power electronics operate within safe temperatures, thus resulting in their durability and enhanced capacity of operation. Moreover, ongoing research in advanced materials extends into specialty uses, particularly in emerging fields like Nuclear Electric Propulsion (NEP) technology. In these applications, thermal management solutions are critical to success in high-performance systems. As envisioned in recent studies, a forward-looking technology advancement strategy for NEP systems is imagined experimenting with power and propulsion systems at kilowatt and megawatt levels, underscoring the need for advanced materials that can withstand extreme thermal environments (Doherty et al., 2025). The research necessitates systematic design of thermal management techniques to not just address immediate cooling demands but also leverage material properties for sustainable operation of NEP systems over the long term. This synergy of high-end materials and research-driven approaches, thus, highlights the necessity for innovation-driven development of thermal management technology, especially in high-risk mission applications like space exploration that necessitate levels of reliability and efficiency unparalleled elsewhere (Figure 10.9).

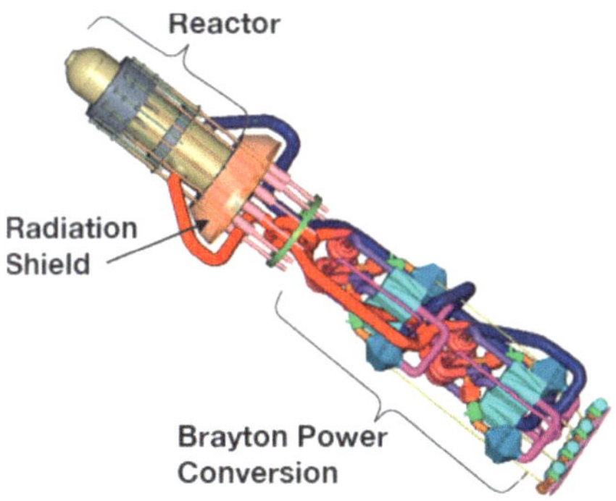

Figure 10.9. Nuclear electric propulsion.

Source: https://nap.nationalacademies.org/openbook/25977/xhtml/images/img-47-1.jpg.

10.4. EMERGING TRENDS IN THERMAL MANAGEMENT

Next-generation trends in thermal management are setting the course for the future of power electronics, driven by the increasing requirement for energy efficiency and reliable performance. One such trend is the integration of next-generation materials such as graphene and carbon nanotubes with increased thermal conductivity. Not only do these materials improve heat dissipation, but they also reduce the weight and size of electronic systems. Moreover, as devices continue to miniaturize, the issue of heat generation control at the nanoscale is becoming more significant. Progress in material science has led to novel TIMs with properties that can enhance thermal conduction among components, thereby enhancing the efficiency of power converters and minimizing thermal-related failures. These advancements hold promise to ultimately result in more efficient and high-performance power electronic devices in applications, from electric cars to renewable power systems. Another recent trend is the research into nanomaterials for phase change materials (PCMs), aimed at optimizing thermal management applications. PCMs have gained popularity because they possess characteristics that enable them to absorb and release thermal energy and thus play a key role in applications where temperature management is unavoidable. The application of nanotechnology to PCMs increases their response rate and thermal energy storage density, thus making them much more efficient at managing heat in power electronic devices. Studies are ongoing in various composite formulations, and research reveals that the modified PCMs can adapt to varying thermal loads efficiently, remaining in the best operating temperatures. The payoff possibilities are obvious, as efficient integration of nanostructured PCMs into electronics is not only able to maximize performance but also extend the lifetimes of devices, minimizing the need for routine maintenance measures. The advent of smart thermal management systems that take advantage of real-time sensing and prognostic analytics is another critical trend in the field. These systems use sensors and adaptive control algorithms for dynamic thermal management optimization; wherein cooling mechanisms are adjusted based on continuous data input regarding device temperatures. The integration of such systems enables the use of more advanced cooling strategies that would once more enhance the reliability of power electronic devices. Smart thermal management also occupies a central role in proactive maintenance because it is able to predict potential thermal failure in advance. Linked with machine learning innovations, such a tendency enables manufacturers and engineers to make their products even more accurate, creating systems not only to prevent overheating but also to meet the highest level of energy efficiency. Ultimately, these innovations in thermal management methods will play a crucial role in pushing the performance capabilities of power electronics in various fields to the next level, making them sustainable for the future (Bao et al., 2018; Guarnieri et al., 2016).

10.4.1. Innovations in Cooling Technologies

Innovations in cooling technology play a vital role in enhancing power electronics thermal management, particularly in applications where heat accumulation can have a significant impact on reliability and performance. With devices being powered at higher power densities, conventional cooling methods become insufficient to maintain optimal working temperatures. The most recent advancements, such as microchannel heat exchangers and two-phase coolers, show how engineered cooling technology can be used to increase heat transfer in small electronics. Microchannel designs, which are highly convoluted flow paths, significantly increase the thermal transfer surface area with very low coolant consumption. This not only enhances thermal efficiency but also helps in space-saving techniques crucial in applications such as electric vehicles and aerospace where thermal management is crucial to system durability and efficiency (Hossain et al., 2024). Another important breakthrough in cooling technologies is the use of advanced materials in thermal management for power electronics. New materials, such as phase change materials (PCMs) and thermally conductive polymers, have also emerged, with significant improvements in thermal interface performance. PCMs, for instance, absorb and release heat upon phase transition, which can effectively buffer temperature change, hence being appropriate for intermittent thermal load conditions. Additionally, integration of these materials into a system can led to a passive cooling solution, reducing active cooling methods to the minimum. This technology aligns with the growing trend of sustainability and energy efficiency in engineering practice that increasingly favors solutions improving performance with decreased energy consumption (Gilman et al., 2025). Finally, bringing artificial intelligence (AI) into cooling technology management presents an exciting new horizon for thermal management of power electronics. Through machine learning algorithms, AI can optimize cooling in real-time, adapting to varying operating conditions and demands. This predictive maintenance requires dynamic behavior, which reduces the threat of overheating and enhances electronic system dependability in such critical applications as automotive and avionics. Furthermore, AI-driven thermal management systems can be integrated with existing enabling technologies such as wide-bandgap semiconductors, delivering still higher thermal performance, and emphasizing the interdependence of advanced materials and intelligent systems. These innovations together lead the way towards a more efficient and sustainable future in the field of electronic design, highlighting how contemporary technology can address conventional thermal management issues (Hossain et al., 2024) (Figure 10.10).

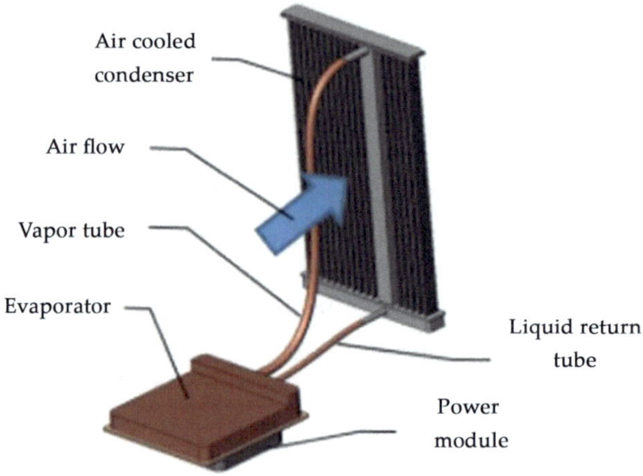

Figure 10.10. Air cooled heat sink using loop thermosiphon.

Source: https://www.electronics-cooling.com/2017/07/advanced-cooling-power-electronics/

10.4.2. Integration of Thermal Management in System Design

In the context of power electronics, the integration of thermal management with system design is the critical consideration to ensure performance and reliability optimization. With electronic components being utilized in increasing power densities, the thermal output becomes a significant concern that must be addressed throughout the entire design process. Effective thermal management practices not only preserve the operational efficiency of power electronic systems but also extend their lifespan and enhance overall system reliability.

This integration involves the choice of thermal materials, in which components are mounted and cooling arrangements planned thoughtfully such that dissipation of heat is aptly controlled. Such integrated design strategies for thermal management are a must in aerospace and automotive applications where performance and heat endurance are not acceptable aspects of system design. The IMD (Integrated Motor Drive) design shows the necessity of power electronic thermal management. With physical integration of converters and electric machines in close proximity to each other, there exists a high likelihood of increased component temperatures, which pose significant problems. Thus, analysis of thermal aspects of the integration becomes increasingly important.

The thermal management methodology must ensure proper heat distribution throughout the IMD to prevent hotspots that can lead to system failure. Further, considering various converter topologies, such as Matrix converters and multi-phase full-bridge converters, provides designers with a range of thermal management schemes adapted to maximize performance as well as thermal stability. Resolution of these factors contributes to the mechanical and structural durability of the system directly, rendering thermal management a cornerstone of effective IMD design (Agarwald et al., 2016) (Figure 10.11).

Figure 10.11. Integrated motor drives (IMD).

Source: https://external-content.duckduckgo.com/iu/?u=https%3A%2F% 2F www07.abb.com%2Fimages%2Flibrariesprovider100%2Fdefault-album%2Ff_00000027.jpg%3FCulture%3Den%26sfvrsn%3D98f3960c_1&f=1&nofb=1&ipt=2cd2 a 3f57314dfd68f66fcbe9a84ccf04908e88edb261c7720774b460991c212.

The utilization of leading-edge materials, like wide bandgap devices GaN and SiC, further depicts developments in thermal management techniques in system integration design. The materials possess higher thermal conductivity as well as high temperature operation capabilities, the primary needs for higher power density as well as efficiency in electronic systems. However, in order to maximize the advantages brought by these technologies, a clear understanding of their packaging and integration into power modules must be known. This guarantees that as system designs progress in incorporating advanced components, the thermal management structures are created parallel to the changing thermal properties. Through the overcoming of unique thermal challenges presented by novel materials and forms, the engineers can enable innovations that drive the next power electronics revolution and thus advance various applications across a number of industries (Hoell et al., 2019).

10.4.3. Future Directions in Thermal Management Research

The future emphasis in thermal management research must take into account increasing miniaturization complexities for power electronics. As devices become smaller, the heat density that is generated in applications such as smartphones and high-end processors becomes even more concentrated. More and more research is focused on innovative cooling schemes and materials that can dissipate heat efficiently with very small design dimensions. Traditional cooling techniques, i.e., heat pipes, have been found to be unsatisfactory at the microscopic scale, and hence new materials like phase change materials and high-performance heat sinks evolved. The TIMs are another field that needs improvement since this decides the efficiency of the heat transfer between the components. Overcoming these issues through systematic thermal analysis, as is highlighted in recent studies, is necessary to make devices safe and functional under different usage conditions, with the threat of catastrophic thermal incidents in small devices (Chen et al., 2019). At the same time, the fast development of data center expansion brings new thermal management problems of unprecedented scale that require immediate redress. The surge in energy consumption in such centers necessitates revolutionary efficiency strategies to neutralize high heat generation. Future research needs to involve incorporating advanced thermal management techniques, such as dynamic voltage and frequency scaling (DVFS) and server consolidation, to improve operation efficiency. Furthermore, understanding the thermal characteristics of data centers will enable the creation of predictive models that will optimize cooling systems in real time. By employing architectural modifications for greater energy efficiency, researchers and engineers are able to significantly reduce the power load and corresponding thermal output. Such a move has the power to bring about a major leap in green computing, highlighting the need for continued research into this pivotal area (Mittal et al., 2014). Finally, interdisciplinary collaboration provides an immense avenue to the creation of thermal management systems through the input of material science, mechanical engineering, and computer science. Through the forging of alliances between researchers, industry participants, and policymakers, all-encompassing solutions with emphasis on successful power electronics thermal management can be developed. The future research should target the integration of emerging technologies such as artificial intelligence and machine learning that can offer predictive solutions in thermal behaviors and management. This may lead to the creation of intelligent cooling systems that dynamically respond to varying loads and weather conditions. Emphasizing an integrated vision encompassing cutting-edge materials, advanced cooling techniques, and intelligent monitoring systems will ensure thermal management of power electronics advances in tandem with the unprecedented rate of technological advancements propelling the sector. Synergistic strategy is key towards achieving enhanced performance and security in increasingly complicated electronic systems.

SUMMARY

The study of thermal management of power electronics underscores the critical significance of adequate thermal solutions in a bid to meet new technology demands. As power electronic devices find themselves increasingly at the core of use cases like electric vehicles and renewable energy systems, their reliability and efficiency need to remain at the forefront. The interface between thermal management strategies and energy optimization is particularly crucial, considering that attempts to enhance device performance need ideally to offset thermal risks. Overcoming this challenge calls for a balanced solution through innovative material, efficient cooling technologies, and robust design mechanisms. Previous research has indicated a relentless pursuit of optimization in this domain, referencing that despite enormous advancements, loopholes still remain in balancing user comfort and energy efficiency (Fayaz et al., 2019). This ongoing research in thermal management not only leads to longer-lasting power electronic systems but also aligns with greater initiatives for sustainability in energy consumption. Novel approaches to energy optimization form central components of effective thermal management strategies in power electronics. Implementing various algorithms for optimization reveals the complexities inherent in balancing increased efficiency against users' comfort. For instance, methods encompassing thermal comfort as well as air and visual comfort have been shown to be critical in smart home systems such that adequate thermal management is more than heat dissipation. Different algorithms have been proposed to address these optimization challenges; however, current literature projects that most of these methods have not yet been utilized to their full potential in real applications (Fayaz et al., 2019). The potential remains vast for employing new and advanced optimization techniques that may require theory breakthrough to reality application in the world. Thus, the quest for solutions to optimizing energy usage will drive future innovation in thermal management solutions. Thermal management and power quality convergence is a large area that merits attention. Power quality issues, as exemplified by voltage continuity, level, and disturbances, can effectively contribute to the performance degradation of power electronic devices. Overcoming these issues requires a multi-faceted approach encompassing electromagnetic compatibility and safety. As power electronic systems become a constituent part of electricity distribution networks, their contribution towards maintaining power quality becomes all the more imperative. The identification and implementation of best practices in thermal management not only enhance device reliability but contribute to overall energy system efficiency. Continued research is necessary, as it reveals the intricate relationship between temperature control and power quality and paves the way for smarter, more reliable electrical infrastructures. The future of power electronics thermal management is therefore set to see big innovation and change, pushing the boundaries both in terms of technology and sustainability initiatives.

In the past few years, the necessity of effective thermal management for power electronics has intensified, particularly with tightening emission regulations worldwide and increasing efficiency demands. One of the secrets to advancing power systems is in the use of advanced thermal management technologies to make them reduce heat generated within energy conversion processes. The use of hybrid-electric systems is an example of such innovation, with new schemes and configurations devised to optimize energy storage as well as the overall efficiency of the system. The reduction of thermal emission is critical, with up to 20% of urban air pollution originating from transportation systems, and thus thermal efficiency needs to be weighed against environmental sustainability (Capata et al., 2018). Such developments not only lead to the enhanced reduction of carbon footprints but also provide opportunities for meeting rigid regulatory requirements. The integration of information and communication technology (ICT) with the thermal management architecture provides additional opportunities for enhancing power electronics. With ICT, researchers can formulate sophisticated algorithms for real-time control and monitoring of electronic device thermal conditions. This capability is crucial in ensuring the consistency and resilience of power systems, since increased operating temperatures may negatively affect component life and efficiency. Additionally, the trend for cleaner energy technologies necessitates a broad knowledge of the rates of energy consumption, particularly in economically emerging nations. The possibility of harnessing various kinds of renewable energy is linked to effective thermal management in power electronics for better performance and minimized operational vulnerability (Nwaobi et al., 2025). Effective thermal management, in its turn, facilitates the possibility of optimal scope for cleaner technologies to the benefit of environmental as well as economic considerations. Moreover, advances in materials science and engineering are key factors in next-generation thermal management technologies for power electronic systems. High-performance TIMs and advanced cooling methods are important in dissipating heat more effectively and attaining optimal operating temperatures of major components. With ongoing developments in power electronics, the focus of research is looking for materials that have higher thermal conductivity and reliability to withstand the environment of high-performance applications. For hybrid technologies, a systematic approach to thermal management not only ensures operational performance but also eliminates reliability challenges as a result of thermal stress. With the market of power electronics increasingly dependent on the pressures of sustainable development and technological advancement, thermal management will remain at the core of achieving the goals of high performance along with lower environmental impact (Capata et al., 2018; Nwaobi et al., 2025).

REFERENCES

1. Abusleme, A., Adam, T., Ahmad, S., Aiello, S., et al. (2020). TAO conceptual design report: A precision measurement of the reactor antineutrino spectrum with sub-percent energy resolution. https://core.ac.uk/download/323310553.pdf (accessed on 23rd October 2025).
2. Agarwald, S. H., Al Ahmar, J., Babb, M., Bai, J. G., Bai, J. G., Bai, J. G., Baliga, B. J., et al. (2016). Integrated motor drives: State of the art and future trends. https://core.ac.uk/download/33576382.pdf (accessed on 23rd October 2025).
3. Anon., Anon., Barlow, M., Barron, D., Braudaway, D., et al. (2007). Mathematical and computer modeling of electro-optic systems using a generic modeling approach. https://core.ac.uk/download/9399574.pdf (accessed on 23rd October 2025).
4. Bao, Y., Feng, W., Jin, M., Liu, H., et al. (2018). A review of microgrid development in the United States–A decade of progress on policies, demonstrations, controls, and software tools. https://core.ac.uk/download/323310623.pdf (accessed on 23rd October 2025).
5. Beck, S. M., Breckenridge, R. A., Clark, L. G., Collier, D., et al. (2025). Technology for the Future: In-Space Technology Experiments Program, Part 2. https://core.ac.uk/download/pdf/42817270.pdf (accessed on 23rd October 2025).
6. Bingham, C., Foster, M. P., Gould, C., Stone, D., et al. (2004). State-variable modelling of CLL resonant converters. https://core.ac.uk/download/54529.pdf (accessed on 23rd October 2025).
7. Brandner, J. J., Brighenti, F., & Kamaruzaman, N. (2011). Experimental investigation on self-similar heat sinks for liquid cooled electronics. 3rd Micro and Nano Flows Conference (MNF2011). https://core.ac.uk/download/9633348.pdf (accessed on 23rd October 2025).
8. Capata, R. (2018). Urban and extra-urban hybrid vehicles: A technological review. https://core.ac.uk/download/188826162.pdf (accessed on 23rd October 2025).
9. Chen, E., Feng, Y., Grimes, C., Gurrum, S., Huang, M., Jaworski, F., et al. (2019). Thermal analysis of lithium-ion battery-equipped smartphone explosions. https://core.ac.uk/download/289095013.pdf (accessed on 23rd October 2025).
10. Davidson, J. L., Esame, İ. O., Gürbüz, Y., Kang, D., et al. (2005). Diamond semiconductor technology for RF device applications. https://core.ac.uk/download/11738207.pdf (accessed on 23rd October 2025).

11. Desmet, J., Heimbach, B., & Renner, H. (2015). Power quality and electromagnetic compatibility: Special report, session 2. https://core.ac.uk/download/55822716.pdf (accessed on 23rd October 2025).
12. Doherty, M. P., & Holcomb, R. S. (2025). Summary and recommendations on nuclear electric propulsion technology for the space exploration initiative. https://core.ac.uk/download/pdf/42807523.pdf (accessed on 23rd October 2025).
13. Fayaz, M., Lajis, A., Nasir, H., Shah, R., et al. (2019). A review on energy consumption optimization techniques in IoT based smart building environments. https://core.ac.uk/download/300473894.pdf (accessed on 23rd October 2025).
14. George, K. (2016). Solid-State Transformers for Interfacing Solar Panels to the Power Grid: An Optimum Design Methodology of a High Frequency Transformer for dc-dc Converter Applications. https://core.ac.uk/download/84119769.pdf (accessed on 23rd October 2025).
15. Gilman, J. A., Paige, J. B., & Schwenk, F. C. (2025). NASA SBIR Abstracts of 1990 Phase 1 Projects. https://core.ac.uk/download/pdf/42808250.pdf (accessed on 23rd October 2025).
16. Guarnieri, M., Mattavelli, P., Petrone, G., Spagnuolo, G., et al. (2016). Vanadium redox flow batteries: Potentials and challenges of an emerging storage technology. https://core.ac.uk/download/80155277.pdf (accessed on 23rd October 2025).
17. Han, X. H., Jacobi, A., Park, Y. G., Sommers, A., et al. (2012). A review of metal foam and metal matrix composites for heat exchangers and heat sinks. https://core.ac.uk/download/55763987.pdf (accessed on 23rd October 2025).
18. Hoell, K., McKay, C., Migneau, M., Mullen, R., et al. (2019). Development of innovative heat sinks for power electronics cooling within the more electrical aircraft. https://core.ac.uk/download/294830160.pdf (accessed on 23rd October 2025).
19. Hossain, M. S., Hussain, M. D., Mia, A., Mosleuzzaman, M., et al. (2024). Electric vehicle powertrain design: Innovations in electrical engineering. https://core.ac.uk/download/624121415.pdf (accessed on 23rd October 2025).
20. Lizy-Destrez, S., & Mimoun, D. (2010). STR: A student developed star tracker for the ESA-led ESMO moon mission. https://core.ac.uk/download/12042098.pdf (accessed on 23rd October 2025).
21. Mittal, S. (2014). Power management techniques for data centers: A survey. http://arxiv.org/abs/1404.6681 (accessed on 23rd October 2025).

22. Mohamed, A. R., Yaacob, K. A., Bahru, R., Yeoh, E., et al. (2017). Electrophoretic deposition of carbon nanotubes on heat spreader for fabrication of thermal interface materials (TIM). https://core.ac.uk/download/144263903.pdf (accessed on 23rd October 2025).
23. N/A. (2025). Publications of the Jet Propulsion Laboratory, 1985. https://core.ac.uk/download/pdf/42838538.pdf (accessed on 23rd October 2025).
24. Nwaobi, G. (2025). Energy power, digital infrastructure and e-learning platforms: African experience. https://core.ac.uk/download/pdf/7301893.pdf (accessed on 23rd October 2025).
25. Poulikakos, D. (2009). On emerging micro- and nanoscale thermofluidic technologies. 2nd Micro and Nano Flows Conference (MNF2009). https://core.ac.uk/download/9633457.pdf (accessed on 23rd October 2025).
26. Rao, K., Song, W., Wardi, Y., Yalamanchili, S., et al. (2015). Temperature regulation in multicore processors using adjustable-gain integral controllers. http://arxiv.org/abs/1507.06357 (accessed on 23rd October 2025).
27. Tappura, K. (2018). A numerical study on the design trade-offs of a thin-film thermoelectric generator for large-area applications. http://arxiv.org/abs/1801.06418 (accessed on 23rd October 2025).

INDEX

D

E

F

G

H

I

J

K

L

M

N

O

P

Q

R

S

T

U

V

W